Science for Engineering

Second Edition

John Bird BSc(Hons), CEng, CMath, MIEE, FIMA, FIIE, FCollP

Newnes

OXFORD   AUCKLAND   BOSTON   JOHANNESBURG   MELBOURNE   NEW DELHI

Newnes
An imprint of Butterworth-Heinemann
Linacre House, Jordan Hill, Oxford OX2 8DP
225 Wildwood Avenue, Woburn, MA 01801-2041
A division of Reed Educational and Professional Publishing Ltd

 A Member of the Reed Elsevier plc Group

First published 1995
Reprinted with revisions 1996
Reprinted 1998
Second edition 2000

British Library Cataloguing in Publication Data
A catalogue record for this book is available from the British Library

Library of Congress Cataloguing in Publication Data
A catalogue record for this book is available from the Library of Congress

ISBN 0 7506 4747 7

FOR EVERY TITLE THAT WE PUBLISH, BUTTERWORTH-HEINEMANN
WILL PAY FOR BTCV TO PLANT AND CARE FOR A TREE.

Typeset by Laser Words, Madras, India
Printed in Great Britain

Contents

Preface

Science for Engineering, Second edition, aims to develop in the reader an understanding of fundamental science concepts developed through energy, electrical and mechanical applications. The aims are to describe engineering systems in terms of basic scientific laws and principles, to investigate the behaviour of simple linear systems in engineering, to calculate the response of engineering systems to changes in variables, and to determine the response of such engineering systems to changes in parameters.

The text covers the following:

 (i) **Applied Science in Engineering** (Advanced GNVQ unit 5)
 (ii) **Applied Science and Mathematics for Engineering** (Intermediate GNVQ unit 4)
 (iii) **Any introductory/access/foundation course** involving Engineering Science and basic Mathematics

This second edition of Science for Engineering is arranged in four sections.

Section 1, Applied Mathematics, Chapters 1 to 10, provides the basic mathematical tools needed to effectively understand the science applications in sections 2, 3 and 4. Basic arithmetic, fractions, decimals, percentages, indices and standard form, calculations and evaluation of formulae, algebra, simple equations, transposition of equations, simultaneous equations, straight line graphs and trigonometry are covered in this first section.

Section 2, Energy Applications, Chapters 11 to 16, covers SI units, work, energy and power, an introduction to electric circuits, chemical effects of electricity, capacitors and inductors, and heat energy.

Section 3, Electrical Applications, Chapters 17 to 23, covers resistance variation, series and parallel circuits, Kirchhoff's laws, electromagnetism, electromagnetic induction, alternating voltages and currents and electrical measuring instruments and measurements.

Section 4, Mechanical Applications, Chapters 24 to 36, covers speed and velocity, acceleration, forces acting at a point, simply supported beams, linear and angular motion, friction, simple machines, the effects of forces on materials, linear momentum and impulse, torque, thermal expansion and the measurement of temperature.

Each topic in the text is presented in a way that assumes in the reader little previous knowledge of that topic. Theory is introduced in each chapter by a reasonably brief outline of essential information, definitions, formulae, laws and procedures. The theory is kept to a minimum, for problem-solving is extensively used to establish and exemplify the theory. It is intended that readers will gain real understanding through seeing problems solved and then through solving similar problems themselves.

Science for Engineering, 2nd Edition contains some **500 worked problems**, together with **400 multi-choice questions**, and some **700 further questions**, arranged in **170 Exercises**, all with answers at the back of the book; the Exercises appear at regular intervals – every 2 or 3 pages – throughout the text. Also included are over **350 short answer questions**, the answers for which can be determined from the preceding material in that particular chapter. **325 line diagrams** further enhance the understanding of the theory. All of the problems – multi-choice, short answer and further questions – mirror where possible practical situations in science and engineering.

At regular intervals throughout the text are thirteen **Assignments** to check understanding. For example, Assignment 1 covers material contained in chapters 1 to 3, Assignment 2 covers the material contained in Chapters 4 to 7, and so on. No answers for the Assignments are contained in the text – lecturers' may obtain a complimentary set of solutions for the assignments when adopting texts for their courses.

A list of relevant **formulae** are included at the end of each of the latter three sections of the book.

'Learning by Example' is at the heart of Science for Engineering.

John Bird
Portsmouth University

Section 1

Applied Mathematics

1

Basic arithmetic

At the end of this chapter you should be able to:

- perform basic arithmetic by addition, subtraction, multiplication and division (without a calculator)
- calculate highest common factors and lowest common multiples
- perform calculations involving brackets
- appreciate the order of precedence when performing calculations

1.1 Arithmetic operations

Whole numbers are called **integers**. +3, +5, +72 are called positive integers; −13, −6, −51 are called negative integers. Between positive and negative integers is the number 0 which is neither positive nor negative.

The four basic arithmetic operators are: add (+), subtract (−), multiply (×) and divide (÷).

For addition and subtraction, when **unlike signs** are together in a calculation, the overall sign is **negative**. Thus, adding minus 4 to 3 is $3 + -4$ and becomes $3 - 4 = -1$. **Like signs** together give an overall **positive sign**. Thus subtracting minus 4 from 3 is $3 - -4$ and becomes $3 + 4 = 7$.

For multiplication and division, when the numbers have **unlike signs**, the answer is **negative**, but when the numbers have **like signs** the answer is **positive**. Thus $3 \times -4 = -12$, whereas $-3 \times -4 = +12$. Similarly,

$$\frac{4}{-3} = -\frac{4}{3} \quad \text{and} \quad \frac{-4}{-3} = +\frac{4}{3}$$

Problem 1. Add 27, −74, 81 and −19

This problem is written as $27 - 74 + 81 - 19$

Adding the positive integers: 27
                                         81
Sum of positive integers is: 108

Adding the negative integers: 74
                                         19
Sum of negative integers is: 93

Taking the sum of the negative integers from the sum of the positive integers gives:

                                        108
                                        −93
                                         15

Thus **$27 - 74 + 81 - 19 = 15$**

Problem 2. Subtract 89 from 123

This is written mathematically as $123 - 89$

                                        123
                                        −89
                                         34

Thus **$123 - 89 = 34$**

Problem 3. Subtract −74 from 377

This problem is written as $377 - -74$. Like signs together give an overall positive sign, hence

$377 - -74 = 377 + 74$

$$
\begin{array}{r}
377 \\
+74 \\
\hline
451
\end{array}
$$

Thus $377 - -74 = 451$

Problem 4.   Multiply 74 by 13

This is written as 74×13

$$
\begin{array}{r}
74 \\
13 \\
\hline
222 \\
740 \\
\hline
962
\end{array}
$$

$\leftarrow 74 \times 3$

$\leftarrow 74 \times 10$

Adding:

Thus $74 \times 13 = 962$

Problem 5.   Divide 1043 by 7

When dividing by the numbers 1 to 12, it is usual to use a method called **short division**.

$$
\begin{array}{r}
1\;4\;9 \\
7\,\overline{)\,10^3 4^6 3}
\end{array}
$$

Step 1. 7 into 10 goes 1, remainder 3. Put 1 above the 0 of 1043 and carry the 3 remainder to the next digit on the right, making it 34;

Step 2. 7 into 34 goes 4, remainder 6. Put 4 above the 4 of 1043 and carry the 6 remainder to the next digit on the right, making it 63;

Step 3. 7 into 63 goes 9, remainder 0. Put 9 above the 3 of 1043

Thus $1043 \div 7 = 149$

Problem 6.   Divide 378 by 14

When dividing by numbers which are larger than 12, it is usual to use a method called **long division**.

$$
\begin{array}{r}
27 \\
14\,\overline{)\,378} \\
28 \\
\hline
98 \\
98 \\
\hline
00
\end{array}
$$

(2) $2 \times 14 \rightarrow$

(4) $7 \times 14 \rightarrow$

(1) 14 into 37 goes twice. Put 2 above the 7 of 378

(3) Subtract. Bring down the 8. 14 into 98 goes 7 times. Put 7 above the 8 of 378

(5) Subtract.

Thus $378 \div 14 = 27$

Problem 7.   Divide 5669 by 46

This problem may be written as $\dfrac{5669}{46}$ or $5669 \div 46$ or 5669/46

Using the long division method shown in Problem 6 gives:

$$
\begin{array}{r}
123 \\
46\,\overline{)\,5669} \\
46 \\
\hline
106 \\
92 \\
\hline
149 \\
138 \\
\hline
11
\end{array}
$$

As there are no more digits to bring down,

$$5669 \div 46 = 123, \text{ remainder } 11 \quad \text{or} \quad 123\frac{11}{46}$$

Now try the following exercise

Exercise 1   Further problems on arithmetic operations (Answers on page 299)

In Problems 1 to 12, determine the values of the expressions given:

1. $67 - 82 + 34$

2. $124 - 273 + 481 - 398$

3. $927 - 114 + 182 - 183 - 247$

4. $2417 - 487 + 2424 - 1778 - 4712$

5. $2715 - 18250 + 11471 - 15093$

6. $813 - (-674)$

7. $647 - 872$

8. $3151 - (-2763)$

9. (a) 261×7 (b) 462×9

10. (a) 783×11 (b) 73×24

11. (a) 27×38 (b) 77×29

12. (a) $288 \div 6$ (b) $979 \div 11$

1.2 Highest common factors and lowest common multiples

When two or more numbers are multiplied together, the individual numbers are called **factors**. Thus a factor is a number which divides into another number exactly. The **highest common factor (HCF)** is the largest number which divides into two or more numbers exactly.

A **multiple** is a number which contains another number an exact number of times. The smallest number which is exactly divisible by each of two or more numbers is called the **lowest common multiple (LCM)**.

Problem 8. Determine the HCF of the numbers 12, 30 and 42

Each number is expressed in terms of its lowest factors. This is achieved by repeatedly dividing by the prime numbers 2, 3, 5, 7, 11, 13, ... (where possible) in turn. Thus

$$12 = \boxed{2} \times 2 \times \boxed{3}$$
$$30 = \boxed{2} \quad \times \boxed{3} \times 5$$
$$42 = \boxed{2} \quad \times \boxed{3} \times 7$$

The factors which are common to each of the numbers are 2 in column 1 and 3 in column 3, shown by the broken lines. Hence the **HCF is 2×3, i.e. 6**. Thus, 6 is the largest number which will divide into 12, 30 and 42 without remainder.

Problem 9. Determine the LCM of the numbers 12, 42 and 90

The LCM is obtained by finding the lowest factors of each of the numbers, as shown in Problems 8 above, and then selecting the largest group of any of the factors present. Thus

$$12 = \boxed{2 \times 2} \times 3$$
$$42 = 2 \quad \times 3 \qquad \times \boxed{7}$$
$$90 = 2 \quad \times \boxed{3 \times 3} \times \boxed{5}$$

The largest group of any of the factors present are shown by the broken lines and are 2×2 in 12, 3×3 in 90, 5 in 90 and 7 in 42.
Hence **the LCM is $2 \times 2 \times 3 \times 3 \times 5 \times 7 = 1260$**, and is the smallest number which 12, 42 and 90 will all divide into exactly.

Problem 10. Determine the LCM of the numbers 150, 210, 735 and 1365

Using the method shown in Problem 9 above:

$$150 = \boxed{2} \times \boxed{3} \times \boxed{5 \times 5}$$
$$210 = 2 \times 3 \times 5 \times 7$$
$$735 = \qquad 3 \times 5 \times \boxed{7 \times 7}$$
$$1365 = \qquad 3 \times 5 \times 7 \qquad \times \boxed{13}$$

The LCM is $2 \times 3 \times 5 \times 5 \times 7 \times 7 \times 13 = 95550$

Now try the following exercise

Exercise 2 Further problems on highest common factors and lowest common multiples (Answers on page 299)

In Problems 1 to 6 find (a) the HCF and (b) the LCM of the numbers given:

1. 6, 10, 14
2. 12, 30, 45
3. 10, 15, 70, 105
4. 90, 105, 300
5. 196, 210, 462, 910
6. 196, 350, 770

1.3 Order of precedence and brackets

When a particular arithmetic operation is to be performed first, the numbers and the operator(s) are placed in brackets. Thus 3 times the result of 6 minus 2 is written as $3 \times (6-2)$. In arithmetic operations, the order in which operations are performed are:

(i) to determine the values of operations contained in brackets,

(ii) multiplication and division (the word 'of' also means multiply), and

(iii) addition and subtraction

This **order of precedence** can be remembered by the word **BODMAS**, indicating **B**rackets, **O**f, **D**ivision, **M**ultiplication, **A**ddition and **S**ubtraction, taken in that order.

The basic laws governing the use of brackets and operators are shown by the following examples:

(i) $2+3 = 3+2$, i.e. the order of numbers when adding does not matter;

(ii) $2 \times 3 = 3 \times 2$, i.e. the order of numbers when multiplying does not matter;

(iii) $2+(3+4) = (2+3)+4$, i.e. the use of brackets when adding does not affect the result;

(iv) $2 \times (3 \times 4) = (2 \times 3) \times 4$, i.e. the use of brackets when multiplying does not affect the result;

(v) $2 \times (3+4) = 2(3+4) = 2 \times 3 + 2 \times 4$, i.e. a number placed outside of a bracket indicates that the whole contents of the bracket must be multiplied by that number;

(vi) $(2+3)(4+5) = (5)(9) = 45$, i.e. adjacent brackets indicate multiplication;

(vii) $2[3 + (4 \times 5)] = 2[3 + 20] = 2 \times 23 = 46$, i.e. when an expression contains inner and outer brackets, the inner brackets are removed first.

Problem 11. Find the value of $6 + 4 \div (5 - 3)$

The order of precedence of operations is remembered by the word BODMAS.

Thus
$$6 + 4 \div (5 - 3) = 6 + 4 \div 2 \qquad \text{(Brackets)}$$
$$= 6 + 2 \qquad \text{(Division)}$$
$$= \mathbf{8} \qquad \text{(Addition)}$$

Problem 12. Determine the value of

$$13 - 2 \times 3 + 14 \div (2 + 5)$$

$$13 - 2 \times 3 + 14 \div (2 + 5) = 13 - 2 \times 3 + 14 \div 7 \qquad \text{(B)}$$
$$= 13 - 2 \times 3 + 2 \qquad \text{(D)}$$
$$= 13 - 6 + 2 \qquad \text{(M)}$$
$$= 15 - 6 \qquad \text{(A)}$$
$$= \mathbf{9} \qquad \text{(S)}$$

Problem 13. Evaluate

$$16 \div (2 + 6) + 18[3 + (4 \times 6) - 21]$$

$$16 \div (2 + 6) + 18[3 + (4 \times 6) - 21]$$
$$= 16 \div (2 + 6) + 18[3 + 24 - 21) \qquad \text{(B)}$$
$$= 16 \div 8 + 18 \times 6 \qquad \text{(B)}$$
$$= 2 + 18 \times 6 \qquad \text{(D)}$$
$$= 2 + 108 \qquad \text{(M)}$$
$$= \mathbf{110} \qquad \text{(A)}$$

Now try the following exercise

Exercise 3 Further problems on order of precedence and brackets (Answers on page 299)

Simplify the expressions given in Problems 1 to 6:

1. $14 + 3 \times 15$

2. $17 - 12 \div 4$

3. $86 + 24 \div (14 - 2)$

4. $7(23 - 18) \div (12 - 5)$

5. $63 - 28(14 \div 2) + 26$

6. $\dfrac{112}{16} - 119 \div 17 + (3 \times 19)$

2

Fractions, decimals and percentages

At the end of this chapter you should be able to:

- add, subtract, multiply and divide fractions
- understand ratio and proportion
- add, subtract, multiply and divide decimals
- perform calculations correct to a certain number of significant figures or decimal places
- perform calculations involving percentages

2.1 Fractions

When 2 is divided by 3, it may be written as $\frac{2}{3}$ or 2/3. $\frac{2}{3}$ is called a **fraction**. The number above the line, i.e. 2, is called the **numerator** and the number below the line, i.e. 3, is called the **denominator**.

When the value of the numerator is less than the value of the denominator, the fraction is called a **proper fraction**; thus $\frac{2}{3}$ is a proper fraction. When the value of the numerator is greater than the denominator, the fraction is called an **improper fraction**. Thus $\frac{7}{3}$ is an improper fraction and can also be expressed as a **mixed number**, that is, an integer and a proper fraction. Thus the improper fraction $\frac{7}{3}$ is equal to the mixed number $2\frac{1}{3}$

When a fraction is simplified by dividing the numerator and denominator by the same number, the process is called **cancelling**. Cancelling by 0 is not permissible.

Problem 1. Simplify $\frac{1}{3} + \frac{2}{7}$

The LCM of the two denominators is 3×7, i.e. 21

Expressing each fraction so that their denominators are 21, gives:

$$\frac{1}{3} + \frac{2}{7} = \frac{1}{3} \times \frac{7}{7} + \frac{2}{7} \times \frac{3}{3} = \frac{7}{21} + \frac{6}{21}$$

$$= \frac{7+6}{21} = \frac{\mathbf{13}}{\mathbf{21}}$$

Alternatively:

$$\begin{array}{cc} \text{Step (2)} & \text{Step (3)} \\ \downarrow & \downarrow \end{array}$$

$$\frac{1}{3} + \frac{2}{7} = \frac{(7 \times 1) + (3 \times 2)}{21}$$

$$\uparrow$$
$$\text{Step (1)}$$

Step 1: the LCM of the two denominators.

Step 2: for the fraction $\frac{1}{3}$, 3 into 21 goes 7 times, 7 × the numerator is 7 × 1

Step 3: for the fraction $\frac{2}{7}$, 7 into 21 goes 3 times, 3 × the numerator is 3 × 2

Thus $\frac{1}{3} + \frac{2}{7} = \frac{7+6}{21} = \frac{\mathbf{13}}{\mathbf{21}}$ as obtained previously.

Problem 2. Find the value of $3\frac{2}{3} - 2\frac{1}{6}$

One method is to split the mixed numbers into integers and their fractional parts. Then

$$3\frac{2}{3} - 2\frac{1}{6} = \left(3 + \frac{2}{3}\right) - \left(2 + \frac{1}{6}\right) = 3 + \frac{2}{3} - 2 - \frac{1}{6}$$

$$= 1 + \frac{4}{6} - \frac{1}{6} = 1\frac{3}{6} = \mathbf{1}\frac{\mathbf{1}}{\mathbf{2}}$$

Another method is to express the mixed numbers as improper fractions.

Since $3 = \dfrac{9}{3}$, then $3\dfrac{2}{3} = \dfrac{9}{3} + \dfrac{2}{3} = \dfrac{11}{3}$

Similarly, $2\dfrac{1}{6} = \dfrac{12}{6} + \dfrac{1}{6} = \dfrac{13}{6}$

Thus $3\dfrac{2}{3} - 2\dfrac{1}{6} = \dfrac{11}{3} - \dfrac{13}{6} = \dfrac{22}{6} - \dfrac{13}{6} = \dfrac{9}{6} = 1\dfrac{1}{2}$ as obtained previously.

Problem 3. Determine the value of $4\dfrac{5}{8} - 3\dfrac{1}{4} + 1\dfrac{2}{5}$

$$4\dfrac{5}{8} - 3\dfrac{1}{4} + 1\dfrac{2}{5} = (4 - 3 + 1) + \left(\dfrac{5}{8} - \dfrac{1}{4} + \dfrac{2}{5}\right)$$

$$= 2 + \dfrac{5 \times 5 - 10 \times 1 + 8 \times 2}{40}$$

$$= 2 + \dfrac{25 - 10 + 16}{40}$$

$$= 2 + \dfrac{31}{40} = 2\dfrac{31}{40}$$

Problem 4. Find the value of $\dfrac{3}{7} \times \dfrac{14}{15}$

Dividing numerator and denominator by 3 gives:

$$\dfrac{1\cancel{3}}{7} \times \dfrac{14}{\cancel{15}\,5} = \dfrac{1}{7} \times \dfrac{14}{5} = \dfrac{1 \times 14}{7 \times 5}$$

Dividing numerator and denominator by 7 gives:

$$\dfrac{1 \times \cancel{14}\,2}{{}_1\cancel{7} \times 5} = \dfrac{1 \times 2}{1 \times 5} = \dfrac{2}{5}$$

This process of dividing both the numerator and denominator of a fraction by the same factor(s) is called **cancelling**.

Problem 5. Evaluate $1\dfrac{3}{5} \times 2\dfrac{1}{3} \times 3\dfrac{3}{7}$

Mixed numbers **must** be expressed as improper fractions before multiplication can be performed. Thus,

$$1\dfrac{3}{5} \times 2\dfrac{1}{3} \times 3\dfrac{3}{7} = \left(\dfrac{5}{5} + \dfrac{3}{5}\right) \times \left(\dfrac{6}{3} + \dfrac{1}{3}\right) \times \left(\dfrac{21}{7} + \dfrac{3}{7}\right)$$

$$= \dfrac{8}{5} \times \dfrac{1\cancel{7}}{{}_1\cancel{3}} \times \dfrac{\cancel{24}\,8}{\cancel{7}\,1} = \dfrac{8 \times 1 \times 8}{5 \times 1 \times 1}$$

$$= \dfrac{64}{5} = 12\dfrac{4}{5}$$

Problem 6. Simplify $\dfrac{3}{7} \div \dfrac{12}{21}$

$$\dfrac{3}{7} \div \dfrac{12}{21} = \dfrac{\dfrac{3}{7}}{\dfrac{12}{21}}$$

Multiplying both numerator and denominator by the reciprocal of the denominator gives:

$$\dfrac{\dfrac{3}{7}}{\dfrac{12}{21}} = \dfrac{\dfrac{1\cancel{3}}{1\cancel{7}} \times \dfrac{\cancel{21}\,3}{\cancel{12}\,4}}{\dfrac{1\cancel{12}}{1\cancel{21}} \times \dfrac{\cancel{21}\,1}{\cancel{12}\,1}} = \dfrac{\dfrac{3}{4}}{\dfrac{1}{1}} = \dfrac{3}{4}$$

This method can be remembered by the rule: invert the second fraction and change the operation from division to multiplication. Thus:

$$\dfrac{3}{7} \div \dfrac{12}{21} = \dfrac{1\cancel{3}}{1\cancel{7}} \times \dfrac{\cancel{21}\,3}{\cancel{12}\,4} = \dfrac{3}{4} \text{ as obtained previously.}$$

Problem 7. Find the value of $5\dfrac{3}{5} \div 7\dfrac{1}{3}$

The mixed numbers must be expressed as improper fractions. Thus,

$$5\dfrac{3}{5} \div 7\dfrac{1}{3} = \dfrac{28}{5} \div \dfrac{22}{3} = \dfrac{14\cancel{28}}{5} \times \dfrac{3}{\cancel{22}\,11} = \dfrac{42}{55}$$

Problem 8. Simplify $\dfrac{1}{3} - \left(\dfrac{2}{5} + \dfrac{1}{4}\right) \div \left(\dfrac{3}{8} \times \dfrac{1}{3}\right)$

The order of precedence of operations for problems containing fractions is the same as that for integers, i.e. remembered by BODMAS (Brackets, Of, Division, Multiplication, Addition and Subtraction). Thus,

$$\dfrac{1}{3} - \left(\dfrac{2}{5} + \dfrac{1}{4}\right) \div \left(\dfrac{3}{8} \times \dfrac{1}{3}\right)$$

$$= \dfrac{1}{3} - \dfrac{4 \times 2 + 5 \times 1}{20} \div \dfrac{\cancel{3}\,1}{\cancel{24}\,8} \quad \text{(B)}$$

$$= \dfrac{1}{3} - \dfrac{13}{5\cancel{20}} \times \dfrac{\cancel{8}\,2}{1} \quad \text{(D)}$$

$$= \dfrac{1}{3} - \dfrac{26}{5} \quad \text{(M)}$$

$$= \dfrac{(5 \times 1) - (3 \times 26)}{15} \quad \text{(S)}$$

$$= \dfrac{-73}{15} = -4\dfrac{13}{15}$$

Now try the following exercise

Exercise 4   Further problems on fractions (Answers on page 299)

Evaluate the following expressions:

1. (a) $\dfrac{1}{2} + \dfrac{2}{5}$ (b) $\dfrac{7}{16} - \dfrac{1}{4}$

2. (a) $\dfrac{2}{7} + \dfrac{3}{11}$ (b) $\dfrac{2}{9} - \dfrac{1}{7} + \dfrac{2}{3}$

3. (a) $5\dfrac{3}{13} + 3\dfrac{3}{4}$ (b) $4\dfrac{5}{8} - 3\dfrac{2}{5}$

4. (a) $\dfrac{3}{4} \times \dfrac{5}{9}$ (b) $\dfrac{17}{35} \times \dfrac{15}{119}$

5. (a) $\dfrac{1}{4} \times \dfrac{3}{11} \times 1\dfrac{5}{39}$ (b) $\dfrac{3}{4} \div 1\dfrac{4}{5}$

6. (a) $\dfrac{3}{8} \div \dfrac{45}{64}$ (b) $1\dfrac{1}{3} \div 2\dfrac{5}{9}$

7. $\dfrac{1}{3} - \dfrac{3}{4} \times \dfrac{16}{27}$

8. $\dfrac{1}{2} + \dfrac{3}{5} \div \dfrac{9}{15} - \dfrac{1}{3}$

9. $\dfrac{7}{15}$ of $\left(15 \times \dfrac{5}{7}\right) + \left(\dfrac{3}{4} \div \dfrac{15}{16}\right)$

10. $\dfrac{1}{4} \times \dfrac{2}{3} - \dfrac{1}{3} \div \dfrac{3}{5} + \dfrac{2}{7}$

2.2   Ratio and proportion

The ratio of one quantity to another is a fraction, and is the number of times one quantity is contained in another quantity **of the same kind**. If one quantity is **directly proportional** to another, then as one quantity doubles, the other quantity also doubles. When a quantity is **inversely proportional** to another, then as one quantity doubles, the other quantity is halved.

Problem 9.   A piece of timber 273 cm long is cut into three pieces in the ratio of 3 to 7 to 11. Determine the lengths of the three pieces.

The total number of parts is $3 + 7 + 11$, that is, 21. Hence 21 parts correspond to 273 cm.

$$1 \text{ part corresponds to } \frac{273}{21} = 13 \text{ cm}$$

$$3 \text{ parts correspond to } 3 \times 13 = 39 \text{ cm}$$

$$7 \text{ parts correspond to } 7 \times 13 = 91 \text{ cm}$$

$$11 \text{ parts correspond to } 11 \times 13 = 143 \text{ cm}$$

i.e. **the lengths of the three pieces are 39 cm, 91 cm and 143 cm**.

(Check: $39 + 91 + 143 = 273$)

Problem 10.   A gear wheel having 80 teeth is in mesh with a 25 tooth gear. What is the gear ratio?

Gear ratio $= 80{:}25 = \dfrac{80}{25} = \dfrac{16}{5} = 3.2$

i.e. gear ratio = **16:5** or **3.2:1**

Problem 11.   An alloy is made up of metals A and B in the ratio 2.5:1 by mass. How much of A has to be added to 6 kg of B to make the alloy?

Ratio A:B::2.5:1 i.e. $\dfrac{A}{B} = \dfrac{2.5}{1} = 2.5$

When $B = 6$ kg, $\dfrac{A}{6} = 2.5$ from which, $A = 6 \times 2.5 =$ **15 kg**.

Problem 12.   If 3 people can complete a task in 4 hours, find how long it will take 5 people to complete the same task, assuming the rate of work remains constant.

The more the number of people, the more quickly the task is done, hence inverse proportion exists.

Thus, if 3 people complete the task in 4 hours,

1 person takes three times as long, i.e. $4 \times 3 = 12$ hours, and 5 people can do it in one fifth of the time that one person takes, that is $\dfrac{12}{5}$ hours or **2 hours 24 minutes**.

Now try the following exercise

Exercise 5   Further problems on ratio and proportion (Answers on page 299)

1. Divide 312 mm in the ratio of 7 to 17

2. Divide 621 cm in the ratio of 3 to 7 to 13

3. When mixing a quantity of paints, dyes of four different colours are used in the ratio of 7:3:19:5. If the mass of the first dye used is $3\dfrac{1}{2}$ g, determine the total mass of the dyes used.

4. Determine how much copper and how much zinc is needed to make a 99 kg brass ingot if they have to be in the proportions copper:zinc :: 8:3 by mass

5. It takes 21 hours for 12 men to resurface a stretch of road. Find how many men it takes to resurface a similar stretch of road in 50 hours 24 minutes, assuming the work rate remains constant.

2.3   Decimals

The decimal system of numbers is based on the **digits** 0 to 9. A number such as 53.17 is called a **decimal fraction**, a decimal point separating the integer part, i.e. 53, from the fractional part, i.e. 0.17

A number which can be expressed exactly as a decimal fraction is called a **terminating decimal** and those which cannot be expressed exactly as a decimal fraction are called **non-terminating decimals**. Thus, $\frac{3}{2} = 1.5$ is a terminating decimal, but $\frac{4}{3} = 1.33333\ldots$ is a non-terminating decimal. $1.33333\ldots$ can be written as $1.\dot{3}$, called 'one point-three recurring'.

The answer to a non-terminating decimal may be expressed in two ways, depending on the accuracy required:

 (i) correct to a number of **significant figures**, that is, figures which signify something, and

 (ii) correct to a number of **decimal places**, that is, the number of figures after the decimal point.

The last digit in the answer is unaltered if the next digit on the right is in the group of numbers 0, 1, 2, 3 or 4, but is increased by 1 if the next digit on the right is in the group of numbers 5, 6, 7, 8 or 9. Thus the non-terminating decimal $7.6183\ldots$ becomes 7.62, correct to 3 significant figures, since the next digit on the right is 8, which is in the group of numbers 5, 6, 7, 8 or 9. Also $7.6183\ldots$ becomes 7.618, correct to 3 decimal places, since the next digit on the right is 3, which is in the group of numbers 0, 1, 2, 3 or 4.

Problem 13.   Evaluate $42.7 + 3.04 + 8.7 + 0.06$

The numbers are written so that the decimal points are under each other. Each column is added, starting from the right.

```
  42.7
   3.04
   8.7
   0.06
 _____
  54.50
```

Thus $42.7 + 3.04 + 8.7 + 0.06 = 54.50$

Problem 14.   Take 81.70 from 87.23

The numbers are written with the decimal points under each other.

```
  8⁶7̷.¹23
 −81. 70
 _____
   5. 53
```

Thus $87.23 - 81.70 = 5.53$

Problem 15.   Determine the value of 74.3×3.8

When multiplying decimal fractions: (i) the numbers are multiplied as if they are integers, and (ii) the position of the decimal point in the answer is such that there are as many digits to the right of it as the sum of the digits to the right of the decimal points of the two numbers being multiplied together. Thus,

(i)
```
     743
      38
  _____
    5944
   22290
  _____
   28234
```

(ii) As there are $(1+1) = 2$ digits to the right of the decimal points of the two numbers being multiplied together, $(74.\underline{3} \times 3.\underline{8})$, then

$$74.3 \times 3.8 = 282.34$$

Problem 16.   Convert (a) 0.4375 to a proper fraction and (b) 4.285 to a mixed number.

(a) 0.4375 can be written as $\dfrac{0.4375 \times 10000}{10000}$ without changing its value, i.e. $0.4375 = \dfrac{4375}{10000}$

By cancelling $\dfrac{4375}{10000} = \dfrac{875}{2000} = \dfrac{175}{400} = \dfrac{35}{80} = \dfrac{7}{16}$

i.e. $\mathbf{0.4375 = \dfrac{7}{16}}$

(b) Similarly, $\mathbf{4.285 = 4\dfrac{285}{1000} = 4\dfrac{57}{200}}$

Now try the following exercise

Exercise 6   Further problems on decimals (Answers on page 299)

In Problems 1 to 5, determine the values of the expressions given:

 1. $23.6 + 14.71 - 18.9 - 7.421$
 2. $73.84 - 113.247 + 8.21 - 0.068$
 3. 5.73×4.2
 4. $3.8 \times 4.1 \times 0.7$
 5. 374.1×0.006
 6. $421.8 \div 17$, (a) correct to 4 significant figures and (b) correct to 3 decimal places.

7. Convert to proper fractions:

    (a) 0.65 (b) 0.84 (c) 0.0125 (d) 0.282 (e) 0.024

8. Convert to mixed numbers:

    (a) 1.82 (b) 4.275 (c) 14.125 (d) 15.35 (e) 16.2125

In Problems 9 to 11, express as decimal fractions to the accuracy stated:

9. $\dfrac{4}{9}$, correct to 5 significant figures.

10. $\dfrac{17}{27}$, correct to 5 decimal place.

11. $1\dfrac{9}{16}$, correct to 4 significant figures.

2.4 Percentages

Percentages are used to give a common standard and are fractions having the number 100 as their denominators. For example, 25 per cent means $\dfrac{25}{100}$ i.e. $\dfrac{1}{4}$ and is written 25%

Problem 17. Express as percentages: (a) 1.875 and (b) 0.0125

A decimal fraction is converted to a percentage by multiplying by 100. Thus,

(a) 1.875 corresponds to $1.875 \times 100\%$, i.e. **187.5%**

(b) 0.0125 corresponds to $0.0125 \times 100\%$, i.e. **1.25%**

Problem 18. Express as percentages:

(a) $\dfrac{5}{16}$ and (b) $1\dfrac{2}{5}$

To convert fractions to percentages, they are (i) converted to decimal fractions and (ii) multiplied by 100

(a) By division, $\dfrac{5}{16} = 0.3125$, hence $\dfrac{5}{16}$ corresponds to $0.3125 \times 100\% = $ **31.25%**

(b) Similarly, $1\dfrac{2}{5} = 1.4$ when expressed as a decimal fraction.

    Hence $1\dfrac{2}{5} = 1.4 \times 100\% = $ **140%**

Problem 19. It takes 50 minutes to machine a certain part. Using a new type of tool, the time can be reduced by 15%. Calculate the new time taken.

15% of 50 minutes $= \dfrac{15}{100} \times 50 = \dfrac{750}{100} = 7.5$ minutes

hence the **new time taken** is $50 - 7.5 = $ **42.5 minutes**.
Alternatively, if the time is reduced by 15%, then it now takes 85% of the original time, i.e.

$$85\% \text{ of } 50 = \dfrac{85}{10} \times 50 = \dfrac{4250}{100}$$

$$= \textbf{42.5 minutes}, \text{ as above.}$$

Problem 20. Find 12.5% of £378

12.5% of £378 means $\dfrac{12.5}{100} \times 378$, since per cent means 'per hundred'.

Hence 12.5% of £378 $= \dfrac{\overset{1}{\cancel{12.5}}}{\underset{8}{\cancel{100}}} \times 378 = \dfrac{378}{8} = $ **£47.25**

Now try the following exercise

Exercise 7 Further problems percentages (Answers on page 299)

1. Convert to percentages:

    (a) 0.057 (b) 0.374 (c) 1.285

2. Express as percentages, correct to 3 significant figures:

    (a) $\dfrac{7}{33}$ (b) $\dfrac{19}{24}$ (c) $1\dfrac{11}{16}$

3. Calculate correct to 4 significant figures:

    (a) 18% of 2758 tonnes (b) 47% of 18.42 grams

    (c) 147% of 14.1 seconds

4. When 1600 bolts are manufactured, 36 are unsatisfactory. Determine the percentage unsatisfactory.

5. Express:

    (a) 140 kg as a percentage of 1 t

    (b) 47 s as a percentage of 5 min

    (c) 13.4 cm as a percentage of 2.5 m

6. A block of monel alloy consists of 70% nickel and 30% copper. If it contains 88.2 g of nickel, determine the mass of copper in the block.

7. A drilling machine should be set to 250 rev/min. The nearest speed available on the machine is 268 rev/min. Calculate the percentage over-speed.

8. 2 kg of a compound contains 30% of element A, 45% of element B and 25% of element C. Determine the masses of the three elements present.

3

Indices and standard form

At the end of this chapter you should be able to:

- understand and use the laws of indices with numbers
- understand and use standard form in calculations

3.1 Indices

The lowest factors of 2000 are $2 \times 2 \times 2 \times 2 \times 5 \times 5 \times 5$. These factors are written as $2^4 \times 5^3$, where 2 and 5 are called **bases** and the numbers 4 and 3 are called **indices**.

When an index is an integer it is called a **power**. Thus, 2^4 is called 'two to the power of four', and has a base of 2 and an index of 4. Similarly, 5^3 is called 'five to the power of 3' and has a base of 5 and an index of 3.

Special names may be used when the indices are 2 and 3, these being called 'squared' and 'cubed', respectively. Thus 7^2 is called 'seven squared' and 9^3 is called 'nine cubed'. When no index is shown, the power is 1, i.e. 2 means 2^1

Reciprocal

The **reciprocal** of a number is when the index is -1 and its value is given by 1 divided by the base. Thus the reciprocal of 2 is 2^{-1} and its value is $\frac{1}{2}$ or 0.5. Similarly, the reciprocal of 5 is 5^{-1} which means $\frac{1}{5}$ or 0.2

Square root

The **square root** of a number is when the index is $\frac{1}{2}$, and the square root of 2 is written as $2^{1/2}$ or $\sqrt{2}$. The value of a square root is the value of the base which when multiplied by itself gives the number. Since $3 \times 3 = 9$, then $\sqrt{9} = 3$. However, $(-3) \times (-3) = 9$, so $\sqrt{9} = -3$. There are always two answers when finding the square root of a number and

this is shown by putting both a $+$ and a $-$ sign in front of the answer to a square root problem. Thus $\sqrt{9} = \pm 3$ and $4^{1/2} = \sqrt{4} = \pm 2$, and so on.

Laws of indices

When simplifying calculations involving indices, certain basic rules or laws can be applied, called the **laws of indices**. These are given below.

(i) When multiplying two or more numbers having the same base, the indices are added. Thus
$$3^2 \times 3^4 = 3^{2+4} = 3^6$$

(ii) When a number is divided by a number having the same base, the indices are subtracted. Thus $\dfrac{3^5}{3^2} = 3^{5-2} = 3^3$

(iii) When a number which is raised to a power is raised to a further power, the indices are multiplied. Thus
$$(3^5)^2 = 3^{5 \times 2} = 3^{10}$$

(iv) When a number has an index of 0, its value is 1. Thus
$$3^0 = 1$$

(v) A number raised to a negative power is the reciprocal of that number raised to a positive power. Thus $3^{-4} = \dfrac{1}{3^4}$

Similarly, $\dfrac{1}{2^{-3}} = 2^3$

(vi) When a number is raised to a fractional power the denominator of the fraction is the root of the number and the numerator is the power.

Thus $\quad 8^{2/3} = \sqrt[3]{8^2} = (2)^2 = 4$

and $\quad 25^{1/2} = \sqrt[2]{25^1} = \sqrt{25^1} = \pm 5$

(Note that $\sqrt{} \equiv \sqrt[2]{}$)

3.2 Worked problems on indices

Problem 1.   Evaluate: (a) $5^2 \times 5^3$ (b) $2 \times 2^2 \times 2^5$

From law (i):

(a) $5^2 \times 5^3 = 5^{(2+3)} = 5^5 = 5 \times 5 \times 5 \times 5 \times 5 = \textbf{3125}$

(b) $2 \times 2^2 \times 2^5 = 2^{(1+2+5)} = 2^8 = \textbf{256}$

Problem 2.   Find the value of: (a) $\dfrac{7^5}{7^3}$ and (b) $\dfrac{5^7}{5^4}$

From law (ii):

(a) $\dfrac{7^5}{7^3} = 7^{(5-3)} = 7^2 = \textbf{49}$

(b) $\dfrac{5^7}{5^4} = 5^{(7-4)} = 5^3 = \textbf{125}$

Problem 3.   Evaluate: (a) $5^2 \times 5^3 \div 5^4$ and
(b) $(3 \times 3^5) \div (3^2 \times 3^3)$

From laws (i) and (ii):

(a) $5^2 \times 5^3 \div 5^4 = \dfrac{5^2 \times 5^3}{5^4} = \dfrac{5^{(2+3)}}{5^4}$

$= \dfrac{5^5}{5^4} = 5^{(5-4)} = 5^1 = \textbf{5}$

(b) $(3 \times 3^5) \div (3^2 \times 3^3) = \dfrac{3 \times 3^5}{3^2 \times 3^3} = \dfrac{3^{(1+5)}}{3^{(2+3)}}$

$= \dfrac{3^6}{3^5} = 3^{6-5} = 3^1 = \textbf{3}$

Problem 4.   Simplify: (a) $(2^3)^4$ (b) $(3^2)^5$, expressing
the answers in index form.

From law (iii):

(a) $(2^3)^4 = 2^{3 \times 4} = \textbf{2}^{\textbf{12}}$ (b) $(3^2)^5 = 3^{2 \times 5} = \textbf{3}^{\textbf{10}}$

Problem 5.   Evaluate: $\dfrac{(10^2)^3}{10^4 \times 10^2}$

From the laws of indices:

$$\dfrac{(10^2)^3}{10^4 \times 10^2} = \dfrac{10^{(2 \times 3)}}{10^{(4+2)}} = \dfrac{10^6}{10^6} = 10^{6-6} = 10^0 = \textbf{1}$$

Problem 6.   Find the value of

(a) $\dfrac{2^3 \times 2^4}{2^7 \times 2^5}$ and   (b) $\dfrac{(3^2)^3}{3 \times 3^9}$

From the laws of indices:

(a) $\dfrac{2^3 \times 2^4}{2^7 \times 2^5} = \dfrac{2^{(3+4)}}{2^{(7+5)}} = \dfrac{2^7}{2^{12}} = 2^{7-12} = 2^{-5} = \dfrac{1}{2^5} = \dfrac{\textbf{1}}{\textbf{32}}$

(b) $\dfrac{(3^2)^3}{3 \times 3^9} = \dfrac{3^{2 \times 3}}{3^{1+9}} = \dfrac{3^6}{3^{10}} = 3^{6-10} = 3^{-4} = \dfrac{1}{3^4} = \dfrac{\textbf{1}}{\textbf{81}}$

Now try the following exercise

**Exercise 8   Further problems on indices (Answers on
page 300)**

Simplify the following expressions, expressing the answers in index form and with positive indices:

1. (a) $3^3 \times 3^4$ (b) $4^2 \times 4^3 \times 4^4$

2. (a) $2^3 \times 2 \times 2^2$ (b) $7^2 \times 7^4 \times 7 \times 7^3$

3. (a) $\dfrac{2^4}{2^3}$ (b) $\dfrac{3^7}{3^2}$

4. (a) $5^6 \div 5^3$ (b) $\dfrac{7^{13}}{7^{10}}$

5. (a) $(7^2)^3$ (b) $(3^3)^2$

6. (a) $(15^3)^5$ (b) $(17^2)^4$

7. (a) $\dfrac{2^2 \times 2^3}{2^4}$ (b) $\dfrac{3^7 \times 3^4}{3^5}$

8. (a) $\dfrac{(9 \times 3^2)^3}{(3 \times 27)^2}$ (b) $\dfrac{(16 \times 4)^2}{(2 \times 8)^3}$

3.3 Further worked problems on indices

Problem 7.   Evaluate (a) $4^{1/2}$ (b) $16^{3/4}$ (c) $27^{2/3}$
(d) $9^{-1/2}$

(a) $4^{1/2} = \sqrt{4} = \pm\textbf{2}$

(b) $16^{3/4} = \sqrt[4]{16^3} = (2)^3 = \textbf{8}$

(Note that it does not matter whether the 4th root of 16 is found first or whether 16 cubed is found first – the same answer will result)

(c) $27^{2/3} = \sqrt[3]{27^2} = (3)^2 = \textbf{9}$

(d) $9^{-1/2} = \dfrac{1}{9^{1/2}} = \dfrac{1}{\sqrt{9}} = \dfrac{1}{\pm 3} = \pm\dfrac{1}{3}$

Problem 8.   Evaluate $\dfrac{3^3 \times 5^7}{5^3 \times 3^4}$

The laws of indices only apply to terms **having the same base**. Grouping terms having the same base, and then applying the laws of indices to each of the groups independently gives:

$$\frac{3^3 \times 5^7}{5^3 \times 3^4} = \frac{3^3}{3^4} \times \frac{5^7}{5^3} = 3^{(3-4)} \times 5^{(7-3)}$$

$$= 3^{-1} \times 5^4 = \frac{5^4}{3^1} = \frac{625}{3} = 208\frac{1}{3}$$

Problem 9.   Find the value of $\dfrac{2^3 \times 3^5 \times (7^2)^2}{7^4 \times 2^4 \times 3^3}$

$$\frac{2^3 \times 3^5 \times (7^2)^2}{7^4 \times 2^4 \times 3^3} = 2^{3-4} \times 3^{5-3} \times 7^{2\times2-4}$$

$$= 2^{-1} \times 3^2 \times 7^0 = \frac{1}{2} \times 3^2 \times 1$$

$$= \frac{9}{2} = 4\frac{1}{2}$$

Problem 10.   Evaluate: $\dfrac{4^{1.5} \times 8^{1/3}}{2^2 \times 32^{-2/5}}$

$4^{1.5} = 4^{3/2} = \sqrt{4^3} = 2^3 = 8$, $8^{1/3} = \sqrt[3]{8} = 2$, $2^2 = 4$

$$32^{-2/5} = \frac{1}{32^{2/5}} = \frac{1}{\sqrt[5]{32^2}} = \frac{1}{2^2} = \frac{1}{4}$$

Hence $\dfrac{4^{1.5} \times 8^{1/3}}{2^2 \times 32^{-2/5}} = \dfrac{8 \times 2}{4 \times \frac{1}{4}} = \dfrac{16}{1} = \mathbf{16}$

Alternatively,

$$\frac{4^{1.5} \times 8^{1/3}}{2^2 \times 32^{-2/5}} = \frac{[(2)^2]^{3/2} \times (2^3)^{1/3}}{2^2 \times (2^5)^{-2/5}} = \frac{2^3 \times 2^1}{2^2 \times 2^{-2}}$$

$$= 2^{3+1-2-(-2)} = 2^4 = \mathbf{16}$$

Problem 11.   Simplify

$$\frac{\left(\dfrac{4}{3}\right)^3 \times \left(\dfrac{3}{5}\right)^{-2}}{\left(\dfrac{2}{5}\right)^{-3}}$$

giving the answer with positive indices.

A fraction raised to a power means that both the numerator and the denominator of the fraction are raised to that power,

i.e. $\left(\dfrac{4}{3}\right)^3 = \dfrac{4^3}{3^3}$

A fraction raised to a negative power has the same value as the inverse of the fraction raised to a positive power.

Thus, $\left(\dfrac{3}{5}\right)^{-2} = \dfrac{1}{\left(\dfrac{3}{5}\right)^2} = \dfrac{1}{\dfrac{3^2}{5^2}} = 1 \times \dfrac{5^2}{3^2} = \dfrac{5^2}{3^2}$

Similarly, $\left(\dfrac{2}{5}\right)^{-3} = \left(\dfrac{5}{2}\right)^3 = \dfrac{5^3}{2^3}$

Thus, $\dfrac{\left(\dfrac{4}{3}\right)^3 \times \left(\dfrac{3}{5}\right)^{-2}}{\left(\dfrac{2}{5}\right)^{-3}} = \dfrac{\dfrac{4^3}{3^3} \times \dfrac{5^2}{3^2}}{\dfrac{5^3}{2^3}}$

$$= \frac{4^3}{3^3} \times \frac{5^2}{3^2} \times \frac{2^3}{5^3}$$

$$= \frac{(2^2)^3 \times 2^3}{3^{(3+2)} \times 5^{(3-2)}} = \frac{2^9}{3^5 \times 5}$$

Now try the following exercise

Exercise 9   Further problems on indices (Answers on page 300)

In Problems 1 and 2, simplify the expressions given, expressing the answers in index form and with positive indices:

1. (a) $\dfrac{3^3 \times 5^2}{5^4 \times 3^4}$ (b) $\dfrac{7^{-2} \times 3^{-2}}{3^5 \times 7^4 \times 7^{-3}}$

2. (a) $\dfrac{4^2 \times 9^3}{8^3 \times 3^4}$ (b) $\dfrac{8^{-2} \times 5^2 \times 3^{-4}}{25^2 \times 2^4 \times 9^{-2}}$

3. Evaluate (a) $\left(\dfrac{1}{3^2}\right)^{-1}$ (b) $81^{0.25}$

   (c) $16^{(-1/4)}$ (d) $\left(\dfrac{4}{9}\right)^{1/2}$

In Problems 4 to 6, evaluate the expressions given.

4. $\dfrac{\left(\dfrac{1}{2}\right)^3 \times \left(\dfrac{2}{3}\right)^{-2}}{\left(\dfrac{3}{5}\right)^2}$ 5. $\dfrac{\left(\dfrac{4}{3}\right)^4}{\left(\dfrac{2}{9}\right)^2}$

6. $\dfrac{(3^2)^{3/2} \times (8^{1/3})^2}{(3)^2 \times (4^3)^{1/2} \times (9)^{-1/2}}$

3.4 Standard form

A number written with one digit to the left of the decimal point and multiplied by 10 raised to some power is said to be written in **standard form**. Thus: 5837 is written as 5.837×10^3 in standard form, and 0.0415 is written as 4.15×10^{-2} in standard form.

When a number is written in standard form, the first factor is called the **mantissa** and the second factor is called the **exponent**. Thus the number 5.8×10^3 has a mantissa of 5.8 and an exponent of 10^3

The laws of indices are used when multiplying or dividing numbers given in standard form. For example,

$$(2.5 \times 10^3) \times (5 \times 10^2) = (2.5 \times 5) \times (10^{3+2})$$
$$= 12.5 \times 10^5 \text{ or } 1.25 \times 10^6$$

Similarly, $\dfrac{6 \times 10^4}{1.5 \times 10^2} = \dfrac{6}{1.5} \times (10^{4-2}) = 4 \times 10^2$

3.5 Worked problems on standard form

> *Problem 12.* Express in standard form:
>
> (a) 38.71 (b) 3746 (c) 0.0124

For a number to be in standard form, it is expressed with only one digit to the left of the decimal point. Thus:

(a) 38.71 must be divided by 10 to achieve one digit to the left of the decimal point and it must also be multiplied by 10 to maintain the equality, i.e.

$$38.71 = \frac{38.71}{10} \times 10 = \mathbf{3.871 \times 10} \text{ in standard form}$$

(b) $3746 = \dfrac{3746}{1000} \times 1000 = \mathbf{3.746 \times 10^3}$ in standard form.

(c) $0.0124 = 0.0124 \times \dfrac{100}{100} = \dfrac{1.24}{100} = \mathbf{1.24 \times 10^{-2}}$ in standard form.

> *Problem 13.* Express the following numbers, which are in standard form, as decimal numbers:
>
> (a) 1.725×10^{-2} (b) 5.491×10^4 (c) 9.84×10^0

(a) $1.725 \times 10^{-2} = \dfrac{1.725}{100} = \mathbf{0.01725}$

(b) $5.491 \times 10^4 = 5.491 \times 10000 = \mathbf{54\,910}$

(c) $9.84 \times 10^0 = 9.84 \times 1 = \mathbf{9.84}$ (since $10^0 = 1$)

> *Problem 14.* Express in standard form, correct to 3 significant figures:
>
> (a) $\dfrac{3}{8}$ (b) $19\dfrac{2}{3}$ (c) $741\dfrac{9}{16}$

(a) $\dfrac{3}{8} = 0.375$, and expressing it in standard form gives:

$$0.375 = \mathbf{3.75 \times 10^{-1}}$$

(b) $19\dfrac{2}{3} = 19.\dot{6} = \mathbf{1.97 \times 10}$ in standard form, correct to 3 significant figures.

(c) $741\dfrac{9}{16} = 741.5625 = \mathbf{7.42 \times 10^2}$ in standard form, correct to 3 significant figures.

> *Problem 15.* Express the following numbers, given in standard form, as fractions or mixed numbers:
>
> (a) 2.5×10^{-1} (b) 6.25×10^{-2} (c) 1.354×10^2

(a) $2.5 \times 10^{-1} = \dfrac{2.5}{10} = \dfrac{25}{100} = \dfrac{\mathbf{1}}{\mathbf{4}}$

(b) $6.25 \times 10^{-2} = \dfrac{6.25}{100} = \dfrac{625}{10\,000} = \dfrac{\mathbf{1}}{\mathbf{16}}$

(c) $1.354 \times 10^2 = 135.4 = 135\dfrac{4}{10} = \mathbf{135\dfrac{2}{5}}$

Now try the following exercise

> **Exercise 10 Further problems on standard form (Answers on page 300)**
>
> In Problems 1 to 5, express in standard form:
>
> 1. (a) 73.9 (b) 197.72
>
> 2. (a) 2748 (b) 33170
>
> 3. (a) 0.2401 (b) 0.0174
>
> 4. (a) 1702.3 (b) 10.04
>
> 5. (a) $\dfrac{1}{2}$ (b) $11\dfrac{7}{8}$ (c) $130\dfrac{3}{5}$ (d) $\dfrac{1}{32}$
>
> In Problems 6 and 7, express the numbers given as integers or decimal fractions:
>
> 6. (a) 1.01×10^3 (b) 9.327×10^2 (c) 5.41×10^4
>   (d) 7×10^0
>
> 7. (a) 3.89×10^{-2} (b) 6.741×10^{-1} (c) 8×10^{-3}

3.6   Further worked problems on standard form

> *Problem 16.*   Find the value of
>
> (a) $7.9 \times 10^{-2} - 5.4 \times 10^{-2}$
>
> (b) $8.3 \times 10^3 + 5.415 \times 10^3$
>
> (c) $9.293 \times 10^2 + 1.3 \times 10^3$ expressing the answers in standard form.

Numbers having the same exponent can be added or subtracted by adding or subtracting the mantissae and keeping the exponent the same. Thus:

(a) $7.9 \times 10^{-2} - 5.4 \times 10^{-2} = (7.9 - 5.4) \times 10^{-2}$

$$= \mathbf{2.5 \times 10^{-2}}$$

(b) $8.3 \times 10^3 + 5.415 \times 10^3 = (8.3 + 5.415) \times 10^3$

$$= 13.715 \times 10^3$$

$$= \mathbf{1.3715 \times 10^4}$$

in standard form.

(c) $9.293 \times 10^2 + 1.3 \times 10^3 = 929.3 + 1300$

$$= 2229.3 = \mathbf{2.2293 \times 10^3}$$

in standard form.

> *Problem 17.*   Evaluate
>
> (a) $(3.75 \times 10^3)(6 \times 10^4)$ and
>
> (b) $\dfrac{3.5 \times 10^5}{7 \times 10^2}$ expressing answers in standard form.

(a) $(3.75 \times 10^3)(6 \times 10^4) = (3.75 \times 6)(10^{3+4})$

$$= 22.50 \times 10^7 = \mathbf{2.25 \times 10^8}$$

(b) $\dfrac{3.5 \times 10^5}{7 \times 10^2} = \dfrac{3.5}{7} \times 10^{5-2}$

$$= 0.5 \times 10^3 = \mathbf{5 \times 10^2}$$

Now try the following exercise

> **Exercise 11   Further problems on standard form (Answers on page 300)**
>
> In Problems 1 to 4, find values of the expressions given, stating the answers in standard form:

1. (a) $3.7 \times 10^2 + 9.81 \times 10^2$

   (b) $8.414 \times 10^{-2} - 2.68 \times 10^{-2}$

2. (a) $4.831 \times 10^2 + 1.24 \times 10^3$

   (b) $3.24 \times 10^{-3} - 1.11 \times 10^{-4}$

3. (a) $(4.5 \times 10^{-2})(3 \times 10^3)$

   (b) $2 \times (5.5 \times 10^4)$

4. (a) $\dfrac{6 \times 10^{-3}}{3 \times 10^{-5}}$

   (b) $\dfrac{(2.4 \times 10^3)(3 \times 10^{-2})}{(4.8 \times 10^4)}$

5. Write the following statements in standard form.

   (a) The density of aluminium is $2710\,\text{kg m}^{-3}$

   (b) Poisson's ratio for gold is 0.44

   (c) The impedance of free space is $376.73\,\Omega$

   (d) The electron rest energy is $0.511\,\text{MeV}$

   (e) The normal volume of a perfect gas is $0.02241\,\text{m}^3\,\text{mol}^{-1}$

Assignment 1

This assignment covers the material contained in Chapters 1 to 3. The marks for each question are shown in brackets at the end of each question.

1. Evaluate the following:

   (a) $\dfrac{120}{15} - 133 \div 19 + (2 \times 17)$

   (b) $\dfrac{2}{5} - \dfrac{1}{15} + \dfrac{5}{6}$

   (c) $49.31 - 97.763 + 9.44 - 0.079$ (8)

2. Evaluate, by long division $\dfrac{4675}{11}$ (3)

3. Find (a) the highest common factor, and (b) the lowest common multiple of the following numbers:
   15   40   75   120   (6)

4. Simplify: $2\dfrac{2}{3} \div 3\dfrac{1}{3}$ (3)

5. A piece of steel, 1.69 m long, is cut into three pieces in the ratio 2 to 5 to 6. Determine, in centimetres, the lengths of the three pieces.   (4)

6. Evaluate $\dfrac{576.29}{19.3}$ (a) correct to 4 significant figures

(b) correct to 1 decimal place (4)

7. Determine, correct to 1 decimal places, 57% of 17.64 g. (2)

8. Express 54.7 mm as a percentage of 1.15 m, correct to 3 significant figures. (3)

9. Evaluate the following:

(a) $\dfrac{2^3 \times 2 \times 2^2}{2^4}$ (b) $\dfrac{(2^3 \times 16)^2}{(8 \times 2)^3}$

(c) $\left(\dfrac{1}{4^2}\right)^{-1}$ (d) $(27)^{-1/3}$ (11)

10. Express the following in standard form:

(a) 1623   (b) 0.076   (c) $145\frac{2}{5}$ (6)

4

Calculations and evaluation of formulae

At the end of this chapter you should be able to:

- appreciate the type of errors in calculations
- determine a sensible approximation of a calculation (i.e. without using a calculator)
- use a calculator in a range of common calculations
- perform calculations from conversion tables
- evaluate given formulae using a calculator

4.1 Errors and approximations

(i) In all problems in which the measurement of distance, time, mass or other quantities occurs, an exact answer cannot be given; only an answer which is correct to a stated degree of accuracy can be given. To take account of this an **error due to measurement** is said to exist.

(ii) To take account of measurement errors it is usual to limit answers so that the result given is **not more than one significant figure greater than the least accurate number given in the data**.

(iii) **Rounding-off errors** can exist with decimal fractions. For example, to state that $\pi = 3.142$ is not strictly correct, but '$\pi = 3.142$ correct to 4 significant figures' is a true statement. (Actually, $\pi = 3.14159265\ldots$).

(iv) It is possible, through an incorrect procedure, to obtain the wrong answer to a calculation. This type of error is known as **a blunder**.

(v) An **order of magnitude error** is said to exist if incorrect positioning of the decimal point occurs after a calculation has been completed.

(vi) Blunders and order of magnitude errors can be reduced by determining **approximate values of calculations**. Answers which do not seem feasible must

be checked and the calculation must be repeated as necessary. An engineer will often need to make a quick mental approximation for a calculation. For example, $\dfrac{49.1 \times 18.4 \times 122.1}{61.2 \times 38.1}$ may be approximated to $\dfrac{50 \times 20 \times 120}{60 \times 40}$ and then, by cancelling,

$$\frac{50 \times \overset{1}{\cancel{20}} \times \cancel{120}\,\cancel{2}^{\,1}}{\underset{1}{\cancel{60}} \times \cancel{40}\,\cancel{2}_{\,1}} = 50.$$ An accurate answer somewhere

between 45 and 55 could therefore be expected. Certainly an answer around 500 or 5 would be a surprise. Actually, by calculator $\dfrac{49.1 \times 18.4 \times 122.1}{61.2 \times 38.1} = 47.31$, correct to 4 significant figures.

Problem 1. The area A of a triangle is given by $A = \frac{1}{2}bh$. The base b when measured is found to be 3.26 cm, and the perpendicular height h is 7.5 cm. Determine the area of the triangle.

Area of triangle $= \frac{1}{2}bh = \frac{1}{2} \times 3.26 \times 7.5 = 12.225\,\text{cm}^2$ (by calculator).

The approximate value is $\frac{1}{2} \times 3 \times 8 = 12\,\text{cm}^2$, so there are no obvious blunder or magnitude errors. However, it is not usual in a measurement type problem to state the answer to an accuracy greater than 1 significant figure more than the least accurate number in the data; this is 7.5 cm, so the result should not have more than 3 significant figures.

Thus **area of triangle $= 12.2\,\text{cm}^2$**, correct to 3 significant figures.

Problem 2. State which type of error has been made in the following statements:

(a) $72 \times 31.429 = 2262.9$

(b) $16 \times 0.08 \times 7 = 89.6$

(c) $11.714 \times 0.0088 = 0.3247$ correct to 4 decimal places

(d) $\dfrac{29.74 \times 0.0512}{11.89} = 0.12$, correct to 2 significant figures

(a) $72 \times 31.429 = 2262.888$ (by calculator), hence a **rounding-off error** has occurred. The answer should have stated: $72 \times 31.429 = 2262.9$, correct to 5 significant figures or 2262.9, correct to 1 decimal place.

(b) $16 \times 0.08 \times 7 = 16 \times \dfrac{\cancel{8}^{\,2}}{\cancel{100}_{\,25}} \times 7 = \dfrac{32 \times 7}{25}$

$$= \dfrac{224}{25} = 8\dfrac{24}{25} = 8.96$$

Hence an **order of magnitude** error has occurred.

(c) 11.714×0.0088 is approximately equal to $12 \times 9 \times 10^{-3}$, i.e. about 108×10^{-3} or 0.108. Thus a **blunder** has been made.

(d) $\dfrac{29.74 \times 0.0512}{11.89} \approx \dfrac{30 \times 5 \times 10^{-2}}{12} = \dfrac{150}{12 \times 10^2}$

$$= \dfrac{15}{120} = \dfrac{1}{8} \quad \text{or} \quad 0.125$$

hence no order of magnitude error has occurred. However, $\dfrac{29.74 \times 0.0512}{11.89} = 0.128$ correct to 3 significant figures, which equals 0.13 correct to 2 significant figures.

Hence a **rounding-off error** has occurred.

Problem 3. Without using a calculator, determine an approximate value of

(a) $\dfrac{11.7 \times 19.1}{9.3 \times 5.7}$ (b) $\dfrac{2.19 \times 203.6 \times 17.91}{12.1 \times 8.76}$

(a) $\dfrac{11.7 \times 19.1}{9.3 \times 5.7}$ is approximately equal to $\dfrac{10 \times 20}{10 \times 5}$, i.e. about **4**

(By calculator, $\dfrac{11.7 \times 19.1}{9.3 \times 5.7} = \mathbf{4.22}$, correct to 3 significant figures).

(b) $\dfrac{2.19 \times 203.6 \times 17.91}{12.1 \times 8.76} \approx \dfrac{2 \times \cancel{200}^{\,2} \times 20}{_1\cancel{10} \times \cancel{10}_{\,1}} = 2 \times 2 \times 20$ after cancelling

i.e. $\dfrac{2.19 \times 203.6 \times 17.91}{12.1 \times 8.76} \approx \mathbf{80}$

(By calculator, $\dfrac{2.19 \times 203.6 \times 17.91}{12.1 \times 8.76} = \mathbf{75.3}$, correct to 3 significant figures.)

Now try the following exercise

Exercise 12 Further problems on errors (Answers on page 300)

In Problems 1 to 5 state which type of error, or errors, have been made:

1. $25 \times 0.06 \times 1.4 = 0.21$ 2. $137 \times 6.842 = 937.4$

3. $\dfrac{24 \times 0.008}{12.6} = 10.42$ 4. $\dfrac{4.6 \times 0.07}{52.3 \times 0.274} = 0.225$

5. For a gas $pV = c$. When pressure $p = 103400\,\text{Pa}$ and $V = 0.54\,\text{m}^3$ then $c = 55836\,\text{Pa}\,\text{m}^3$

In Problems 6 to 8, evaluate the expressions approximately, without using a calculator.

6. 4.7×6.3 7. $\dfrac{2.87 \times 4.07}{6.12 \times 0.96}$ 8. $\dfrac{72.1 \times 1.96 \times 48.6}{139.3 \times 5.2}$

4.2 Use of calculator

The most modern aid to calculations is the pocket-sized electronic calculator. With one of these, calculations can be quickly and accurately performed, correct to about 9 significant figures. The scientific type of calculator has made the use of tables and logarithms largely redundant.

To help you to become competent at using your calculator check that you agree with the answers to the following problems:

Problem 4. Evaluate the following, correct to 4 significant figures:

(a) $4.7826 + 0.02713$ (b) $17.6941 - 11.8762$

(a) $4.7826 + 0.02713 = 4.80973 = \mathbf{4.810}$, correct to 4 significant figures.

(b) $17.6941 - 11.8762 = 5.8179 = \mathbf{5.818}$, correct to 4 significant figures.

Problem 5. Evaluate the following, correct to 4 decimal places:

(a) $46.32 \times 97.17 \times 0.01258$ (b) $\dfrac{4.621}{23.76}$

(a) $46.32 \times 97.17 \times 0.01258 = 56.6215031\ldots = \mathbf{56.6215}$, correct to 4 decimal places.

(b) $\dfrac{4.621}{23.76} = 0.19448653\ldots = \mathbf{0.1945}$, correct to 4 decimal places.

Problem 6.   Evaluate the following, correct to 3 decimal places:

(a) $\dfrac{1}{0.0275}$ (b) $\dfrac{1}{4.92} + \dfrac{1}{1.97}$

(a) $\dfrac{1}{0.0275} = 36.3636363\ldots = \mathbf{36.364}$, correct to 3 decimal places.

(b) $\dfrac{1}{4.92} + \dfrac{1}{1.97} = 0.71086624\ldots = \mathbf{0.711}$, correct to 3 decimal places.

Problem 7.   Evaluate the following, expressing the answers in standard form, correct to 4 significant figures:

(a) $(0.00451)^2$ (b) $631.7 - (6.21 + 2.95)^2$

(a) $(0.00451)^2 = 2.03401 \times 10^{-5} = \mathbf{2.034 \times 10^{-5}}$, correct to 4 significant figures.

(b) $631.7 - (6.21 + 2.95)^2 = 547.7944 = 5.477944 \times 10^2 = \mathbf{5.478 \times 10^2}$, correct to 4 significant figures.

Problem 8.   Evaluate the following, correct to 4 significant figures:

(a) $\sqrt{5.462}$ (b) $\sqrt{54.62}$

(a) $\sqrt{5.462} = 2.3370922\ldots = \mathbf{2.337}$, correct to 4 significant figures.

(b) $\sqrt{54.62} = 7.39053448\ldots = \mathbf{7.391}$, correct to 4 significant figures.

Problem 9.   Evaluate the following, correct to 3 decimal places:

(a) $\sqrt{52.91} - \sqrt{31.76}$ (b) $\sqrt{(1.6291 \times 10^4)}$

(a) $\sqrt{52.91} - \sqrt{31.76} = 1.63832491\ldots = \mathbf{1.638}$, correct to 3 decimal places.

(b) $\sqrt{(1.6291 \times 10^4)} = \sqrt{16291} = 127.636201\ldots = \mathbf{127.636}$, correct to 3 decimal places.

Problem 10.   Evaluate the following, correct to 4 significant figures:

(a) 4.72^3 (b) $(0.8316)^4$

(a) $4.72^3 = 105.15404\ldots = \mathbf{105.2}$, correct to 4 significant figures.

(b) $(0.8316)^4 = 0.47825324\ldots = \mathbf{0.4783}$, correct to 4 significant figures.

Now try the following exercise

Exercise 13   Further problems on use of calculator (Answers on page 300)

In Problems 1 to 8, use a calculator to evaluate correct to 4 significant figures:

1. (a) 43.27×12.91 (b) 54.31×0.5724

2. (a) $127.8 \times 0.0431 \times 19.8$ (b) $15.76 \div 4.329$

3. (a) $\dfrac{137.6}{552.9}$ (b) $\dfrac{11.82 \times 1.736}{0.041}$

4. (a) $\dfrac{1}{17.31}$ (b) $\dfrac{1}{0.0346}$

5. (a) 13.6^3 (b) 3.476^4

6. (a) $\sqrt{347.1}$ (b) $\sqrt{0.027}$

7. (a) $\left(\dfrac{24.68 \times 0.0532}{7.412}\right)^3$ (b) $\left(\dfrac{0.2681 \times 41.2^2}{32.6 \times 11.89}\right)^4$

8. (a) $\dfrac{14.32^3}{21.68^2}$ (b) $\dfrac{4.821^3}{17.33^2 - 15.86 \times 11.6}$

9. Evaluate correct to 3 decimal places:

   (a) $\dfrac{29.12}{(5.81)^2 - (2.96)^2}$ (b) $\sqrt{53.98} - \sqrt{21.78}$

4.3   Conversion tables

It is often necessary to make calculations from various conversion tables and charts. Examples include currency exchange rates, imperial to metric unit conversions, train or bus timetables, production schedules and so on.

Problem 11.   Some approximate imperial to metric conversions are shown in Table 4.1.

Table 4.1

length	1 inch = 2.54 cm
	1 mile = 1.61 km
weight	2.2 lb = 1 kg
	(1 lb = 16 oz)
capacity	1.76 pints = 1 litre
	(8 pints = 1 gallon)

Use the table to determine:

(a) the number of millimetres in 9.5 inches,

(b) a speed of 50 miles per hour in kilometres per hour,

(c) the number of miles in 300 km,

(d) the number of kilograms in 30 pounds weight,

(e) the number of pounds and ounces in 42 kilograms (correct to the nearest ounce),

(f) the number of litres in 15 gallons, and

(g) the number of gallons in 40 litres.

(a) 9.5 inches = 9.5 × 2.54 cm = 24.13 cm

   24.13 cm = 24.13 × 10 mm = **241.3 mm**

(b) 50 m.p.h. = 50 × 1.61 km/h = **80.5 km/h**

(c) 300 km = $\dfrac{300}{1.61}$ miles = **186.3 miles**

(d) 30 lb = $\dfrac{30}{2.2}$ kg = **13.64 kg**

(e) 42 kg = 42 × 2.2 lb = 92.4 lb

   0.4 lb = 0.4 × 16 oz = 6.4 oz = 6 oz, correct to the nearest ounce.

   Thus 42 kg = **92 lb 6 oz**, correct to the nearest ounce.

(f) 15 gallons = 15 × 8 pints = 120 pints

   120 pints = $\dfrac{120}{1.76}$ litres = **68.18 litres**

(g) 40 litres = 40 × 1.76 pints = 70.4 pints

   70.4 pints = $\dfrac{70.4}{8}$ gallons = **8.8 gallons**

Now try the following exercise

Exercise 14 Further problems conversion tables and charts (Answers on page 300)

1. Currency exchange rates listed in a newspaper included the following:

France	£1 = 9.60 francs
Japan	£1 = 205 yen
Germany	£1 = 2.85 Deutschmarks
U.S.A.	£1 = $1.65
Spain	£1 = 250 pesetas

   Calculate (a) how many French francs £32.50 will buy, (b) the number of American dollars that can be purchased for £74.80, (c) the pounds sterling which can be exchanged for 14000 yen, (d) the pounds sterling which can be exchanged for 1750 pesetas, and (e) the German Deutschmarks which can be bought for £55.

2. Use Table 4.1 to determine (a) the number of millimetres in 15 inches, (b) a speed of 35 mph in km/h, (c) the number of kilometres in 235 miles, (d) the number of pounds and ounces in 24 kg (correct to the nearest ounce), (e) the number of kilograms in 15 lb, (f) the number of litres in 12 gallons and (g) the number of gallons in 25 litres.

4.4 Evaluation of formulae

The statement $v = u + at$ is said to be a **formula** for v in terms of u, a and t.

v, u, a and t are called **symbols**.

The single term on the left-hand side of the equation, v, is called the **subject of the formulae**.

Provided values are given for all the symbols in a formula except one, the remaining symbol can be made the subject of the formula and may be evaluated by using a calculator.

Problem 12. In an electrical circuit the voltage V is given by Ohm's law, i.e. $V = IR$. Find, correct to 4 significant figures, the voltage when $I = 5.36$ A and $R = 14.76\,\Omega$.

$$V = IR = (5.36)(14.76)$$

Hence **voltage $V = 79.11$ V, correct to 4 significant figures**.

Problem 13. Velocity v is given by $v = u + at$. If $u = 9.86$ m/s, $a = 4.25$ m/s² and $t = 6.84$ s, find v, correct to 3 significant figures.

$$v = u + at = 9.86 + (4.25)(6.84)$$
$$= 9.86 + 29.07$$
$$= 38.93$$

Hence **velocity $v = 38.9$ m/s, correct to 3 significant figures.**

Problem 14. The volume V cm³ of a right circular cone is given by $V = \frac{1}{3}\pi r^2 h$. Given that $r = 4.321$ cm and $h = 18.35$ cm, find the volume, correct to 4 significant figures.

$$V = \tfrac{1}{3}\pi r^2 h = \tfrac{1}{3}\pi (4.321)^2 (18.35)$$
$$= \tfrac{1}{3}\pi (18.671041)(18.35)$$

Hence **volume, $V = 358.8$ cm³, correct to 4 significant figures**.

Problem 15. Force F newtons is given by the formula $F = \dfrac{Gm_1m_2}{d^2}$, where m_1 and m_2 are masses, d their distance apart and G is a constant. Find the value of the force given that $G = 6.67 \times 10^{-11}$, $m_1 = 7.36$, $m_2 = 15.5$ and $d = 22.6$. Express the answer in standard form, correct to 3 significant figures.

$$F = \frac{Gm_1m_2}{d^2} = \frac{(6.67 \times 10^{-11})(7.36)(15.5)}{(22.6)^2}$$

$$= \frac{(6.67)(7.36)(15.5)}{(10^{11})(510.76)} = \frac{1.490}{10^{11}}$$

Hence force $F = 1.49 \times 10^{-11}$ newtons, correct to 3 significant figures.

Problem 16. The time of swing t seconds, of a simple pendulum is given by $t = 2\pi\sqrt{\dfrac{l}{g}}$. Determine the time, correct to 3 decimal places, given that $l = 12.0$ and $g = 9.81$

$$t = 2\pi\sqrt{\frac{l}{g}} = (2)\pi\sqrt{\frac{12.0}{9.81}}$$

$$= (2)\pi\sqrt{1.22324159}$$

$$= (2)\pi(1.106002527)$$

Hence time $t = 6.950$ seconds, correct to 3 decimal places.

Now try the following exercise

Exercise 15   Further problems on evaluation of formulae (Answers on page 300)

1. The circumference C of a circle is given by the formula $C = 2\pi r$. Determine the circumference given $\pi = 3.14$ and $r = 8.40$ mm.

2. The velocity of a body is given by $v = u + at$. The initial velocity u is measured when time t is 15 seconds and found to be 12 m/s. If the acceleration a is 9.81 m/s^2 calculate the final velocity v.

3. Find the distance s, given that $s = \frac{1}{2} gt^2$. Time $t = 0.032$ seconds and acceleration due to gravity $g = 9.81$ m/s^2.

4. The energy stored in a capacitor is given by $E = \frac{1}{2}CV^2$ joules. Determine the energy when capacitance $C = 5 \times 10^{-6}$ farads and voltage $V = 240$ V.

5. Resistance R_2 is given by $R_2 = R_1(1 + \alpha t)$. Find R_2, correct to 4 significant figures, when $R_1 = 220$, $\alpha = 0.00027$ and $t = 75.6$

6. Density $= \dfrac{\text{mass}}{\text{volume}}$. Find the density when the mass is 2.462 kg and the volume is 173 cm^3. Give the answer in units of kg/m^3.

7. Velocity $=$ frequency $\times$ wavelength. Find the velocity when the frequency is 1825 Hz and the wavelength is 0.154 m.

8. Evaluate resistance R_T, given $\dfrac{1}{R_T} = \dfrac{1}{R_1} + \dfrac{1}{R_2} + \dfrac{1}{R_3}$ when $R_1 = 5.5\,\Omega$, $R_2 = 7.42\,\Omega$ and $R_3 = 12.6\,\Omega$.

9. Power $= \dfrac{\text{force} \times \text{distance}}{\text{time}}$. Find the power when a force of 3760 N raises an object a distance of 4.73 m in 35 s.

10. The potential difference, V volts, available at battery terminals is given by $V = E - Ir$. Evaluate V when $E = 5.62$, $I = 0.70$ and $R = 4.30$

11. Given force $F = \frac{1}{2}m(v^2 - u^2)$, find F when $m = 18.3$, $v = 12.7$ and $u = 8.24$

12. The time, t seconds, of oscillation for a simple pendulum is given by $t = 2\pi\sqrt{\dfrac{l}{g}}$. Determine the time when $\pi = 3.142$, $l = 54.32$ and $g = 9.81$

13. Energy, E joules, is given by the formula $E = \frac{1}{2}LI^2$. Evaluate the energy when $L = 5.5$ and $I = 1.2$

14. Distance s metres is given by the formula $s = ut + \frac{1}{2}at^2$. If $u = 9.50$, $t = 4.60$ and $a = -2.50$, evaluate the distance.

5

Algebra

At the end of this chapter you should be able to:

- use the basic operations of addition, subtraction, multiplication and division with algebra
- use the laws of indices to simplify algebraic expressions
- understand the use of brackets in simple algebraic expressions
- factorise simple algebraic expressions
- appreciate and apply the laws of precedence

5.1 Basic operations

Algebra is that part of mathematics in which the relations and properties of numbers are investigated by means of general symbols. For example, the area of a rectangle is found by multiplying the length by the breadth; this is expressed algebraically as $A = l \times b$, where A represents the area, l the length and b the breadth.

The basic laws introduced in arithmetic are generalized in algebra.

Let a, b, c and d represent any four numbers. Then:

(i) $a + (b + c) = (a + b) + c$

(ii) $a(bc) = (ab)c$

(iii) $a + b = b + a$

(iv) $ab = ba$

(v) $a(b + c) = ab + ac$

(vi) $\dfrac{a + b}{c} = \dfrac{a}{c} + \dfrac{b}{c}$

(vii) $(a + b)(c + d) = ac + ad + bc + bd$

> **Problem 1.** Evaluate $3ab - 2bc + abc$ when $a = 1$, $b = 3$ and $c = 5$

Replacing a, b and c with their numerical values gives:

$$3ab - 2bc + abc = 3 \times 1 \times 3 - 2 \times 3 \times 5 + 1 \times 3 \times 5$$
$$= 9 - 30 + 15 = -6$$

> **Problem 2.** Find the value of $4p^2qr^3$, given that
> $$p = 2, q = \tfrac{1}{2} \text{ and } r = 1\tfrac{1}{2}$$

Replacing p, q and r with their numerical values gives:

$$4p^2qr^3 = 4(2)^2 \left(\tfrac{1}{2}\right) \left(\tfrac{3}{2}\right)^3$$
$$= 4 \times 2 \times 2 \times \tfrac{1}{2} \times \tfrac{3}{2} \times \tfrac{3}{2} \times \tfrac{3}{2} = 27$$

> **Problem 3.** Find the sum of $3x$, $2x$, $-x$ and $-7x$

The sum of the positive terms is: $3x + 2x = 5x$

The sum of the negative terms is: $x + 7x = 8x$

Taking the sum of the negative terms from the sum of the positive terms gives:

$$5x - 8x = -3x$$

Alternatively,

$$3x + 2x + (-x) + (-7x) = 3x + 2x - x - 7x = -3x$$

> **Problem 4.** Find the sum of $4a$, $3b$, c, $-2a$, $-5b$ and $6c$

Each symbol must be dealt with individually.

For the '*a*' terms: $+4a - 2a = 2a$

For the '*b*' terms: $+3b - 5b = -2b$

For the '*c*' terms: $+c + 6c = 7c$

Thus $4a + 3b + c + (-2a) + (-5b) + 6c$

$$= 4a + 3b + c - 2a - 5b + 6c$$

$$= \mathbf{2a - 2b + 7c}$$

Problem 5.   Subtract $2x + 3y - 4z$ from $x - 2y + 5z$

$$
\begin{array}{r}
x - 2y + 5z \\
2x + 3y - 4z \\
\hline
\end{array}
$$

Subtracting gives: $\mathbf{-x - 5y + 9z}$

(Note that $+5z - -4z = +5z + 4z = 9z$)

An alternative method of subtracting algebraic expressions is to 'change the signs of the bottom line and add'. Hence:

$$
\begin{array}{r}
x - 2y + 5z \\
-2x - 3y + 4z \\
\hline
\end{array}
$$

Adding gives: $\mathbf{-x - 5y + 9z}$

Problem 6.   Multiply $(2a + 3b)$ by $(a + b)$

Each term in the first expression is multiplied by a, then each term in the first expression is multiplied by b, and the two results are added. The usual layout is shown below.

$$
\begin{array}{r}
2a + 3b \\
a + b \\
\hline
\end{array}
$$

Multiplying by $a \rightarrow$ $2a^2 + 3ab$

Multiplying by $b \rightarrow$ $+ 2ab + 3b^2$

Adding gives: $\mathbf{2a^2 + 5ab + 3b^2}$

Problem 7.   Simplify $2p \div 8pq$

$2p \div 8pq$ means $\dfrac{2p}{8pq}$. This can be reduced by cancelling as in arithmetic.

Thus: $\dfrac{2p}{8pq} = \dfrac{\overset{1}{\cancel{2}} \times \overset{1}{\cancel{p}}}{\underset{4}{\cancel{8}} \times \underset{1}{\cancel{p}} \times q} = \dfrac{\mathbf{1}}{\mathbf{4q}}$

Now try the following exercise

**Exercise 16   Further problems on basic operations
(Answers on page 300)**

1. Find the value of $2xy + 3yz - xyz$, when $x = 2$, $y = -2$ and $z = 4$

2. Evaluate $3pq^2r^3$ when $p = \frac{2}{3}$, $q = -2$ and $r = -1$

3. Find the sum of $3a$, $-2a$, $-6a$, $5a$ and $4a$

4. Add together $2a + 3b + 4c$, $-5a - 2b + c$, $4a - 5b - 6c$

5. Add together $3d + 4e$, $-2e + f$, $2d - 3f$, $4d - e + 2f - 3e$

6. From $4x - 3y + 2z$ subtract $x + 2y - 3z$

7. Multiply $3x + 2y$ by $x - y$

8. Multiply $2a - 5b + c$ by $3a + b$

9. Simplify (i) $3a \div 9ab$ (ii) $4a^2b \div 2a$

5.2   Laws of Indices

The laws of indices are:

(i) $a^m \times a^n = a^{m+n}$ (ii) $\dfrac{a^m}{a^n} = a^{m-n}$

(iii) $(a^m)^n = a^{mn}$ (iv) $a^{m/n} = \sqrt[n]{a^m}$

(v) $a^{-n} = \dfrac{1}{a^n}$ (vi) $a^0 = 1$

Problem 8.   Simplify $a^3b^2c \times ab^3c^5$

Grouping like terms gives: $a^3 \times a \times b^2 \times b^3 \times c \times c^5$

Using the first law of indices gives: $a^{3+1} \times b^{2+3} \times c^{1+5}$

i.e. $a^4 \times b^5 \times c^6 = \mathbf{a^4b^5c^6}$

Problem 9.   Simplify $\dfrac{a^3b^2c^4}{abc^{-2}}$ and evaluate when $a = 3$, $b = \dfrac{1}{8}$ and $c = 2$

Using the second law of indices,

$$\frac{a^3}{a} = a^{3-1} = a^2, \quad \frac{b^2}{b} = b^{2-1} = b$$

and $\dfrac{c^4}{c^{-2}} = c^{4--2} = c^6$

Thus $\dfrac{a^3b^2c^4}{abc^{-2}} = \mathbf{a^2bc^6}$

When $a = 3$, $b = \frac{1}{8}$ and $c = 2$,

$$a^2bc^6 = (3)^2 \left(\tfrac{1}{8}\right)(2)^6 = (9)\left(\tfrac{1}{8}\right)(64) = \mathbf{72}$$

Problem 10. Simplify $(p^3)^{1/2}(q^2)^4$

Using the third law of indices gives:

$$p^{3\times(1/2)}q^{2\times4} = \boldsymbol{p}^{(3/2)}\boldsymbol{q}^{\mathbf{8}} \quad \text{or} \quad \sqrt{\boldsymbol{p^3}}\,\boldsymbol{q}^{\mathbf{8}}$$

Problem 11. Simplify $\dfrac{(mn^2)^3}{(m^{1/2}n^{1/4})^4}$

The brackets indicate that each letter in the bracket must be raised to the power outside.

Using the third law of indices gives:

$$\frac{(mn^2)^3}{(m^{1/2}n^{1/4})^4} = \frac{m^{1\times3}n^{2\times3}}{m^{(1/2)\times4}n^{(1/4)\times4}} = \frac{m^3n^6}{m^2n^1}$$

Using the second law of indices gives:

$$\frac{m^3n^6}{m^2n^1} = m^{3-2}n^{6-1} = \boldsymbol{mn^5}$$

Problem 12. Simplify $(a^3b)(a^{-4}b^{-2})$, expressing the answer with positive indices only.

Using the first law of indices gives: $a^{3+-4}b^{1+-2} = a^{-1}b^{-1}$

Using the fifth law of indices gives: $a^{-1}b^{-1} = \dfrac{1}{a^{+1}b^{+1}} = \dfrac{1}{\boldsymbol{ab}}$

Now try the following exercise

Exercise 17   Further problems on laws of indices (Answers on page 300)

1. Simplify $(x^2y^3z)(x^3yz^2)$ and evaluate when $x = \frac{1}{2}$, $y = 2$ and $z = 3$

2. Simplify $(a^{3/2}bc^{-3})(a^{1/2}b^{-1/2}c)$ and evaluate when $a = 3$, $b = 4$ and $c = 2$

3. Simplify $\dfrac{a^5bc^3}{a^2b^3c^2}$ and evaluate when $a = \dfrac{3}{2}$, $b = \dfrac{1}{2}$ and $c = \dfrac{2}{3}$

In Problems 4 to 5, simplify the given expressions:

4. $(a^2)^{1/2}(b^2)^3(c^{1/2})^3$ 5. $\dfrac{(abc)^2}{(a^2b^{-1}c^{-3})^3}$

5.3 Brackets and factorisation

When two or more terms in an algebraic expression contain a common factor, then this factor can be shown outside of a **bracket**. For example

$$ab + ac = a(b + c)$$

which is simply the reverse of law (v) of algebra on page 23, and

$$6px + 2py - 4pz = 2p(3x + y - 2z)$$

This process is called **factorisation**.

Problem 13. Remove the brackets and simplify the expression $(3a + b) + 2(b + c) - 4(c + d)$

Both b and c in the second bracket have to be multiplied by 2, and c and d in the third bracket by -4 when the brackets are removed. Thus:

$$(3a + b) + 2(b + c) - 4(c + d)$$
$$= 3a + b + 2b + 2c - 4c - 4d$$

Collecting similar terms together gives: $\mathbf{3a + 3b - 2c - 4d}$

Problem 14. Simplify $a^2 - (2a - ab) - a(3b + a)$

When the brackets are removed, both $2a$ and $-ab$ in the first bracket must be multiplied by -1 and both $3b$ and a in the second bracket by $-a$. Thus

$$a^2 - (2a - ab) - a(3b + a)$$
$$= a^2 - 2a + ab - 3ab - a^2$$

Collecting similar terms together gives: $-2a - 2ab$

Since $-2a$ is a common factor, the answer can be expressed as $-\mathbf{2a\,(1 + b)}$

Problem 15. Simplify $(a + b)(a - b)$

Each term in the second bracket has to be multiplied by each term in the first bracket. Thus:

$$(a + b)(a - b) = a(a - b) + b(a - b)$$
$$= a^2 - ab + ab - b^2 = \boldsymbol{a^2 - b^2}$$

Alternatively

$$\begin{array}{rr} a & +\ b \\ a & -\ b \\ \hline \end{array}$$

Multiplying by $a \rightarrow \quad a^2 + ab$

Multiplying by $-b \rightarrow \quad -ab - b^2$

Adding gives: $a^2 \qquad - b^2$

Problem 16. Remove the brackets from the expression $(x - 2y)(3x + y^2)$

$$(x - 2y)(3x + y^2) = x(3x + y^2) - 2y(3x + y^2)$$
$$= 3x^2 + xy^2 - 6xy - 2y^3$$

Problem 17. Simplify $(3x - 3y)^2$

$$(2x - 3y)^2 = (2x - 3y)(2x - 3y)$$
$$= 2x(2x - 3y) - 3y(2x - 3y)$$
$$= 4x^2 - 6xy - 6xy + 9y^2$$
$$= 4x^2 - 12xy + 9y^2$$

Alternatively,

$$\begin{array}{r} 2x - 3y \\ 2x - 3y \\ \hline \end{array}$$

Multiplying by $2x \rightarrow$ $\qquad 4x^2 - 6xy$

Multiplying by $-3y \rightarrow$ $\qquad - 6xy + 9y^2$

Adding gives: $\qquad \dfrac{4x^2 - 12xy + 9y^2}{}$

Problem 18. Remove the brackets from the expression

$$2[p^2 - 3(q + r) + q^2]$$

In this problem there are two brackets and the 'inner' one is removed first.

Hence $2[p^2 - 3(q + r) + q^2] = 2[p^2 - 3q - 3r + q^2]$
$$= 2p^2 - 6q - 6r + 2q^2$$

Problem 19. Remove the brackets and simplify the expression

$$2a - [3\{2(4a - b) - 5(a + 2b)\} + 4a]$$

Removing the innermost brackets gives:

$$2a - [3\{8a - 2b - 5a - 10b\} + 4a]$$

Collecting together similar terms gives:

$$2a - [3\{3a - 12b\} + 4a]$$

Removing the 'curly' brackets gives:

$$2a - [9a - 36b + 4a]$$

Collecting together similar terms gives:

$$2a - [13a - 36b]$$

Removing the outer brackets gives:

$$2a - 13a + 36b$$

i.e. $-11a + 36b$ or $36b - 11a$ (see law (iii), page 23).

Problem 20. Factorise (a) $xy - 3xz$ (b) $4a^2 + 16ab^3$ (c) $3a^2b - 6ab^2 + 15ab$

For each part of this problem, the HCF of the terms will become one of the factors. Thus:

(a) $xy - 3xz = x(y - 3z)$

(b) $4a^2 + 16ab^3 = 4a(a + 4b^3)$

(c) $3a^2b - 6ab^2 + 15ab = 3ab(a - 2b + 5)$

Now try the following exercise

Exercise 18 Further problems on brackets and factorization (Answers on page 301)

In Problems 1 to 11, remove the brackets and simplify where possible:

1. $(x + 2y) + (2x - y)$
2. $(4a + 3y) - (a - 2y)$
3. $2(x - y) - 3(y - x)$
4. $2x^2 - 3(x - xy) - x(2y - x)$
5. $2(p + 3q - r) - 4(r - q + 2p) + p$
6. $(a + b)(a + 2b)$
7. $(p + q)(3p - 2q)$
8. (i) $(x - 2y)^2$ (ii) $(3a - b)^2$
9. $3a(b + c) + 4c(a - b)$
10. $3a + 2[a - (3a - 2)]$
11. $24p - [2(3(5p - q) - 2(p + 2q)) + 3q]$

In Problems 12 to 13, factorise:

12. (i) $pb + 2pc$ (ii) $2q^2 + 8qn$
13. (i) $21a^2b^2 - 28ab$ (ii) $2xy^2 + 6x^2y + 8x^3y$

5.4 Fundamental laws and precedence

The **laws of precedence** which apply to arithmetic also apply to algebraic expressions. The order is Brackets, Of, Division, Multiplication, Addition and Subtraction (i.e. **BODMAS**).

Problem 21. Simplify $2a + 5a \times 3a - a$

Multiplication is performed before addition and subtraction thus:

$$2a + 5a \times 3a - a = 2a + 15a^2 - a$$
$$= a + 15a^2 = \mathbf{a(1 + 15a)}$$

Problem 22. Simplify $a + 5a \times (2a - 3a)$

The order of precedence is brackets, multiplication, then subtraction. Hence

$$a + 5a \times (2a - 3a) = a + 5a \times -a = a + -5a^2$$
$$= a - 5a^2 = \mathbf{a(1 - 5a)}$$

Problem 23. Simplify $a \div 5a + 2a - 3a$

The order of precedence is division, then addition and subtraction. Hence

$$a \div 5a + 2a - 3a = \frac{a}{5a} + 2a - 3a$$
$$= \frac{1}{5} + 2a - 3a = \frac{1}{5} - a$$

Problem 24. Simplify $a \div (5a + 2a) - 3a$

The order of precedence is brackets, division and subtraction. Hence

$$a \div (5a + 2a) - 3a = a \div 7a - 3a$$
$$= \frac{a}{7a} - 3a = \frac{1}{7} - 3a$$

Problem 25. Simplify $a \div (5a + 2a - 3a)$

The order of precedence is brackets, then division. Hence:

$$a \div (5a + 2a - 3a) = a \div 4a = \frac{a}{4a} = \frac{1}{4}$$

Problem 26. Simplify $3c + 2c \times 4c + c \div (5c - 8c)$

The order of precedence is division, multiplication and addition. Hence:

$$3c + 2c \times 4c + c \div (5c - 8c) = 3c + 2c \times 4c + c \div -3c$$
$$= 3c + 2c \times 4c + \frac{c}{-3c}$$

Now $\dfrac{c}{-3c} = \dfrac{1}{-3}$

Multiplying numerator and denominator by -1 gives

$$\frac{1 \times -1}{-3 \times -1} \text{ i.e. } -\frac{1}{3}$$

Hence: $3c + 2c \times 4c + \dfrac{c}{-3c} = 3c + 2c \times 4c - \dfrac{1}{3}$

$$= 3c + 8c^2 - \frac{1}{3} \text{ or } c(3 + 8c) - \frac{1}{3}$$

Problem 27. Simplify $(2a - 3) \div 4a + 5 \times 6 - 3a$

The bracket around the $(2a - 3)$ shows that both $2a$ and -3 have to be divided by $4a$, and to remove the bracket the expression is written in fraction form. Hence:

$$(2a - 3) \div 4a + 5 \times 6 - 3a = \frac{2a - 3}{4a} + 5 \times 6 - 3a$$
$$= \frac{2a - 3}{4a} + 30 - 3a$$
$$= \frac{2a}{4a} - \frac{3}{4a} + 30 - 3a$$
$$= \frac{1}{2} - \frac{3}{4a} + 30 - 3a$$
$$= 30\frac{1}{2} - \frac{3}{4a} - 3a$$

Now try the following exercise

Exercise 19 Further problems on fundamental laws and precedence (Answers on page 301)

In Problems 1 to 10, simplify:

1. $2x \div 4x + 6x$

2. $2x \div (4x + 6x)$

3. $3a - 2a \times 4a + a$

4. $(3a - 2a)4a + a$

5. $3a - 2a(4a + a)$

6. $2y + 4 \div 6y + 3 \times 4 - 5y$

7. $(2y + 4) \div 6y + 3 \times 4 - 5y$

8. $2y + 4 \div 6y + 3(4 - 5y)$

9. $3 \div y + 2 \div y + 1$

10. $(x + 1)(x - 4) \div (2x + 2)$

6

Simple equations

At the end of this chapter you should be able to:

- solve simple equations

- apply the solution of simple equations to practical science and engineering applications

6.1 Expressions, equations and identities

$(3x - 5)$ is an example of an **algebraic expression**, whereas $3x - 5 = 1$ is an example of an **equation** (i.e. it contains an 'equals' sign).

An equation is simply a statement that two quantities are equal. For example,

$$1\,\text{m} = 1000\,\text{mm} \quad \text{or} \quad F = \tfrac{9}{5}C + 32 \quad \text{or} \quad y = mx + c$$

An **identity** is a relationship which is true for all values of the unknown, whereas an equation is only true for particular values of the unknown. For example, $3x - 5 = 1$ is an equation, since it is only true when $x = 2$, whereas $3x \equiv 8x - 5x$ is an identity since it is true for all values of x.

(Note '$\equiv$' means 'is identical to')

Simple linear equations (or equations of the first degree) are those in which an unknown quantity is raised only to the power 1.

To **'solve an equation'** means 'to find the value of the unknown'.

Any arithmetic operation may be applied to an equation **as long as the equality of the equation is maintained**.

6.2 Worked problems on simple equations

Problem 1. Solve the equation $4x = 20$

Dividing each side of the equation by 4 gives: $\dfrac{4x}{4} = \dfrac{20}{4}$

(Note that the same operation has been applied to both the left-hand side (LHS) and the right-hand side (RHS) of the equation so the equality has been maintained.)

Cancelling gives: $x = 5$, which is the solution to the equation.

Solutions to simple equations should always be checked and this is accomplished by substituting the solution into the original equation. In this case, LHS $= 4(5) = 20 =$ RHS.

Problem 2. Solve $\dfrac{2x}{5} = 6$

The LHS is a fraction and this can be removed by multiplying both sides of the equation by 5.

Hence $\qquad\qquad 5\left(\dfrac{2x}{5}\right) = 5(6)$

Cancelling gives: $\qquad 2x = 30$

Dividing both sides of the equation by 2 gives:

$$\dfrac{2x}{2} = \dfrac{30}{2} \quad \text{i.e.} \quad x = 15$$

Problem 3. Solve $a - 5 = 8$

Adding 5 to both sides of the equation gives:

$$a - 5 + 5 = 8 + 5$$

i.e. $\qquad\qquad a = 13$

The result of the above procedure is to move the '−5' from the LHS of the original equation, across the equals sign, to the RHS, but the sign is changed to +

Problem 4. Solve $x + 3 = 7$

Subtracting 3 from both sides of the equation gives:

$$x + 3 - 3 = 7 - 3$$

i.e. $\qquad x = 4$

The result of the above procedure is to move the '+3' from the LHS of the original equation, across the equals sign, to the RHS, but the sign is changed to −. Thus a term can be moved from one side of an equation to the other as long as a change in sign is made.

Problem 5. Solve $6x + 1 = 2x + 9$

In such equations the terms containing x are grouped on one side of the equation and the remaining terms grouped on the other side of the equation. As in Problems 3 and 4, changing from one side of an equation to the other must be accompanied by a change of sign.

Thus since $\quad 6x + 1 = 2x + 9$

then $\qquad 6x - 2x = 9 - 1$

$$4x = 8$$

$$\frac{4x}{4} = \frac{8}{4}$$

i.e. $\qquad x = 2$

Check: LHS of original equation $= 6x + 1 = 6(2) + 1 = 13$

RHS of original equation $= 2x + 9 = 2(2) + 9 = 13$

Hence the solution $x = 2$ is correct.

Problem 6. Solve $4 - 3p = 2p - 11$

In order to keep the p term positive the terms in p are moved to the RHS and the constant terms to the LHS.

Hence $\quad 4 + 11 = 2p + 3p$

$$15 = 5p$$

$$\frac{15}{5} = \frac{5p}{5}$$

Hence $\qquad 3 = p \quad$ or $\quad p = 3$

Check: LHS $= 4 - 3(3) = 4 - 9 = -5$

RHS $= 2(3) - 11 = 6 - 11 = -5$

Hence the solution $p = 3$ is correct.

If, in this example, the unknown quantities had been grouped initially on the LHS instead of the RHS then:

$$-3p - 2p = -11 - 4$$

i.e. $\qquad -5p = -15$

$$\frac{-5p}{-5} = \frac{-15}{-5}$$

and $\qquad p = 3$, as before

It is often easier, however, to work with positive values where possible.

Problem 7. Solve $3(x - 2) = 9$

Removing the bracket gives: $\quad 3x - 6 = 9$

Rearranging gives: $\qquad\qquad 3x = 9 + 6$

$$3x = 15$$

$$\frac{3x}{3} = \frac{15}{3}$$

i.e. $\qquad x = 5$

Check: LHS $= 3(5 - 2) = 3(3) = 9 =$ RHS.

Hence the solution $x = 5$ is correct.

Problem 8. Solve $4(2r - 3) - 2(r - 4) = 3(r - 3) - 1$

Removing brackets gives:

$$8r - 12 - 2r + 8 = 3r - 9 - 1$$

Rearranging gives: $\quad 8r - 2r - 3r = -9 - 1 + 12 - 8$

i.e. $\qquad\qquad\qquad\qquad 3r = -6$

$$r = \frac{-6}{3} = -2$$

Check: LHS $= 4(-4 - 3) - 2(-2 - 4) = -28 + 12 = -16$

RHS $= 3(-2 - 3) - 1 = -15 - 1 = -16$

Hence the solution $r = -2$ is correct.

Now try the following exercise

Exercise 20 Further problems on simple equations (Answers on page 301)

Solve the following equations:

1. $2x + 5 = 7$ 2. $8 - 3t = 2$

3. $\frac{2}{3}c - 1 = 3$ 4. $2x - 1 = 5x + 11$

5. $7 - 4p = 2p - 3$ 6. $2a + 6 - 5a = 0$

7. $3x - 2 - 5x = 2x - 4$

8. $20d - 3 + 3d = 11d + 5 - 8$

9. $2(x - 1) = 4$

10. $16 = 4(t + 2)$

11. $5(f - 2) - 3(2f + 5) + 15 = 0$

12. $2x = 4(x - 3)$

13. $6(2 - 3y) - 42 = -2(y - 1)$

14. $2(3g - 5) - 5 = 0$

15. $4(3x + 1) = 7(x + 4) - 2(x + 5)$

6.3 Further worked problems on simple equations

> *Problem 9.* Solve $\dfrac{3}{x} = \dfrac{4}{5}$

The lowest common multiple (LCM) of the denominators, i.e. the lowest algebraic expression that both x and 5 will divide into, is $5x$

Multiplying both sides by $5x$ gives:

$$5x \left(\frac{3}{x} \right) = 5x \left(\frac{4}{5} \right)$$

Cancelling gives: $15 = 4x$ $\qquad\qquad$ (1)

$$\frac{15}{4} = \frac{4x}{4}$$

i.e. $\qquad\qquad x = \dfrac{15}{4}$ or $3\dfrac{3}{4}$

Check: LHS $= \dfrac{3}{3\frac{3}{4}} = \dfrac{3}{\frac{15}{4}} = 3 \left(\dfrac{4}{15} \right) = \dfrac{12}{15} = \dfrac{4}{5} =$ RHS.

(Note that when there is only one fraction on each side of an equation, 'cross-multiplication' can be applied. In this example, if $\dfrac{3}{x} = \dfrac{4}{5}$ then $(3)(5) = 4x$, which is a quicker way of arriving at equation (1) above.)

> *Problem 10.* Solve $\dfrac{2y}{5} + \dfrac{3}{4} + 5 = \dfrac{1}{20} - \dfrac{3y}{2}$

The LCM of the denominators is 20. Multiplying each term by 20 gives:

$$20 \left(\frac{2y}{5} \right) + 20 \left(\frac{3}{4} \right) + 20(5) = 20 \left(\frac{1}{20} \right) - 20 \left(\frac{3y}{2} \right)$$

Cancelling gives: $4(2y) + 5(3) + 100 = 1 - 10(3y)$

i.e. $\qquad\qquad 8y + 15 + 100 = 1 - 30y$

Rearranging gives: $\qquad 8y + 30y = 1 - 15 - 100$

$$38y = -114$$

$$y = \frac{-114}{38} = -3$$

Check: LHS $= \dfrac{2(-3)}{5} + \dfrac{3}{4} + 5 = \dfrac{-6}{5} + \dfrac{3}{4} + 5$

$$= \frac{-9}{20} + 5 = 4\frac{11}{20}$$

RHS $= \dfrac{1}{20} - \dfrac{3(-3)}{2} = \dfrac{1}{20} + \dfrac{9}{2} = 4\dfrac{11}{20}$

Hence the solution $y = -3$ is correct.

> *Problem 11.* Solve $\sqrt{x} = 2$

[$\sqrt{x} = 2$ is not a 'simple equation' since the power of x is $\frac{1}{2}$ i.e. $\sqrt{x} = x^{1/2}$; however, it is included here since it occurs often in practise.]

Wherever square root signs are involved with the unknown quantity, both sides of the equation must be squared. Hence

$$(\sqrt{x})^2 = (2)^2$$

i.e. $\qquad x = 4$

> *Problem 12.* Solve $2\sqrt{2} = 8$

To avoid possible errors it is usually best to arrange the term containing the square root on its own. Thus

$$\frac{2\sqrt{d}}{2} = \frac{8}{2}$$

i.e. $\qquad \sqrt{d} = 4$

Squaring both sides gives: $d = 16$, which may be checked in the original equation.

> *Problem 13.* Solve $x^2 = 25$

This problem involves a square term and thus is not a simple equation (it is, in fact, a quadratic equation). However the solution of such an equation is often required and is therefore included here for completeness.

Whenever a square of the unknown is involved, the square root of both sides of the equation is taken. Hence

$$\sqrt{x^2} = \sqrt{25}$$

i.e. $\qquad x = 5$

However, $x = -5$ is also a solution of the equation because $(-5) \times (-5) = +25$. Therefore, whenever the square root of a number is required there are always two answers, one positive, the other negative.

The solution of $x^2 = 25$ is thus written as $x = \pm 5$

Problem 14. Solve $\dfrac{15}{4t^2} = \dfrac{2}{3}$

'Cross-multiplying' gives: $15(3) = 2(4t^2)$

i.e. $$45 = 8t^2$$
$$\frac{45}{8} = t^2$$

i.e. $$t^2 = 5.625$$

Hence $t = \sqrt{5.625} = \pm 2.372$, correct to 4 significant figures.

Now try the following exercise

**Exercise 21   Further problems on simple equations
(Answers on page 301)**

Solve the following equations:

1. $\dfrac{1}{5}d + 3 = 4$ 2. $2 + \dfrac{3}{4}y = 1 + \dfrac{2}{3}y + \dfrac{5}{6}$

3. $\dfrac{1}{4}(2x - 1) + 3 = \dfrac{1}{2}$ 4. $\dfrac{x}{3} - \dfrac{x}{5} = 2$

5. $1 - \dfrac{y}{3} = 3 + \dfrac{y}{3} - \dfrac{y}{6}$ 6. $\dfrac{2}{a} = \dfrac{3}{8}$

7. $\dfrac{x}{4} - \dfrac{x+6}{5} = \dfrac{x+3}{2}$ 8. $3\sqrt{t} = 9$

9. $2\sqrt{y} = 5$ 10. $\dfrac{3\sqrt{x}}{1 - \sqrt{x}} = -6$

11. $10 = 5\sqrt{\left(\dfrac{x}{2} - 1\right)}$ 12. $16 = \dfrac{t^2}{9}$

13. $\dfrac{6}{a} = \dfrac{2a}{3}$ 14. $\dfrac{11}{2} = 5 + \dfrac{8}{x^2}$

6.4   Practical problems involving simple equations

Problem 15. The temperature coefficient of resistance α may be calculated from the formula $R_t = R_0(1 + \alpha t)$. Find α given $R_t = 0.928$, $R_0 = 0.8$ and $t = 40$

Since $R_t = R_0(1 + \alpha t)$

then $$0.928 = 0.8[1 + \alpha(40)]$$
$$0.928 = 0.8 + (0.8)(\alpha)(40)$$
$$0.928 - 0.8 = 32\alpha$$
$$0.128 = 32\alpha$$

Hence $$\alpha = \frac{0.128}{32} = 0.004$$

Problem 16. The distance s metres travelled in time t seconds is given by the formula $s = ut + \frac{1}{2}at^2$, where u is the initial velocity in m/s and a is the acceleration in m/s². Find the acceleration of the body if it travels 168 m in 6 s, with an initial velocity of 10 m/s.

$s = ut + \dfrac{1}{2}at^2$, and $s = 168$, $u = 10$ and $t = 6$

Hence $$168 = (10)(6) + \frac{1}{2}a(6)^2$$
$$168 = 60 + 18a$$
$$168 - 60 = 18a$$
$$108 = 18a$$
$$a = \frac{108}{18} = 6$$

Hence the acceleration of the body is 6 m/s²

Problem 17. When three resistors in an electrical circuit are connected in parallel the total resistance R_T is given by: $\dfrac{1}{R_T} = \dfrac{1}{R_1} + \dfrac{1}{R_2} + \dfrac{1}{R_3}$. Find the total resistance when $R_1 = 5\,\Omega$, $R_2 = 10\,\Omega$ and $R_3 = 30\,\Omega$.

$$\frac{1}{R_T} = \frac{1}{5} + \frac{1}{10} + \frac{1}{30} = \frac{6+3+1}{30} = \frac{10}{30} = \frac{1}{3}$$

Taking the reciprocal of both sides gives: $R_T = 3\,\Omega$.

Alternatively, if $\dfrac{1}{R_T} = \dfrac{1}{5} + \dfrac{1}{10} + \dfrac{1}{30}$ the LCM of the denominators is $30\,R_T$

Hence $30R_T\left(\dfrac{1}{R_T}\right) = 30R_T\left(\dfrac{1}{5}\right) + 30R_T\left(\dfrac{1}{10}\right)$

$$+ 30R_T\left(\frac{1}{30}\right)$$

Cancelling gives: $30 = 6R_T + 3R_T + R_T$

$$30 = 10R_T$$

$$R_T = \frac{30}{10} = 3\,\Omega, \text{ as above}$$

Now try the following exercise

Exercise 22   Practical problems involving simple equations (Answers on page 301)

1. Force F newtons is given by $F = ma$, where m is the mass in kilograms and a is the acceleration in metres per second squared. Find the acceleration when a force of 4 kN is applied to a mass of 500 kg.

2. $PV = mRT$ is the characteristic gas equation. Find the value of m when $P = 100 \times 10^3$, $V = 3.00$, $R = 288$ and $T = 300$

3. When three resistors R_1, R_2 and R_3 are connected in parallel the total resistance R_T is determined from $\dfrac{1}{R_T} = \dfrac{1}{R_1} + \dfrac{1}{R_2} + \dfrac{1}{R_3}$

   (a) Find the total resistance when $R_1 = 3\,\Omega$, $R_2 = 6\,\Omega$ and $R_3 = 18\,\Omega$.

   (b) Find the value of R_3 given that $R_T = 3\,\Omega$, $R_1 = 5\,\Omega$ and $R_2 = 10\,\Omega$.

4. Ohm's law may be represented by $I = V/R$, where I is the current in amperes, V is the voltage in volts and R is the resistance in ohms. A soldering iron takes a current of 0.30 A from a 240 V supply. Find the resistance of the element.

6.5   Further practical problems involving simple equations

Problem 18.   The extension x m of an aluminium tie bar of length l m and cross-sectional area A m² when carrying a load of F newtons is given by the modulus of elasticity $E = Fl/Ax$. Find the extension of the tie bar (in mm) if $E = 70 \times 10^9$ N/m², $F = 20 \times 10^6$ N, $A = 0.1$ m² and $l = 1.4$ m.

$E = Fl/Ax$, hence $70 \times 10^9\,\dfrac{\text{N}}{\text{m}^2} = \dfrac{(20 \times 10^6\,\text{N})(1.4\,\text{m})}{(0.1\,\text{m}^2)(x)}$

(the unit of x is thus metres)

$$(70 \times 10^9)(0.1)(x) = (20 \times 10^6)(1.4)$$

$$x = \frac{20 \times 10^6 \times 1.4}{70 \times 10^9 \times 0.1}$$

Cancelling gives: $x = \dfrac{2 \times 1.4}{7 \times 100}\,\text{m} = \dfrac{2 \times 1.4}{7 \times 100} \times 1000\,\text{mm}$

Hence the extension of the tie bar, $x = 4$ mm

Problem 19.   Power in a d.c. circuit is given by $P = \dfrac{V^2}{R}$ where V is the supply voltage and R is the circuit resistance. Find the supply voltage if the circuit resistance is 1.25 Ω and the power measured is 320 W.

Since $P = \dfrac{V^2}{R}$ then $320 = \dfrac{V^2}{1.25}$

$$(320)(1.25) = V^2$$

i.e. $V^2 = 400$

Supply voltage, $V = \sqrt{400} = \pm 20$ V

Problem 20.   The stress f in a material of a thick cylinder can be obtained from $\dfrac{D}{d} = \sqrt{\left(\dfrac{f+p}{f-p}\right)}$ Calculate the stress, given that $D = 21.5$, $d = 10.75$ and $p = 1800$

Since $\dfrac{D}{d} = \sqrt{\left(\dfrac{f+p}{f-p}\right)}$

then $\dfrac{21.5}{10.75} = \sqrt{\left(\dfrac{f+1800}{f-1800}\right)}$

i.e. $2 = \sqrt{\left(\dfrac{f+1800}{f-1800}\right)}$

Squaring both sides gives:

$$4 = \frac{f+1800}{f-1800}$$

$$4(f - 1800) = f + 1800$$

$$4f - 7200 = f + 1800$$

$$4f - f = 1800 + 7200$$

$$3f = 9000$$

$$f = \frac{9000}{3} = 3000$$

Hence   **stress, $f = 3000$**

Now try the following exercise

Exercise 23   Practical problems involving simple equations (Answers on page 301)

1. Given $R_2 = R_1(1 + \alpha t)$, find α given $R_1 = 5.0$, $R_2 = 6.03$ and $t = 51.5$

2. If $v^2 = u^2 + 2as$, find u given $v = 24$, $a = -40$ and $s = 4.05$

3. The relationship between the temperature on a Fahrenheit scale and that on a Celsius scale is given by $F = \dfrac{9}{5}C + 32$. Express $113°F$ in degrees Celsius.

4. If $t = 2\pi\sqrt{\dfrac{w}{Sg}}$, find the value of S given $w = 1.219$, $g = 9.81$ and $t = 0.3132$

5. A rectangular laboratory has a length equal to one and a half times its width and a perimeter of $40\,\text{m}$. Find its length and width.

7

Transposition of formulae

At the end of this chapter you should be able to:

- transpose formulae applicable to engineering and science applications

7.1 Introduction to transposition of formulae

When a symbol other than the subject is required to be calculated it is usual to rearrange the formula to make a new subject. This rearranging process is called **transposing the formula** or **transposition**.

The rules used for transposition of formulae are the same as those used for the solution of simple equations (see Chapter 6)–basically, **that the equality of an equation must be maintained**.

7.2 Worked problems on transposition of formulae

Problem 1. Transpose $p = q + r + s$ to make r the subject.

The aim is to obtain r on its own on the left-hand side (LHS) of the equation. Changing the equation around so that r is on the LHS gives:

$$q + r + s = p \tag{1}$$

Subtracting $(q + s)$ from both sides of the equation gives:

$$q + r + s - (q + s) = p - (q + s)$$

Thus $\quad q + r + s - q - s = p - q - s$

i.e. $\qquad\qquad\quad r = p - q - s \tag{2}$

It is shown with simple equations, that a quantity can be moved from one side of an equation to the other with an appropriate change of sign. Thus equation (2) follows immediately from equation (1) above.

Problem 2. If $a + b = w - x + y$, express x as the subject.

Rearranging gives:

$$w - x + y = a + b \quad \text{and} \quad -x = a + b - w - y$$

Multiplying both sides by -1 gives:

$$(-1)(-x) = (-1)(a + b - w - y)$$

i.e. $\qquad\qquad x = -a - b + w + y$

The result of multiplying each side of the equation by -1 is to change all the signs in the equation.

It is conventional to express answers with positive quantities first. Hence rather than $x = -a - b + w + y$, $x = w + y - a - b$, since the order of terms connected by $+$ and $-$ signs is immaterial.

Problem 3. Transpose $M = Fd$ to make d the subject.

Rearranging gives: $Fd = M$

Dividing both sides by F gives: $\dfrac{Fd}{F} = \dfrac{M}{F}$, i.e. $d = \dfrac{M}{F}$

Problem 4. When a body falls freely through a height h, the velocity v is given by $v^2 = 2gh$. Express this formula with h as the subject.

Rearranging gives: $2gh = v^2$

Dividing both sides by $2g$ gives: $\dfrac{2gh}{2g} = \dfrac{v^2}{2g}$, i.e. $\boldsymbol{h = \dfrac{v^2}{2g}}$

Problem 5. If $I = \dfrac{V}{R}$, rearrange to make V the subject.

Rearranging gives: $\dfrac{V}{R} = I$

Multiplying both sides by R gives:

$$R\left(\dfrac{V}{R}\right) = R(I)$$

Hence $\qquad \boldsymbol{V = IR}$

Problem 6. Transpose $a = \dfrac{F}{m}$ for m

Rearranging gives: $\dfrac{F}{m} = a$

Multiplying both sides by m gives:

$$m\left(\dfrac{F}{m}\right) = m(a) \quad \text{i.e.} \quad F = ma$$

Rearranging gives: $\qquad ma = F$

Dividing both sides by a gives: $\dfrac{ma}{a} = \dfrac{F}{a}$

i.e. $\qquad\qquad \boldsymbol{m = \dfrac{F}{a}}$

Problem 7. Rearrange the formula $R = \dfrac{\rho l}{a}$ to make (i) a the subject, and (ii) l the subject.

(i) Rearranging gives: $\dfrac{\rho l}{a} = R$

Multiplying both sides by a gives:

$$a\left(\dfrac{\rho l}{a}\right) = a(R) \quad \text{i.e.} \quad \rho l = aR$$

Rearranging gives: $aR = \rho l$

Dividing both sides by R gives:

$$\dfrac{aR}{R} = \dfrac{\rho l}{R}$$

i.e. $\quad \boldsymbol{a = \dfrac{\rho l}{R}}$

(ii) Multiplying both sides of $\dfrac{\rho l}{a} = R$ by a gives:

$$\rho l = aR$$

Dividing both sides by ρ gives:

$$\dfrac{\rho l}{\rho} = \dfrac{aR}{\rho}$$

i.e. $\quad \boldsymbol{l = \dfrac{aR}{\rho}}$

Now try the following exercise

Exercise 24 Further problems on transposition of formulae (Answers on page 301)

Make the symbol indicated the subject of each of the formulae shown and express each in its simplest form.

1. $a + b = c - d - e$ $\qquad$ (d)

2. $x + 3y = t$ $\qquad$ (y)

3. $c = 2\pi r$ $\qquad$ (r)

4. $y = mx + c$ $\qquad$ (x)

5. $I = PRT$ $\qquad$ (T)

6. $I = \dfrac{E}{R}$ $\qquad$ (R)

7. $S = \dfrac{a}{1 - r}$ $\qquad$ (r)

8. $F = \dfrac{9}{5}C + 32$ $\qquad$ (C)

7.3 Further worked problems on transposition of formulae

Problem 8. Transpose the formula $v = u + \dfrac{ft}{m}$ to make f the subject.

Rearranging gives: $u + \dfrac{ft}{m} = v$ and $\dfrac{ft}{m} = v - u$

Multiplying each side by m gives:

$$m\left(\dfrac{ft}{m}\right) = m(v - u) \quad \text{i.e.} \quad ft = m(v - u)$$

Dividing both sides by t gives:

$$\dfrac{ft}{t} = \dfrac{m}{t}(v - u) \quad \text{i.e.} \quad \boldsymbol{f = \dfrac{m}{t}(v - u)}$$

Problem 9. The final length, l_2 of a piece of wire heated through $\theta°C$ is given by the formula $l_2 = l_1(1 + \alpha\theta)$. Make the coefficient of expansion, α, the subject.

Rearranging gives: $\qquad l_1(1 + \alpha\theta) = l_2$

Removing the bracket gives: $l_1 + l_1\alpha\theta = l_2$

Rearranging gives: $\qquad l_1\alpha\theta = l_2 - l_1$

Dividing both sides by $l_1\theta$ gives:

$$\frac{l_1\alpha\theta}{l_1\theta} = \frac{l_2 - l_1}{l_1\theta} \quad \text{i.e.} \quad \boldsymbol{\alpha = \frac{l_2 - l_1}{l_1\theta}}$$

Problem 10.   A formula for the distance moved by a body is given by: $s = \frac{1}{2}(v + u)t$. Rearrange the formula to make u the subject.

Rearranging gives: $\qquad\qquad \dfrac{1}{2}(v + u)t = s$

Multiplying both sides by 2 gives: $(v + u)t = 2s$

Dividing both sides by t gives: $\qquad \dfrac{(v + u)t}{t} = \dfrac{2s}{t}$

i.e. $\qquad\qquad\qquad\qquad v + u = \dfrac{2s}{t}$

Hence $\qquad \boldsymbol{u = \dfrac{2s}{t} - v} \quad \text{or} \quad \boldsymbol{u = \dfrac{2s - vt}{t}}$

Problem 11.   A formula for kinetic energy is $k = \frac{1}{2}mv^2$. Transpose the formula to make v the subject.

Rearranging gives: $\frac{1}{2}mv^2 = k$

Whenever the prospective new subject is a squared term, that term is isolated on the LHS, and then the square root of both sides of the equation is taken.

Multiplying both sides by 2 gives: $mv^2 = 2k$

Dividing both sides by m gives: $\qquad \dfrac{mv^2}{m} = \dfrac{2k}{m}$

i.e. $\qquad\qquad\qquad\qquad v^2 = \dfrac{2k}{m}$

Taking the square root of both sides gives:

$$\sqrt{v^2} = \sqrt{\left(\frac{2k}{m}\right)}$$

i.e. $\qquad v = \sqrt{\left(\dfrac{2k}{m}\right)}$

Problem 12.   Given $t = 2\pi\sqrt{\dfrac{l}{g}}$, find g in terms of t, l and π

Whenever the prospective new subject is within a square root sign, it is best to isolate that term on the LHS and then to square both sides of the equation.

Rearranging gives: $2\pi\sqrt{\dfrac{l}{g}} = t$

Dividing both sides by 2π gives: $\sqrt{\dfrac{l}{g}} = \dfrac{t}{2\pi}$

Squaring both sides gives: $\dfrac{l}{g} = \left(\dfrac{t}{2\pi}\right)^2 = \dfrac{t^2}{4\pi^2}$

Cross-multiplying, i.e. multiplying each term by $4\pi^2 g$, gives:

$$4\pi^2 l = gt^2$$

or $\qquad\qquad\qquad gt^2 = 4\pi^2 l$

Dividing both sides by t^2 gives: $\dfrac{gt^2}{t^2} = \dfrac{4\pi^2 l}{t^2}$

i.e. $\qquad\qquad\qquad \boldsymbol{g = \dfrac{4\pi^2 l}{t^2}}$

Now try the following exercise

Exercise 25   Further problems on transposition of formulae (Answers on page 301)

Make the symbol indicated the subject of each of the formulae shown and express each in its simplest form.

1. $y = \dfrac{\lambda(x - d)}{d}$ $\qquad\qquad$ (x)

2. $A = \dfrac{3(F - f)}{L}$ $\qquad\qquad$ (f)

3. $y = \dfrac{Ml^2}{8EI}$ $\qquad\qquad$ (E)

4. $R = R_0(1 + \alpha t)$ $\qquad\qquad$ (t)

5. $\dfrac{1}{R} = \dfrac{1}{R_1} + \dfrac{1}{R_2}$ $\qquad\qquad$ (R_2)

6. $I = \dfrac{E - e}{R + r}$ $\qquad\qquad$ (R)

7. $t = 2\pi\sqrt{\dfrac{l}{g}}$ $\qquad\qquad$ (l)

8. $v^2 = u^2 + 2as$ $\qquad\qquad$ (u)

7.4 Harder worked problems on transposition of formulae

> *Problem 13.* If $cd = 3d + e - ad$, express d in terms of a, c and e

Rearranging to obtain the terms in d on the LHS gives:

$$cd - 3d + ad = e$$

Factorising the LHS gives:

$$d(c - 3 + a) = e$$

Dividing both sides by $(c - 3 + a)$ gives:

$$d = \frac{e}{c - 3 + a}$$

> *Problem 14.* If $a = \dfrac{b}{1+b}$, make b the subject of the formula.

Rearranging gives:
$$\frac{b}{1+b} = a$$

Multiplying both sides by $(1 + b)$ gives: $\quad b = a(1 + b)$

Removing the bracket gives: $\quad b = a + ab$

Rearranging to obtain terms in b on the LHS gives:
$$b - ab = a$$

Factorising the LHS gives: $\quad b(1 - a) = a$

Dividing both sides by $(1 - a)$ gives: $\quad b = \dfrac{a}{1-a}$

> *Problem 15.* Transpose the formula $V = \dfrac{Er}{R+r}$ to make r the subject.

Rearranging gives:
$$\frac{Er}{R+r} = V$$

Multiplying both sides by $(R + r)$ gives: $\quad Er = V(R + r)$

Removing the bracket gives: $\quad Er = VR + Vr$

Rearranging to obtain terms in r on the LHS gives:
$$Er - Vr = VR$$

Factorising gives: $\quad r(E - V) = VR$

Dividing both sides by $(E - V)$ gives: $\quad r = \dfrac{VR}{E-V}$

Now try the following exercise

> **Exercise 26 Further problems on transposition of formulae (Answers on page 301)**
>
> Make the symbol indicated the subject of each of the formulae shown in Problems 1 to 4, and express each in its simplest form.
>
> 1. $y = \dfrac{a^2m - a^2n}{x}$ (a)
> 2. $x + y = \dfrac{r}{3+r}$ (r)
> 3. $m = \dfrac{\mu L}{L + rCR}$ (L)
> 4. $a^2 = \dfrac{b^2 - c^2}{b^2}$ (b)
> 5. A formula for the focal length, f, of a convex lens is $\dfrac{1}{f} = \dfrac{1}{u} + \dfrac{1}{v}$. Transpose the formula to make v the subject and evaluate v when $f = 5$ and $u = 6$
> 6. The quantity of heat, Q, is given by the formula $Q = mc(t_2 - t_1)$. Make t_2 the subject of the formula and evaluate t_2 when $m = 10$, $t_1 = 15$, $c = 4$ and $Q = 1600$
> 7. The velocity, v, of water in a pipe appears in the formula $h = \dfrac{0.03Lv^2}{2dg}$. Express v as the subject of the formula and evaluate v when $h = 0.712$, $L = 150$, $d = 0.30$ and $g = 9.81$
> 8. The sag S at the centre of a wire is given by the formula: $S = \sqrt{\left(\dfrac{3d(l-d)}{8}\right)}$. Make l the subject of the formula and evaluate l when $d = 1.75$ and $S = 0.80$

Assignment 2

This assignment covers the material contained in Chapters 4 to 7. The marks for each question are shown in brackets at the end of each question.

1. Evaluate the following, each correct to 4 significant figures:

(a) $\dfrac{1}{0.0419}$ (b) $\sqrt{0.0527}$ (c) $\left(\dfrac{36.2^2 \times 0.561}{27.8 \times 12.83}\right)^3$

(7)

2. If $1.6\,\text{km} = 1$ mile, determine the speed of 45 miles/hour in kilometres per hour (3)

3. The area A of a circle is given by $A = \pi r^2$. Find the area of a circle of radius $r = 3.73\,\text{cm}$, correct to 2 decimal places. (3)

4. Evaluate $3xy^2z^3 - 2yz$ when $x = \frac{8}{3}$, $y = 2$ and $z = \frac{1}{2}$ (3)

5. Simplify the following:

  (a) $(2a + 3b)(x - y)$ (b) $3x + 4 \div 2x + 5 \times 2 - 4x$ (5)

6. Remove the brackets in the following expressions and simplify:

  (a) $(2x - y)^2$ (b) $4ab - [3\{2(4a - b) + b(2 - a)\}]$ (4)

7. Factorise $3x^2y + 9xy^2 + 6xy^3$ (2)

8. Solve the following equations:

  (a) $3t - 2 = 5t + 4$

  (b) $4(k - 1) - 2(3k + 2) + 14 = 0$

  (c) $3\sqrt{y} = 2$ (9)

9. Transpose the following equations:

  (a) $y = mx + c$ for m (b) $x = \dfrac{2(y - z)}{t}$ for z

  (c) $\dfrac{1}{R_T} = \dfrac{1}{R_A} + \dfrac{1}{R_B}$ for R_A (10)

10. The passage of sound waves through walls is governed by the equation:

$$v = \sqrt{\dfrac{K + \dfrac{4}{3}G}{\rho}}$$

Make the shear modulus G the subject of the formula. (4)

8

Simultaneous equations

At the end of this chapter you should be able to:

- solve simultaneous equations by substitution and by elimination
- solve practical engineering problems involving simultaneous equations

8.1 Introduction to simultaneous equations

Only one equation is necessary when finding the value of a **single unknown quantity** (as with simple equations in Chapter 6). However, when an equation contains **two unknown quantities** it has an infinite number of solutions. When two equations are available connecting the same two unknown values then a unique solution is possible. Similarly, for three unknown quantities it is necessary to have three equations in order to solve for a particular value of each of the unknown quantities, and so on.

Equations which have to be solved together to find the unique values of the unknown quantities, which are true for each of the equations, are called **simultaneous equations**.

Two methods of solving simultaneous equations analytically are:

(a) by **substitution**, and (b) by **elimination**.

8.2 Worked problems on simultaneous equations

Problem 1. Solve the following equations for x and y, (a) by substitution, and (b) by elimination:

$$x + 2y = -1 \qquad (1)$$
$$4x - 3y = 18 \qquad (2)$$

(a) **By substitution**

From equation (1): $x = -1 - 2y$

Substituting this expression for x into equation (2) gives:

$$4(-1 - 2y) - 3y = 18$$

This is now a simple equation in y.

Removing the bracket gives:

$$-4 - 8y - 3y = 18$$
$$-11y = 18 + 4 = 22$$
$$y = \frac{22}{-11} = -2$$

Substituting $y = -2$ into equation (1) gives:

$$x + 2(-2) = -1$$
$$x - 4 = -1$$
$$x = -1 + 4 = 3$$

Thus $x = 3$ and $y = -2$ is the solution to the simultaneous equations.

(Check: In equation (2), since $x = 3$ and $y = -2$, LHS $= 4(3) - 3(-2) = 12 + 6 = 18 =$ RHS.)

(b) **By elimination**

$$x + 2y = -1 \qquad (1)$$
$$4x - 3y = 18 \qquad (2)$$

If equation (1) is multiplied throughout by 4 the coefficient of x will be the same as in equation (2), giving:

$$4x + 8y = -4 \qquad (3)$$

Subtracting equation (3) from equation (2) gives:

$$4x - 3y = 18 \qquad (2)$$
$$4x + 8y = -4 \qquad (3)$$
$$\overline{0 - 11y = 22}$$

Hence $y = \dfrac{22}{-11} = -2$

(Note, in the above subtraction, $18 - -4 = 18 + 4 = 22$)

Substituting $y = -2$ into either equation (1) or equation (2) will give $x = 3$ as in method (a). The solution $x = 3$, $y = -2$ is the only pair of values that satisfies both of the original equations.

Problem 2. Solve, by a substitution method, the simultaneous equations

$$3x - 2y = 12 \qquad (1)$$
$$x + 3y = -7 \qquad (2)$$

From equation (2), $x = -7 - 3y$

Substituting for x in equation (1) gives:

$$3(-7 - 3y) - 2y = 12$$

i.e. $\qquad -21 - 9y - 2y = 12$

$$-11y = 12 + 21 = 33$$

Hence $\qquad y = \dfrac{33}{-11} = -3$

Substituting $y = -3$ in equation (2) gives:

$$x + 3(-3) = -7$$

i.e. $\qquad x - 9 = -7$

Hence $\qquad x = -7 + 9 = 2$

Thus $x = 2$, $y = -3$ is the solution of the simultaneous equations.

(Such solutions should always be checked by substituting values into each of the original two equations.)

Problem 3. Use an elimination method to solve the simultaneous equations

$$3x + 4y = 5 \qquad (1)$$
$$2x - 5y = -12 \qquad (2)$$

If equation (1) is multiplied throughout by 2 and equation (2) by 3, then the coefficient of x will be the same in the newly formed equations. Thus,

2 × equation (1) gives: $\qquad 6x + 8y = 10 \qquad (3)$

3 × equation (2) gives: $\qquad 6x - 15y = -36 \qquad (4)$

Equation (3) − equation (4) gives: $\quad \overline{0 + 23y = 46}$

i.e. $\qquad y = \dfrac{46}{23} = 2$

(Note $+8y - -15y = 8y + 15y = 23y$ and $10 - -36 = 10 + 36 = 46$. Alternatively, 'change the signs of the bottom line and add'.)

Substituting $y = 2$ in equation (1) gives:

$$3x + 4(2) = 5$$

from which $\quad 3x = 5 - 8 = -3$

and $\qquad x = -1$

Checking in equation (2), left-hand side $= 2(-1) - 5(2) = -2 - 10 = -12 =$ right-hand side.

Hence $x = -1$ and $y = 2$ is the solution of the simultaneous equations.

The elimination method is the most common method of solving simultaneous equations.

Problem 4. Solve

$$7x - 2y = 26 \qquad (1)$$
$$6x + 5y = 29 \qquad (2)$$

When equation (1) is multiplied by 5 and equation (2) by 2 the coefficients of y in each equation are numerically the same, i.e. 10, but are of opposite sign.

5 × equation (1) gives: $\qquad 35x - 10y = 130 \quad (3)$

2 × equation (2) gives: $\qquad 12x + 10y = 58 \quad (4)$

Adding equation (3) and (4) gives: $\quad \overline{47x + 0 = 188}$

Hence $x = \dfrac{188}{47} = 4$

[Note that when the signs of common coefficients are **different** the two equations are **added**, and when the signs of common coefficients are the **same** the two equations are **subtracted** (as in Problems 1 and 3)]

Substituting $x = 4$ in equation (1) gives:

$$7(4) - 2y = 26$$
$$28 - 2y = 26$$
$$28 - 26 = 2y$$
$$2 = 2y$$

Hence $\qquad y = 1$

Checking, by substituting $x = 4$ and $y = 1$ in equation (2), gives:

$$\text{LHS} = 6(4) + 5(1) = 24 + 5 = 29 = \text{RHS}$$

Thus the solution is $x = 4$, $y = 1$, since these values maintain the equality when substituted in both equations.

Now try the following exercise

Exercise 27 Further problems on simultaneous equations (Answers on page 301)

Solve the following simultaneous equations and verify the results.

1. $a + b = 7$
   $a - b = 3$

2. $2x + 5y = 7$
   $x + 3y = 4$

3. $3s + 2t = 12$
   $4s - t = 5$

4. $3x - 2y = 13$
   $2x + 5y = -4$

5. $5m - 3n = 11$
   $3m + n = 8$

6. $8a - 3b = 51$
   $3a + 4b = 14$

8.3 Further worked problems on simultaneous equations

Problem 5. Solve

$$3p = 2q \tag{1}$$

$$4p + q + 11 = 0 \tag{2}$$

Rearranging gives:

$$3p - 2q = 0 \tag{3}$$

$$4p + q = -11 \tag{4}$$

Multiplying equation (4) by 2 gives:

$$8p + 2q = -22 \tag{5}$$

Adding equations (3) and (5) gives:

$$11p + 0 = -22$$

$$p = \frac{-22}{11} = -2$$

Substituting $p = -2$ into equation (1) gives:

$$3(-2) = 2q$$

$$-6 = 2q$$

$$q = \frac{-6}{2} = -3$$

Checking, by substituting $p = -2$ and $q = -3$ into equation (2) gives:

$$\text{LHS} = 4(-2) + (-3) + 11 = -8 - 3 + 11 = 0 = \text{RHS}$$

Hence the solution is $p = -2$, $q = -3$

Problem 6. Solve

$$\frac{x}{8} + \frac{5}{2} = y \tag{1}$$

$$13 - \frac{y}{3} = 3x \tag{2}$$

Whenever fractions are involved in simultaneous equations it is usual to firstly remove them. Thus, multiplying equation (1) by 8 gives:

$$8\left(\frac{x}{8}\right) + 8\left(\frac{5}{2}\right) = 8y$$

i.e. $\qquad x + 20 = 8y \tag{3}$

Multiplying equation (2) by 3 gives:

$$39 - y = 9x \tag{4}$$

Rearranging equations (3) and (4) gives:

$$x - 8y = -20 \tag{5}$$

$$9x + y = 39 \tag{6}$$

Multiplying equation (6) by 8 gives:

$$72x + 8y = 312 \tag{7}$$

Adding equations (5) and (7) gives:

$$73x + 0 = 292$$

$$x = \frac{292}{73} = 4$$

Substituting $x = 4$ into equation (5) gives:

$$4 - 8y = -20$$

$$4 + 20 = 8y$$

$$24 = 8y$$

$$y = \frac{24}{8} = 3$$

Checking: substituting $x = 4$, $y = 3$ in the original equations, gives:

Equation (1): $\text{LHS} = \frac{4}{8} + \frac{5}{2} = \frac{1}{2} + 2\frac{1}{2} = 3 = y = \text{RHS}$

Equation (2): $\text{LHS} = 13 - \frac{3}{3} = 13 - 1 = 12$

$$\text{RHS} = 3x = 3(4) = 12$$

Hence the solution is $x = 4$, $y = 3$

Now try the following exercise

Exercise 28   Further problems on simultaneous equations (Answers on page 302)

Solve the following simultaneous equations and verify the results.

1. $7p + 11 + 2q = 0$

   $-1 = 3q - 5p$

2. $\dfrac{x}{2} + \dfrac{y}{3} = 4$

   $\dfrac{x}{6} - \dfrac{y}{9} = 0$

3. $\dfrac{a}{2} - 7 = -2b$

   $12 = 5a + \dfrac{2}{3}b$

4. $\dfrac{3}{2}s - 2t = 8$

   $\dfrac{s}{4} + 3t = -2$

8.4   Practical problems involving simultaneous equations

There are a number of situations in engineering and science where the solution of simultaneous equations is required. Some are demonstrated in the following worked problems.

Problem 7. The law connecting friction F and load L for an experiment is of the form $F = aL + b$, where a and b are constants. When $F = 5.6$, $L = 8.0$ and when $F = 4.4$, $L = 2.0$. Find the values of a and b and the value of F when $L = 6.5$

Substituting $F = 5.6$, $L = 8.0$ into $F = aL + b$ gives:

$$5.6 = 8.0a + b \tag{1}$$

Substituting $F = 4.4$, $L = 2.0$ into $F = aL + b$ gives:

$$4.4 = 2.0a + b \tag{2}$$

Subtracting equation (2) from equation (1) gives:

$$1.2 = 6.0a$$

$$a = \frac{1.2}{6.0} = \mathbf{\frac{1}{5}}$$

Substituting $a = \frac{1}{5}$ into equation (1) gives:

$$5.6 = 8.0\left(\frac{1}{5}\right) + b$$

$$5.6 = 1.6 + b$$

$$5.6 - 1.6 = b$$

i.e. $\mathbf{b = 4}$

Checking, substituting $a = \frac{1}{5}$ and $b = 4$ in equation (2), gives:

$$\text{RHS} = 2.0\left(\frac{1}{5}\right) + 4 = 0.4 + 4 = 4.4 = \text{LHS}$$

Hence $a = \frac{1}{5}$ and $b = 4$

When $L = 6.5$, $F = aL + b = \frac{1}{5}(6.5) + 4 = 1.3 + 4$, i.e. $\mathbf{F = 5.30}$

Problem 8. The distance s metres from a fixed point of a vehicle travelling in a straight line with constant acceleration, a m/s^2, is given by $s = ut + \frac{1}{2}at^2$, where u is the initial velocity in m/s and t the time in seconds. Determine the initial velocity and the acceleration given that $s = 42$ m when $t = 2$ s and $s = 144$ m when $t = 4$ s. Find also the distance travelled after 3 s

Substituting $s = 42$, $t = 2$ into $s = ut + \frac{1}{2}at^2$ gives:

$$42 = 2u + \frac{1}{2}a(2)^2$$

i.e. $42 = 2u + 2a \tag{1}$

Substituting $s = 144$, $t = 4$ into $s = ut + \frac{1}{2}at^2$ gives:

$$144 = 4u + \frac{1}{2}a(4)^2$$

i.e. $144 = 4u + 8a \tag{2}$

Multiplying equation (1) by 2 gives:

$$84 = 4u + 4a \tag{3}$$

Subtracting equation (3) from equation (2) gives:

$$60 = 0 + 4a$$

$$a = \frac{60}{4} = \mathbf{15}$$

Substituting $a = 15$ into equation (1) gives:

$$42 = 2u + 2(15)$$

$$42 - 30 = 2u$$

$$u = \frac{12}{2} = \mathbf{6}$$

Substituting $a = 15$, $u = 6$ in equation (2) gives:

$$\text{RHS} = 4(6) + 8(15) = 24 + 120 = 144 = \text{LHS}$$

Hence the initial velocity, $u = 6$ m/s and the acceleration, $a = 15$ m/s^2

Distance travelled after 3 s is given by $s = ut + \frac{1}{2}at^2$ where $t = 3$, $u = 6$ and $a = 15$

Hence $s = (6)(3) + \frac{1}{2}(15)(3)^2 = 18 + 67.5$

i.e. distance travelled after 3 s = 85.5 m

Problem 9. The molar heat capacity of a solid compound is given by the equation $c = a + bT$, where a and b are constants. When $c = 52$, $T = 100$ and when $c = 172$, $T = 400$. Determine the values of a and b.

When $c = 52$, $T = 100$, hence

$$52 = a + 100b \qquad (1)$$

When $c = 172$, $T = 400$, hence

$$172 = a + 400b \qquad (2)$$

Equation (2) − equation (1) gives:

$$120 = 300b$$

from which, $b = \dfrac{120}{300} = \mathbf{0.4}$

Substituting $b = 0.4$ in equation (1) gives:

$$52 = a + 100(0.4)$$

$$a = 52 - 40 = \mathbf{12}$$

Hence $a = 12$ and $b = 0.4$

Now try the following exercise

Exercise 29 Further practical problems involving simultaneous equations (Answers on page 302)

1. In a system of pulleys, the effort P required to raise a load W is given by $P = aW + b$, where a and b are constants. If $W = 40$ when $P = 12$ and $W = 90$ when $P = 22$, find the values of a and b.

2. Velocity v is given by the formula $v = u + at$. If $v = 20$ when $t = 2$ and $v = 40$ when $t = 7$ find the values of u and a. Hence find the velocity when $t = 3.5$

3. $y = mx + c$ is the equation of a straight line of gradient m and y-axis intercept c. If the line passes through the point where $x = 2$ and $y = 2$, and also through the point where $x = 5$ and $y = \frac{1}{2}$, find the gradient and y-axis intercept of the straight line.

4. The molar heat capacity of a solid compound is given by the equation $c = a + bT$. When $c = 52$, $T = 100$ and when $c = 172$, $T = 400$. Find the values of a and b.

9

Straight line graphs

At the end of this chapter you should be able to:

- plot a straight line from given co-ordinates
- determine the gradient and the vertical axis intercept of a straight line graph
- produce a table of co-ordinates from a given equation to plot a straight line graph
- interpolate and extrapolate values from a straight line graph
- determine the law of a graph
- appreciate where typical practical engineering examples of straight line graphs occur

9.1   Introduction to graphs

A **graph** is a pictorial representation of information showing how one quantity varies with another related quantity.

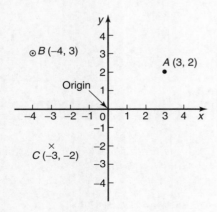

Fig. 9.1

The most common method of showing a relationship between two sets of data is to use **Cartesian** or **rectangular axes** as shown in Fig. 9.1.

The points on a graph are called **co-ordinates**. Point A in Fig. 9.1 has the co-ordinates $(3, 2)$, i.e. 3 units in the x direction and 2 units in the y direction. Similarly, point B has co-ordinates $(-4, 3)$ and C has co-ordinates $(-3, -2)$. The origin has co-ordinates $(0, 0)$.

9.2   The straight line graph

Let a relationship between two variables x and y be

$$y = 3x + 2$$

When $x = 0$,

$$y = 3(0) + 2 = 2,$$

when $x = 1$,

$$y = 3(1) + 2 = 5,$$

when $x = 2$,

$$y = 3(2) + 2 = 8, \text{ and so on.}$$

Thus co-ordinates $(0, 2)$, $(1, 5)$ and $(2, 8)$ have been produced from the equation by selecting arbitrary values of x, and are shown plotted in Fig. 9.2. When the points are joined together a **straight-line graph** results.

The **gradient** or **slope** of a straight line is the ratio of the change in the value of y to the change in the value of x between any two points on the line. If, as x increases, ($\rightarrow$), y also increases ($\uparrow$), then the gradient is positive.

In Fig. 9.3(a), the gradient of AC

$$= \frac{\text{change in } y}{\text{change in } x} = \frac{CB}{BA} = \frac{7 - 3}{3 - 1} = \frac{4}{2} = 2$$

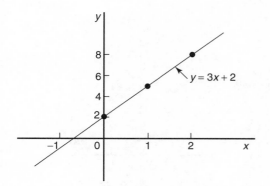

Fig. 9.2

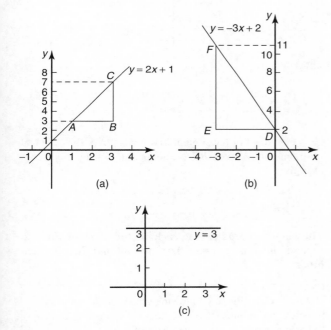

(a)

(b)

(c)

Fig. 9.3

If as x increases ($\rightarrow$), y decreases ($\downarrow$), then the gradient is negative.

In Fig. 9.3(b), the gradient of DF

$$= \frac{\text{change in } y}{\text{change in } x} = \frac{FE}{ED} = \frac{11-2}{-3-0} = \frac{9}{-3} = -3$$

Figure 9.3(c) shows a straight line graph $y = 3$. Since the straight line is horizontal the gradient is zero.

The value of y when $x = 0$ is called the **y-axis intercept**. In Fig. 9.3(a) the y-axis intercept is 1 and in Fig. 9.3(b) is 2.

If the equation of a graph is of the form $y = mx + c$, where m and c are constants, **the graph will always be a straight line**, m representing the gradient and c the y-axis intercept. Thus $y = 5x + 2$ represents a straight line of gradient 5 and y-axis intercept 2. Similarly, $y = -3x - 4$

represents a straight line of gradient -3 and y-axis intercept -4.

Summary of general rules to be applied when drawing graphs

(i) Give the graph a title clearly explaining what is being illustrated.

(ii) Choose scales such that the graph occupies as much space as possible on the graph paper being used.

(iii) Choose scales so that interpolation is made as easy as possible. Usually scales such as 1 cm = 1 unit, or 1 cm = 2 units, or 1 cm = 10 units are used. Awkward scales such as 1 cm = 3 units or 1 cm = 7 units should not be used.

(iv) The scales need not start at zero, particularly when starting at zero produces an accumulation of points within a small area of the graph paper.

(v) The co-ordinates, or points, should be clearly marked. This may be done either by a cross, or a dot and circle, or just by a dot (see Fig. 9.1).

(vi) A statement should be made next to each axis explaining the numbers represented with their appropriate units.

(vii) Sufficient numbers should be written next to each axis without cramping.

Problem 1. Plot the graph $y = 4x + 3$ in the range $x = -3$ to $x = +4$. From the graph, find (a) the value of y when $x = 2.2$, and (b) the value of x when $y = -3$

Whenever an equation is given and a graph is required, a table giving corresponding values of the variable is necessary. The table is achieved as follows:

When $x = -3$,

$$y = 4x + 3 = 4(-3) + 3 = -12 + 3 = -9,$$

when $x = -2$,

$$y = 4(-2) + 3 = -8 + 3 = -5, \text{ and so on.}$$

Such a table is shown below:

x	-3	-2	-1	0	1	2	3	4
y	-9	-5	-1	3	7	11	15	19

The co-ordinates $(-3, -9)$, $(-2, -5)$, $(-1, -1)$, and so on, are plotted and joined together to produce the straight line shown in Fig. 9.4. (Note that the scales used on the x and y axes do not have to be the same) From the graph:

(a) when $x = 2.2$, $y = \mathbf{11.8}$, and
(b) when $y = -3$, $x = \mathbf{-1.5}$

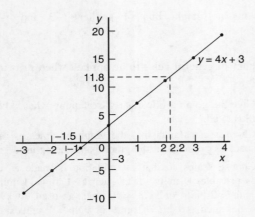

Fig. 9.4

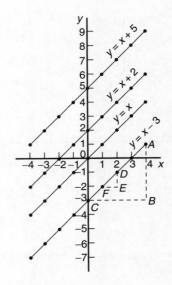

Fig. 9.5

$$= \frac{AB}{BC} = \frac{1-(-3)}{4-0} = \frac{4}{4} = 1$$

i.e. the gradient of the straight line $y = x - 3$ is 1. The actual positioning of AB and BC is unimportant for the gradient is also given by

$$\frac{DE}{EF} = \frac{-1-(-2)}{2-1} = \frac{1}{1} = 1$$

The slope or gradient of each of the straight lines in Fig. 9.5 is thus 1 since they are all parallel to each other.

Problem 2. Plot the following graphs on the same axes between the range $x = -4$ to $x = +4$, and determine the gradient of each.

(a) $y = x$ (b) $y = x + 2$

(c) $y = x + 5$ (d) $y = x - 3$

A table of co-ordinates is produced for each graph.

(a) $y = x$

x	−4	−3	−2	−1	0	1	2	3	4
y	−4	−3	−2	−1	0	1	2	3	4

(b) $y = x + 2$

x	−4	−3	−2	−1	0	1	2	3	4
y	−2	−1	0	1	2	3	4	5	6

(c) $y = x + 5$

x	−4	−3	−2	−1	0	1	2	3	4
y	1	2	3	4	5	6	7	8	9

(d) $y = x - 3$

x	−4	−3	−2	−1	0	1	2	3	4
y	−7	−6	−5	−4	−3	−2	−1	0	1

The co-ordinates are plotted and joined for each graph. The results are shown in Fig. 9.5. Each of the straight lines produced are parallel to each other, i.e. the slope or gradient is the same for each.

To find the gradient of any straight line, say, $y = x - 3$ a horizontal and vertical component needs to be constructed. In Fig. 9.5, AB is constructed vertically at $x = 4$ and BC constructed horizontally at $y = -3$. The gradient of AC

Problem 3. Plot the following graphs on the same axes between the values $x = -3$ to $x = +3$ and determine the gradient and y-axis intercept of each.

(a) $y = 3x$ (b) $y = 3x + 7$

(c) $y = -4x + 4$ (d) $y = -4x - 5$

A table of co-ordinates is drawn up for each equation.

(a) $y = 3x$

x	−3	−2	−1	0	1	2	3
y	−9	−6	−3	0	3	6	9

(b) $y = 3x + 7$

x	−3	−2	−1	0	1	2	3
y	−2	1	4	7	10	13	16

(c) $y = -4x + 4$

x	-3	-2	-1	0	1	2	3
y	16	12	8	4	0	-4	-8

(d) $y = -4x - 5$

x	-3	-2	-1	0	1	2	3
y	7	3	-1	-5	-9	-13	-17

Each of the graphs is plotted as shown in Fig. 9.6, and each is a straight line. $y = 3x$ and $y = 3x + 7$ are parallel to each other and thus have the same gradient. The gradient of AC is given by

$$\frac{CB}{BA} = \frac{16 - 7}{3 - 0} = \frac{9}{3} = 3$$

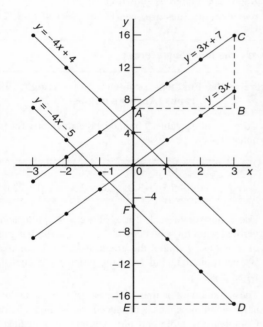

Fig. 9.6

Hence the gradient of both $y = 3x$ and $y = 3x + 7$ is 3.
$y = -4x + 4$ and $y = -4x - 5$ are parallel to each other and thus have the same gradient. The gradient of DF is given by

$$\frac{FE}{ED} = \frac{-5 - (-17)}{0 - 3} = \frac{12}{-3} = -4$$

Hence the gradient of both $y = -4x + 4$ and $y = -4x - 5$ is -4

The y-axis intercept means the value of y where the straight line cuts the y-axis. From Fig. 9.6,

$y = 3x$ cuts the y-axis at $y = 0$

$y = 3x + 7$ cuts the y-axis at $y = +7$

$y = -4x + 4$ cuts the y-axis at $y = +4$

and $y = -4x - 5$ cuts the y-axis at $y = -5$

Some general conclusions can be drawn from the graphs shown in Figs. 9.4, 9.5 and 9.6

When an equation is of the form $y = mx + c$, where m and c are constants, then

(i) a graph of y against x produces a straight line,

(ii) m represents the slope or gradient of the line, and

(iii) c represents the y-axis intercept.

Thus, given an equation such as $y = 3x + 7$, it may be deduced 'on sight' that its gradient is $+3$ and its y-axis intercept is $+7$, as shown in Fig. 9.6. Similarly, if $y = -4x - 5$, then the gradient is -4 and the y-axis intercept is -5, as shown in Fig. 9.6.

When plotting a graph of the form $y = mx + c$, only two co-ordinates need be determined. When the co-ordinates are plotted a straight line is drawn between the two points. Normally, three co-ordinates are determined, the third one acting as a check.

Problem 4. The following equations represent straight lines. Determine, without plotting graphs, the gradient and y-axis intercept for each.

(a) $y = 4$ (b) $y = 2x$

(c) $y = 5x - 1$ (d) $2x + 3y = 3$

(a) $y = 4$ (which is of the form $y = 0x + 4$) represents a horizontal straight line intercepting the y-axis at **4**. Since the line is horizontal its **gradient is zero**.

(b) $y = 2x$ is of the form $y = mx + c$, where c is zero. Hence **gradient $= 2$ and y-axis intercept $= 0$** (i.e. the origin).

(c) $y = 5x - 1$ is of the form $y = mx + c$. Hence **gradient $= 5$ and y-axis intercept $= -1$**

(d) $2x + 3y = 3$ is not in the form $y = mx + c$ as it stands. Transposing to make y the subject gives $3y = 3 - 2x$, i.e.

$$y = \frac{3 - 2x}{3} = \frac{3}{3} - \frac{2x}{3}$$

i.e. $y = -\frac{2x}{3} + 1$

which is of the form $y = mx + c$.

Hence **gradient $= -\frac{2}{3}$ and y-axis intercept $= +1$**

Problem 5. Determine the gradient of the straight line graph passing through the co-ordinates (a) $(-2, 5)$ and $(3, 4)$ (b) $(-2, -3)$ and $(-1, 3)$

A straight line graph passing through co-ordinates (x_1, y_1) and (x_2, y_2) has, from Fig. 9.7, a gradient given by:

$$m = \frac{y_2 - y_1}{x_2 - x_1}$$

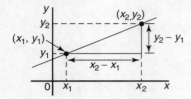

Fig. 9.7

(a) A straight line passes through $(-2, 5)$ and $(3, 4)$, hence $x_1 = -2$, $y_1 = 5$, $x_2 = 3$ and $y_2 = 4$,

hence gradient, $m = \dfrac{y_2 - y_1}{x_2 - x_1}$

$$= \frac{4 - 5}{3 - (-2)} = -\frac{1}{5}$$

(b) A straight line passes through $(-2, -3)$ and $(-1, 3)$, hence $x_1 = -2$, $y_1 = -3$, $x_2 = -1$ and $y_2 = 3$,

hence gradient, $m = \dfrac{y_2 - y_1}{x_2 - x_1}$

$$= \frac{3 - (-3)}{-1 - (-2)}$$

$$= \frac{3 + 3}{-1 + 2} = \frac{6}{1} = 6$$

Problem 6.   Plot the graph $3x + y + 1 = 0$ and $2y - 5 = x$ on the same axes and find their point of intersection.

Rearranging $3x + y + 1 = 0$ gives $y = -3x - 1$

Rearranging $2y - 5 = x$ gives $2y = x + 5$ and $y = \frac{1}{2}x + 2\frac{1}{2}$

Since both equations are of the form $y = mx + c$ both are straight lines. Knowing an equation is a straight line means that only two co-ordinates need to be plotted and a straight line drawn through them. A third co-ordinate is usually determined to act as a check. A table of values is produced for each equation as shown below.

x	1	0	-1
$y = -3x - 1$	-4	-1	2

x	2	0	-3
$y = \frac{1}{2}x + 2\frac{1}{2}$	$3\frac{1}{2}$	$2\frac{1}{2}$	1

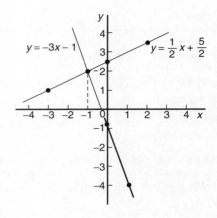

Fig. 9.8

The graphs are plotted as shown in Fig. 9.8.
The two straight lines are seen to intersect at $(-1, 2)$

Now try the following exercise

Exercise 30   Further problems on straight line graphs (Answers on page 302)

1. Corresponding values obtained experimentally for two quantities are:

x	-2.0	-0.5	0	1.0	2.5	3.0	5.0
y	-13.0	-5.5	-3.0	2.0	9.5	12.0	22.0

   Use a horizontal scale for x of $1\,\text{cm} = \frac{1}{2}$ unit and a vertical scale for y of $1\,\text{cm} = 2$ units and draw a graph of x against y. Label the graph and each of its axes. By interpolation, find from the graph the value of y when x is 3.5

2. The equation of a line is $4y = 2x + 5$. A table of corresponding values is produced and is shown below. Complete the table and plot a graph of y against x. Find the gradient of the graph.

x	-4	-3	-2	-1	0	1	2	3	4
y		-0.25			1.25				3.25

3. Determine the gradient and intercept on the y-axis for each of the following equations:

   (a) $y = 4x - 2$ (b) $y = -x$

   (c) $y = -3x - 4$ (d) $y = 7$

4. Determine the gradient of the straight line graphs passing through the co-ordinates:

   (a) $(2, 7)$ and $(-3, 4)$ (b) $(-4, -1)$ and $(-5, 3)$

5. State which of the following equations will produce graphs which are parallel to one another:

  (a) $y - 4 = 2x$ (b) $4x = -(y + 1)$

  (c) $x = \frac{1}{2}(y + 5)$ (d) $1 + \frac{1}{2}y = \frac{3}{2}x$

  (e) $2x = \frac{1}{2}(7 - y)$

6. Draw a graph of $y - 3x + 5 = 0$ over a range of $x = -3$ to $x = 4$. Hence determine (a) the value of y when $x = 1.3$ and (b) the value of x when $y = -9.2$

7. Draw on the same axes the graphs of $y = 3x - 5$ and $3y + 2x = 7$. Find the co-ordinates of the point of intersection. Check the result obtained by solving the two simultaneous equations algebraically.

9.3 Practical problems involving straight line graphs

When a set of co-ordinate values are given or are obtained experimentally and it is believed that they follow a law of the form $y = mx + c$, then if a straight line can be drawn reasonably close to most of the co-ordinate values when plotted, this verifies that a law of the form $y = mx + c$ exists. From the graph, constants m (i.e. gradient) and c (i.e. y-axis intercept) can be determined.

Problem 7. The temperature in degrees Celsius and the corresponding values in degrees Fahrenheit are shown in the table below. Construct rectangular axes, choose a suitable scale and plot a graph of degrees Celsius (on the horizontal axis) against degrees Fahrenheit (on the vertical scale).

°C	10	20	40	60	80	100
°F	50	68	104	140	176	212

From the graph find (a) the temperature in degrees Fahrenheit at 55°C, (b) the temperature in degrees Celsius at 167°F, (c) the Fahrenheit temperature at 0°C, and (d) the Celsius temperature at 230°F

The co-ordinates (10, 50), (20, 68), (40, 104), and so on, are plotted as shown in Fig. 9.9. When the co-ordinates are joined, a straight line is produced. Since a straight line results there is a linear relationship between degrees Celsius and degrees Fahrenheit.

(a) To find the Fahrenheit temperature at 55°C a vertical line AB is constructed from the horizontal axis to meet the straight line at B. The point where the horizontal

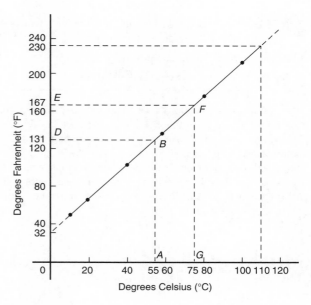

Fig. 9.9

line BD meets the vertical axis indicates the equivalent Fahrenheit temperature.

Hence 55°C is equivalent to 131°F

This process of finding an equivalent value in between the given information in the above table is called **interpolation**.

(b) To find the Celsius temperature at 167°F, a horizontal line EF is constructed as shown in Fig. 9.9. The point where the vertical line FG cuts the horizontal axis indicates the equivalent Celsius temperature.

Hence 167°F is equivalent to 75°C

(c) If the graph is assumed to be linear even outside of the given data, then the graph may be extended at both ends (shown by broken lines in Fig. 9.9)

From Fig. 9.9, **0°C corresponds to 32°F**

(d) **230°F is seen to correspond to 110°C.**

The process of finding equivalent values outside of the given range is called **extrapolation**.

Problem 8. In an experiment demonstrating Hooke's law, the strain in an aluminium wire was measured for various stresses. The results were:

Stress N/mm²	4.9	8.7	15.0
Strain	0.00007	0.00013	0.00021
Stress N/mm²	18.4	24.2	27.3
Strain	0.00027	0.00034	0.00039

Plot a graph of stress (vertically) against strain (horizontally). Find:

(a) Young's Modulus of Elasticity for aluminium which is given by the gradient of the graph,

(b) the value of the strain at a stress of 20 N/mm², and

(c) the value of the stress when the strain is 0.00020

The co-ordinates (0.00007, 4.9), (0.00013, 8.7), and so on, are plotted as shown in Fig. 9.10. The graph produced is the best straight line which can be drawn corresponding to these points. (With experimental results it is unlikely that all the points will lie exactly on a straight line.) The graph, and each of its axes, are labelled. Since the straight line passes through the origin, then stress is directly proportional to strain for the given range of values.

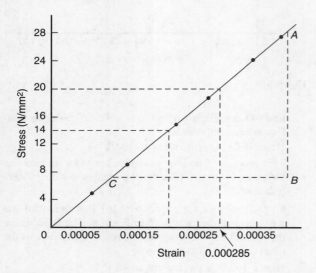

Fig. 9.10

(a) The gradient of the straight line AC is given by

$$\frac{AB}{BC} = \frac{28 - 7}{0.00040 - 0.00010} = \frac{21}{0.00030}$$

$$= \frac{21}{3 \times 10^{-4}} = \frac{7}{10^{-4}}$$

$$= 7 \times 10^4 = 70\,000 \text{ N/mm}^2$$

Thus Young's Modulus of Elasticity for aluminium is 70 000 N/mm²

Since $1 \text{ m}^2 = 10^6 \text{ mm}^2$, $70\,000 \text{ N/mm}^2$ is equivalent to $70\,000 \times 10^6 \text{ N/m}^2$, i.e. **70 × 10⁹ N/m² (or Pascals)**.
From Fig. 9.10:

(b) the value of the strain at a stress of 20 N/mm² is **0.000285**, and

(c) the value of the stress when the strain is 0.00020 is **14 N/mm²**

Problem 9. The following values of resistance R ohms and corresponding voltage V volts are obtained from a test on a filament lamp.

R ohms	30	48.5	73	107	128
V volts	16	29	52	76	94

Choose suitable scales and plot a graph with R representing the vertical axis and V the horizontal axis. Determine (a) the gradient of the graph, (b) the R axis intercept value, (c) the equation of the graph, (d) the value of resistance when the voltage is 60 V, and (e) the value of the voltage when the resistance is 40 ohms. (f) If the graph were to continue in the same manner, what value of resistance would be obtained at 110 V?

The co-ordinates (16, 30), (29, 48.5), and so on, are shown plotted in Fig. 9.11 where the best straight line is drawn through the points.

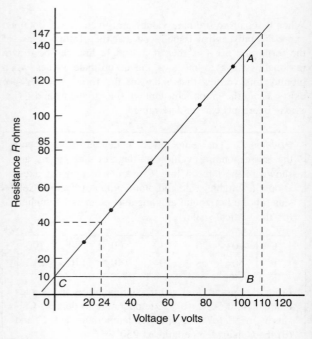

Fig. 9.11

(a) The slope or gradient of the straight line AC is given by

$$\frac{AB}{BC} = \frac{135 - 10}{100 - 0} = \frac{125}{100} = \mathbf{1.25}$$

(Note that the vertical line AB and the horizontal line BC may be constructed anywhere along the length of the straight line. However, calculations are made easier if the horizontal line *BC* is carefully chosen, in this case, 100).

(b) The *R*-axis intercept is at $R = 10\,\text{ohms}$ (by extrapolation).

(c) The equation of a straight line is $y = mx + c$, when y is plotted on the vertical axis and x on the horizontal axis. m represents the gradient and c the y-axis intercept. In this case, R corresponds to y, V corresponds to x, $m = 1.25$ and $c = 10$. Hence the equation of the graph is $R = (1.25\,V + 10)\,\Omega$.

From Fig. 9.11,

(d) when the voltage is 60 V, the resistance is **85 Ω**,

(e) when the resistance is 40 ohms, the voltage is **24 V**, and

(f) by extrapolation, when the voltage is 110 V, the resistance is **147 Ω**.

Now try the following exercise

Exercise 31 Further practical problems involving straight line graphs (Answers on page 302)

1. The resistance *R* ohms of a copper winding is measured at various temperatures *t*°C and the results are as follows:

R ohms	112	120	126	131	134
t°C	20	36	48	58	64

   Plot a graph of *R* (vertically) against *t* (horizontally) and find from it (a) the temperature when the resistance is 122 Ω and (b) the resistance when the temperature is 52°C

2. The following table gives the force *F* newtons which, when applied to a lifting machine, overcomes a corresponding load of *L* newtons.

Force *F* newtons	25	47	64	120	149	187
Load *L* newtons	50	140	210	430	550	700

   Choose suitable scales and plot a graph of *F* (vertically) against *L* (horizontally). Draw the best straight line through the points. Determine from the graph (a) the gradient, (b) the *F*-axis intercept, (c) the equation of the graph, (d) the force applied when the load is 310 N, and (e) the load that a force of 160 N will overcome. (f) If the graph were to continue in the same manner, what value of force will be needed to overcome a 800 N load?

3. The velocity *v* of a body after varying time intervals *t* was measured as follows:

t (seconds)	2	5	8	11	15	18
v (m/s)	16.9	19.0	21.1	23.2	26.0	28.1

   Plot *v* vertically and *t* horizontally and draw a graph of velocity against time. Determine from the graph (a) the velocity after 10 s, (b) the time at 20 m/s and (c) the equation of the graph.

4. An experiment with a set of pulley blocks gave the following results:

Effort, *E* (newtons)	9.0	11.0	13.6	17.4	20.8	23.6
Load, *L* (newtons)	15	25	38	57	74	88

   Plot a graph of effort (vertically) against load (horizontally) and determine (a) the gradient, (b) the vertical axis intercept, (c) the law of the graph, (d) the effort when the load is 30 N and (e) the load when the effort is 19 N.

10

Trigonometry

At the end of this chapter you should be able to:

- state and use Pythagoras' theorem
- define the sine, cosine and tangent of an acute angle
- solve right-angled triangles
- evaluate trigonometric ratios for angles of any magnitude
- plot graphs of sine, cosine and tangent over one cycle
- use the sine and cosine rules to solve any triangle
- calculate the area of any triangle
- appreciate typical practical science and engineering applications where trigonometry is needed

10.1 Introduction

Trigonometry is the branch of mathematics which deals with the measurement of sides and angles of triangles, and their relationship with each other. There are many applications in engineering where a knowledge of trigonometry is needed.

10.2 The theorem of Pythagoras

With reference to Fig. 10.1, the side opposite the right angle (i.e. side b) is called the **hypotenuse**. The **theorem of Pythagoras** states:

'In any right-angled triangle, the square on the hypotenuse is equal to the sum of the squares on the other two sides.'

Hence $b^2 = a^2 + c^2$

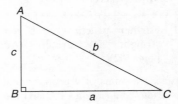

Fig. 10.1

Problem 1. In Fig. 10.2, find the length of BC

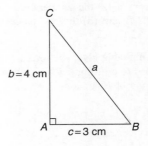

Fig. 10.2

By Pythagoras' theorem,

$$a^2 = b^2 + c^2$$

i.e. $a^2 = 4^2 + 3^2 = 16 + 9 = 25$

Hence $a = \sqrt{25} = \pm 5$

(-5 has no meaning in this context and is thus ignored)

Thus $BC = 5\,\text{cm}$

Problem 2. In Fig. 10.3, find the length of *EF*

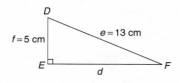

Fig. 10.3

By Pythagoras' theorem: $\quad e^2 = d^2 + f^2$

Hence $\qquad\qquad\qquad\quad 13^2 = d^2 + 5^2$

$$169 = d^2 + 25$$

$$d^2 = 169 - 25 = 144$$

Thus $\qquad\qquad\qquad\quad d = \sqrt{144} = 12\,\text{cm}$

i.e. $\qquad\qquad\qquad\quad$ **$EF = 12\,\text{cm}$**

Problem 3. Two aircraft leave an airfield at the same time. One travels due north at an average speed of 300 km/h and the other due west at an average speed of 220 km/h. Calculate their distance apart after 4 hours.

After 4 hours, the first aircraft has travelled $4 \times 300 = 1200$ km, due north, and the second aircraft has travelled $4 \times 220 = 880$ km due west, as shown in Fig. 10.4. Distance apart after 4 hours $= BC$

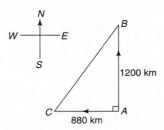

Fig. 10.4

From Pythagoras' theorem:

$$BC^2 = 1200^2 + 880^2 = 1\,440\,000 + 774\,400$$

and $\quad BC = \sqrt{2\,214\,400}$

Hence distance apart after 4 hours = 1488 km.

Now try the following exercise

Exercise 32   Further problems on the theorem of Pythagoras (Answers on page 302)

1. In a triangle *ABC*, $\angle B$ is a right angle, $AB = 6.92$ cm and $BC = 8.78$ cm. Find the length of the hypotenuse.

2. In a triangle *CDE*, $D = 90°$, $CD = 14.83$ mm and $CE = 28.31$ mm. Determine the length of *DE*.

3. A man cycles 24 km due south and then 20 km due east. Another man, starting at the same time as the first man, cycles 32 km due east and then 7 km due south. Find the distance between the two men.

4. A ladder 3.5 m long is placed against a perpendicular wall with its foot 1.0 m from the wall. How far up the wall (to the nearest centimetre) does the ladder reach? If the foot of the ladder is now moved 30 cm further away from the wall, how far does the top of the ladder fall?

10.3   Trigonometric ratios of acute angles

(a) With reference to the right-angled triangle shown in Fig. 10.5:

(i) $\quad$ sine $\theta = \dfrac{\text{opposite side}}{\text{hypotenuse}}$, $\qquad$ i.e. $\mathbf{\sin\theta = \dfrac{b}{c}}$

(ii) $\quad$ cosine $\theta = \dfrac{\text{adjacent side}}{\text{hypotenuse}}$, $\qquad$ i.e. $\mathbf{\cos\theta = \dfrac{a}{c}}$

(iii) $\quad$ tangent $\theta = \dfrac{\text{opposite side}}{\text{adjacent side}}$, $\qquad$ i.e. $\mathbf{\tan\theta = \dfrac{b}{a}}$

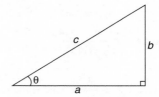

Fig. 10.5

Problem 4. From Fig. 10.6, find sin *D*, cos *D* and tan *F*

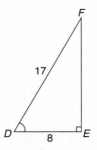

Fig. 10.6

By Pythagoras' theorem,

$$17^2 = 8^2 + EF^2$$

from which, $EF = \sqrt{17^2 - 8^2} = 15$

$$\sin D = \frac{EF}{DF} = \frac{15}{17} \quad \text{or} \quad \mathbf{0.8824}$$

$$\cos D = \frac{DE}{DF} = \frac{8}{17} \quad \text{or} \quad \mathbf{0.4706}$$

and $\tan F = \dfrac{DE}{EF} = \dfrac{8}{15} \quad \text{or} \quad \mathbf{0.5333}$

Problem 5.   Determine the values of $\sin\theta$, $\cos\theta$ and $\tan\theta$ for the right-angled triangle *ABC* shown in Fig. 10.7.

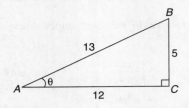

Fig. 10.7

By definition: $\sin\theta = \dfrac{\text{opposite side}}{\text{hypotenuse}} = \dfrac{5}{13} = \mathbf{0.3846}$

$$\cos\theta = \frac{\text{adjacent side}}{\text{hypotenuse}} = \frac{12}{13} = \mathbf{0.9231}$$

and $\tan\theta = \dfrac{\text{opposite side}}{\text{adjacent side}} = \dfrac{5}{12} = \mathbf{0.4167}$

Problem 6.   If $\cos X = \dfrac{9}{41}$ determine the value of $\sin X$ and $\tan X$.

Figure 10.8 shows a right-angled triangle *XYZ*.

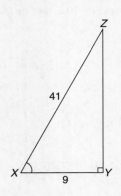

Fig. 10.8

Since $\cos X = \dfrac{9}{41}$, then $XY = 9$ units and $XZ = 41$ units.

Using Pythagoras' theorem:

$$41^2 = 9^2 + YZ^2$$

from which $YZ = \sqrt{41^2 - 9^2} = 40$ units

Thus $\sin X = \dfrac{40}{41}$ and $\tan X = \dfrac{40}{9} = 4\dfrac{4}{9}$

Problem 7.   Point *A* lies at co-ordinate (2,3) and point *B* at (8,7). Determine (a) the distance *AB*, (b) the gradient of the straight line *AB*, and (c) the angle *AB* makes with the horizontal.

(a) Points *A* and *B* are shown in Fig. 10.9(a).

In Fig. 10.9(b), the horizontal and vertical lines *AC* and *BC* are constructed.

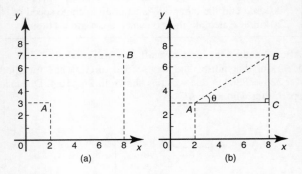

Fig. 10.9

Since *ABC* is a right-angled triangle, and $AC = (8 - 2) = 6$ and $BC = (7 - 3) = 4$, then by Pythagoras' theorem

$$AB^2 = AC^2 + BC^2 = 6^2 + 4^2$$

and $AB = \sqrt{6^2 + 4^2} = \sqrt{52}$

$$= \mathbf{7.211}, \text{ correct to 3 decimal places}$$

(b) The gradient of *AB* is given by $\tan\theta$,

i.e. **gradient** $= \tan\theta = \dfrac{BC}{AC}$

$$= \frac{4}{6} = \frac{2}{3}$$

(c) **The angle *AB* makes with the horizontal** is given by $\tan^{-1}\frac{2}{3} = \mathbf{33.69°}$ (see Problem 12, page 56).

Now try the following exercise

Exercise 33 Further problems on trigonometric ratios of acute angles (Answers on page 302)

1. Sketch a triangle XYZ such that $\angle Y = 90°$, $XY = 9$ cm and $YZ = 40$ cm. Determine $\sin Z$, $\cos Z$, $\tan X$ and $\cos X$.

2. In triangle ABC shown in Fig. 10.10, find $\sin A$, $\cos A$, $\tan A$, $\sin B$, $\cos B$ and $\tan B$.

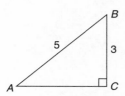

Fig. 10.10

3. If $\tan X = \dfrac{15}{112}$, find $\sin X$ and $\cos X$, in fraction form.

4. For the right-angled triangle shown in Fig. 10.11, find:
   (a) $\sin \alpha$ (b) $\cos \theta$ (c) $\tan \theta$

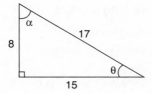

Fig. 10.11

5. Point P lies at co-ordinate $(-3,1)$ and point Q at $(5,-4)$. Determine
   (a) the distance PQ,
   (b) the gradient of the straight line PQ, and
   (c) the angle PQ makes with the horizontal.

10.4 Evaluating trigonometric ratios of any angles

The easiest method of evaluating trigonometric functions of any angle is by using a **calculator.**

The following values, correct to 4 decimal places, may be checked:

sine $18° = 0.3090$ cosine $56° = 0.5592$

tangent $29° = 0.5543$,

sine $172° = 0.1392$ cosine $115° = -0.4226$

tangent $178° = -0.0349$,

sine $241.63° = -0.8799$ cosine $331.78° = 0.8811$

tangent $296.42° = -2.0127$

To evaluate, say, sine $42°23'$ using a calculator means finding sine $42\dfrac{23°}{60}$ since there are 60 minutes in 1 degree.

$\dfrac{23}{60} = 0.383\dot{3}$,   thus $42°23' = 42.383\dot{3}°$

Thus sine $42°23' = $ sine $42.383\dot{3}° = 0.674$, correct to 3 decimal places.

Similarly, cosine $72°38' = $ cosine $72\dfrac{38°}{60} = 0.29849$, correct to 5 significant figures.

Radians

Another method of measuring angles is by using **radians**. The relationship between degrees and radians is:

$$360° = 2\pi \text{ radians}$$

or $\boxed{180° = \pi \text{ radians}}$

Hence 1 rad $= \dfrac{180°}{\pi} = 57.3°$ correct to 3 significant figures.

Most calculators enable a change of mode between degrees and radians. If your calculator is in radian mode then you can check the following values: $\sin 1.23$ rad $= 0.9425$, $\cos 0.356$ rad $= 0.9373$ and $\tan(-3.42$ rad$) = -0.2858$ each correct to 4 decimal places.

Problem 8. Evaluate correct to 4 decimal places:
(a) $\sin 121.68°$ (b) $\sin 259°10'$

(a) $\sin 121.68° = \mathbf{0.8510}$

(c) $\sin 259°10' = \sin 259\dfrac{10°}{60} = \mathbf{-0.9822}$

Problem 9. Evaluate correct to 4 decimal places:
(a) $\cos 159.32°$ (c) $\cos 321°41'$

(a) $\cos 159.32° = \mathbf{-0.9356}$

(b) $\cos 321°41' = $ cosine $321\dfrac{41°}{60} = \mathbf{0.7846}$

Problem 10. Evaluate, correct to 4 significant figures:
(a) $\tan 131.29°$ (b) $\tan 76°58'$

(a) $\tan 131.29° = \mathbf{-1.139}$

(b) $\tan 76°58' = \tan 76\dfrac{58°}{60} = \mathbf{4.320}$

Problem 11.   Evaluate, correct to 4 significant figures:

(a) $\sin 1.481$ (b) $\cos(3\pi/5)$ (c) $\tan 2.93$

If no degrees sign is shown then the angle is assumed to be in radians

(a) $\sin 1.481$ means the sine of 1.481 radians. Hence a calculator needs to be on the radian function.

Hence $\sin 1.481 = \mathbf{0.9960}$

(b) $\cos(3\pi/5) = \cos 1.884955\ldots = \mathbf{-0.3090}$

(c) $\tan 2.93 = \mathbf{-0.2148}$

Problem 12.   Determine the acute angles:

(a) $\sin^{-1} 0.7321$ (b) $\cos^{-1} 0.4174$

(c) $\tan^{-1} 1.4695$

(a) $\sin^{-1} \theta$ is an abbreviation for 'the angle whose sine is equal to θ'.

By calculator, $\sin^{-1} 0.7321 = 47.06273\ldots°$.

Subtracting 47 leaves $0.06273\ldots°$ and multiplying this by 60 gives $4'$ to the nearest minute.

Hence $\sin^{-1} 0.7321 = \mathbf{47.06°}$ or $\mathbf{47°4'}$.

Alternatively, in radians,

$$\sin^{-1} 0.7321 = \mathbf{0.821\ radians}$$

(b) $\cos^{-1} 0.4174 = \mathbf{65.33°}$ or $\mathbf{65°20'}$ or $\mathbf{1.140\ radians}$.

(c) $\tan^{-1} 1.4695 = \mathbf{55.76°}$ or $\mathbf{55°46'}$ or $\mathbf{0.973\ radians}$.

Problem 13.   Evaluate correct to 4 decimal places:

(a) $\sin(-112°)$ (b) $\operatorname{cosine}(-93°16')$

(c) $\operatorname{tangent}(-217.29°)$

(a) Positive angles are considered by convention to be anticlockwise and negative angles as clockwise. From Fig. 10.12, $-112°$ is actually the same as $+248°$ (i.e. $360° - 112°$)

Hence, by calculator,

$$\sin(-112°) = \sin 248° = \mathbf{-0.9272}$$

(b) $\operatorname{cosine}(-93°6') = \operatorname{cosine}\left(-93\dfrac{16}{60}\right)° = \mathbf{-0.0570}$

(c) $\operatorname{tangent}(-217.29°) = \mathbf{-0.7615}$ (which is the same as $\tan(360° - 217.29°)$, i.e. $\tan 141.71°$)

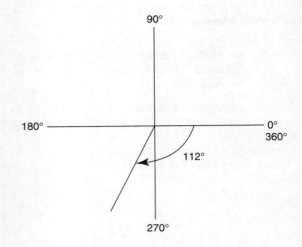

Fig. 10.12

Now try the following exercise

Exercise 34   Further problems on evaluating trigonometric ratios of any angle (Answers on page 302)

In Problems 1 to 4, evaluate correct to 4 decimal places:

1. (a) $\sin 172.41°$ (b) $\sin 302°52'$

2. (a) $\cos 21.46°$ (b) $\cos 284°10'$

3. (a) $\tan 310.59°$ (b) $\tan 49°16'$

4. (a) $\sin \dfrac{2\pi}{3}$,   (b) $\cos 1.681$ (c) $\tan 3.672$

In Problems 5 to 7, determine the acute angle in degrees (correct to 2 decimal places), degrees and minutes, and in radians (correct to 3 decimal places).

5. $\sin^{-1} 0.2341$ 6. $\cos^{-1} 0.8271$ 7. $\tan^{-1} 0.8106$

8. Evaluate correct to 4 decimal places:

   (a) $\sin(-125°)$ (b) $\tan(-241°)$ (c) $\cos(-49°15')$

10.5   Solution of right-angled triangles

To 'solve a right-angled triangle' means 'to find the unknown sides and angles'. This is achieved by using (i) the theorem of Pythagoras, and/or (ii) trigonometric ratios. This is demonstrated in the following problems.

Problem 14.   In triangle PQR shown in Fig. 10.13, find the lengths of PQ and PR.

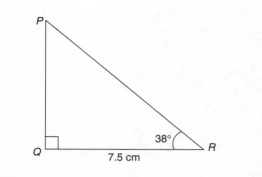

Fig. 10.13

$$\tan 38° = \frac{PQ}{QR} = \frac{PQ}{7.5}$$

hence $PQ = 7.5\tan 38° = 7.5(0.7813) = \mathbf{5.860\,cm}$

$$\cos 38° = \frac{QR}{PR} = \frac{7.5}{PR},$$

hence $PR = \frac{7.5}{\cos 38°} = \frac{7.5}{0.7880} = \mathbf{9.518\,cm}$

[Check: Using Pythagoras' theorem $(7.5)^2 + (5.860)^2 = 90.59 = (9.518)^2$]

Problem 15. Solve the triangle ABC shown in Fig. 10.14.

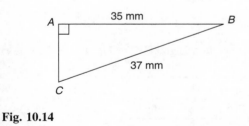

Fig. 10.14

To 'solve triangle ABC' means 'to find the length AC and angles B and C'

$$\sin C = \frac{35}{37} = 0.94595$$

hence $C = \sin^{-1} 0.94595 = 71.08°$

$$B = 180° - 90° - 71.08° = \mathbf{18.92°}$$

(since angles in a triangle add up to 180°)

$$\sin B = \frac{AC}{37}$$

hence $AC = 37\sin 18.92° = 37(0.3242) = \mathbf{12.0\,mm}$,

or, using Pythagoras' theorem,

$$37^2 = 35^2 + AC^2,$$

from which $AC = \sqrt{(37^2 - 35^2)} = \mathbf{12.0\,mm}$

Problem 16. Solve triangle XYZ given $\angle X = 90°$, $\angle Y = 23°$ and $YZ = 20.0$ mm. Determine also its area.

It is always advisable to make a reasonably accurate sketch so as to visualise the expected magnitudes of unknown sides and angles. Such a sketch is shown in Fig. 10.15.

$$\angle Z = 180° - 90° - 23° = \mathbf{67°}$$

$$\sin 23° = \frac{XZ}{20.0} \quad \text{hence} \quad XZ = 20.0\sin 23°$$

$$= 20.0(0.3907) = \mathbf{7.814\,mm}$$

$$\cos 23° = \frac{XY}{20.0} \quad \text{hence} \quad XY = 20.0\cos 23°$$

$$= 20.0(0.9205) = \mathbf{18.41\,mm}$$

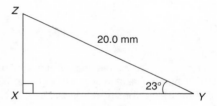

Fig. 10.15

(Check: Using Pythagoras' theorem $(18.41)^2 + (7.814)^2 = 400.0 = (20.0)^2$)

$$\text{Area of triangle } XYZ = \frac{1}{2}(\text{base})(\text{perpendicular height})$$

$$= \frac{1}{2}(XY)(XZ) = \frac{1}{2}(18.41)(7.814)$$

$$= \mathbf{71.93\,mm^2}$$

Now try the following exercise

Exercise 35 Further problems on the solution of right-angled triangles (Answers on page 302)

1. Solve triangle ABC in Fig. 10.16(i).
2. Solve triangle DEF in Fig. 10.16(ii).
3. Solve the triangle JKL in Fig. 10.17(i) and find its area.
4. Solve the triangle MNO in Fig. 10.17(ii) and find its area.

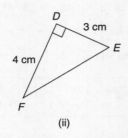

Fig. 10.16

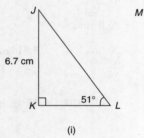
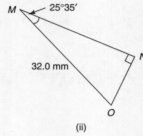

Fig. 10.17

5. A ladder rests against the top of the perpendicular wall of a building and makes an angle of 73° with the ground. If the foot of the ladder is 2 m from the wall, calculate the height of the building.

10.6   Graphs of trigonometric functions

By drawing up tables of values from 0° to 360°, graphs of $y = \sin A$, $y = \cos A$ and $y = \tan A$ may be plotted. Values obtained with a calculator (correct to 3 decimal places – which is more than sufficient for plotting graphs), using 30° intervals, are shown below, with the respective graphs shown in Fig. 10.18.

(a) $y = \sin A$

A	0	30°	60°	90°	120°	150°	180°
sin A	0	0.500	0.866	1.000	0.866	0.500	0

A	210°	240°	270°	300°	330°	360°
sin A	−0.500	−0.866	−1.000	−0.866	−0.500	0

(b) $y = \cos A$

A	0	30°	60°	90°	120°	150°	180°
cos A	1.000	0.866	0.500	0	−0.500	−0.866	−1.000

A	210°	240°	270°	300°	330°	360°
cos A	−0.866	−0.500	0	0.500	0.866	1.000

(c) $y = \tan A$

A	0	30°	60°	90°	120°	150°	180°
tan A	0	0.577	1.732	∞	−1.732	−0.577	0

A	210°	240°	270°	300°	330°	360°
tan A	0.577	1.732	∞	−1.732	−0.577	0

From Fig. 10.19 it is seen that:

(i) Sine and cosine graphs oscillate between peak values of ±1

(ii) The cosine curve is the same shape as the sine curve but displaced by 90°

(iii) The sine and cosine curves are continuous and they repeat at intervals of 360°; the tangent curve appears to be discontinuous and repeats at intervals of 180°

10.7   Sine and cosine rules

To **'solve a triangle'** means 'to find the values of unknown sides and angles'. If a triangle is **right angled**, trigonometric ratios and the theorem of Pythagoras may be used for its solution, as shown earlier. However, for a **non-right-angled triangle**, trigonometric ratios and Pythagoras' theorem **cannot** be used. Instead, two rules, called the **sine rule** and the **cosine rule**, are used.

Sine rule

With reference to triangle *ABC* of Fig. 10.19, the **sine rule** states:

$$\frac{a}{\sin A} = \frac{b}{\sin B} = \frac{c}{\sin C}$$

The rule may be used only when:

(i) 1 side and any 2 angles are initially given, or

(ii) 2 sides and an angle (not the included angle) are initially given.

Cosine rule

With reference to triangle *ABC* of Fig. 10.19, the **cosine rule** states:

$$a^2 = b^2 + c^2 - 2bc\cos A$$
$$\text{or}\quad b^2 = a^2 + c^2 - 2ac\cos B$$
$$\text{or}\quad c^2 = a^2 + b^2 - 2ab\cos C$$

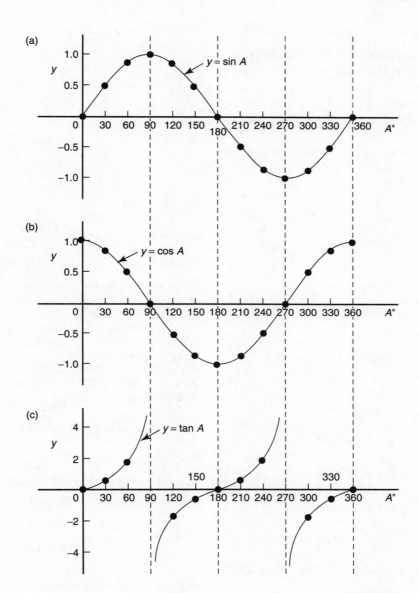

Fig. 10.18

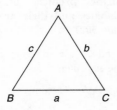

Fig. 10.19

The rule may be used only when:

(i) 2 sides and the included angle are initially given, or

(ii) 3 sides are initially given.

10.8 Area of any triangle

The **area of any triangle** such as *ABC* of Fig. 10.19 is given by:

(i) $\frac{1}{2} \times$ **base** $\times$ **perpendicular height**, or

(ii) $\frac{1}{2}ab \sin C$ or $\frac{1}{2}ac \sin B$ or $\frac{1}{2}bc \sin A$, or

(iii) $\sqrt{[s(s-a)(s-b)(s-c)]}$

where $s = \dfrac{a+b+c}{2}$

10.9 Worked problems on the solution of triangles and their areas

> *Problem 17.* In a triangle XYZ, $\angle X = 51°$, $\angle Y = 67°$ and $YZ = 15.2$ cm. Solve the triangle and find its area.

The triangle XYZ is shown in Fig. 10.20. Since the angles in a triangle add up to 180°, then $Z = 180° - 51° - 67° = \mathbf{62°}$

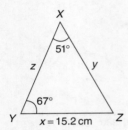

Fig. 10.20

Applying the sine rule:

$$\frac{15.2}{\sin 51°} = \frac{y}{\sin 67°} = \frac{z}{\sin 62°}$$

Using $\dfrac{15.2}{\sin 51°} = \dfrac{y}{\sin 67°}$ and transposing gives:

$$y = \frac{15.2 \sin 67°}{\sin 51°} = \mathbf{18.00\,cm} = XZ$$

Using $\dfrac{15.2}{\sin 51°} = \dfrac{z}{\sin 62°}$ and transposing gives:

$$z = \frac{15.2 \sin 62°}{\sin 51°} = \mathbf{17.27\,cm} = XY$$

Area of triangle $XYZ = \frac{1}{2}xy \sin Z = \frac{1}{2}(15.2)(18.00) \sin 62°$

$$= \mathbf{120.8\,cm^2}$$

(or area $= \frac{1}{2}xz \sin Y = \frac{1}{2}(15.2)(17.27) \sin 67° = \mathbf{120.8\,cm^2}$)

It is always worth checking with triangle problems that the longest side is opposite the largest angle, and vice-versa. In this problem, Y is the largest angle and XZ is the longest of the three sides.

> *Problem 18.* Solve the triangle ABC given $B = 78.85°$, $AC = 22.31$ mm and $AB = 17.92$ mm. Find also its area.

Triangle ABC is shown in Fig. 10.21.
Applying the sine rule:

$$\frac{22.31}{\sin 78.85°} = \frac{17.92}{\sin C}$$

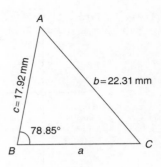

Fig. 10.21

from which, $\qquad \sin C = \dfrac{17.92 \sin 78.85°}{22.31} = 0.7881$

Hence $\qquad C = \sin^{-1} 0.7881 = 52.0°$

[It can be seen from Fig. 10.18, that if $y = 0.7881$ two values of A are possible, i.e. 52.0° and $(180° - 52.0°) = 128.0°$ by symmetry. Your calculator does not give you this second value; it has to be deduced from Figure 10.18.]

Since $B = 78.85°$, C cannot be $128°0'$, since $128°0' + 78.85°$ is greater than 180°.

Thus only $C = 52.0°$ is valid.
Angle $A = 180° - 78.85° - 52.0° = 49.15°$
Applying the sine rule:

$$\frac{a}{\sin 49.15°} = \frac{22.31}{\sin 78.85°}$$

from which, $\qquad a = \dfrac{22.31 \sin 49.15°}{\sin 78.85°} = 17.20$ mm

Hence $A = \mathbf{49.15°}$, $C = \mathbf{52.0°}$ and $BC = \mathbf{17.20\,mm}$.
Area of triangle $ABC = \frac{1}{2}ac \sin B$

$$= \frac{1}{2}(17.20)(17.92) \sin 78.85°$$

$$= \mathbf{151.2\,mm^2}$$

Now try the following exercise

> **Exercise 36 Further problems on the solution of triangles and their areas (Answers on page 302)**
>
> In Problems 1 and 2, use the sine rule to solve the triangles ABC and find their areas:
>
> 1. $A = 29°$, $B = 68°$, $b = 27$ mm.
> 2. $B = 71.43°$, $C = 56.53°$, $b = 8.60$ cm.
>
> In Problems 3 and 4, use the sine rule to solve the triangles DEF and find their areas:
>
> 3. $d = 17$ cm, $f = 22$ cm, $F = 26°$
> 4. $d = 32.6$ mm, $e = 25.4$ mm, $D = 104.37°$

10.10 Further worked problems on the solution of triangles and their areas

> *Problem 19.* Solve triangle *DEF* and find its area given that $EF = 35.0$ mm, $DE = 25.0$ mm and $\angle E = 64°$

Triangle *DEF* is shown in Fig. 10.22.

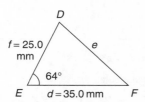

Fig. 10.22

Applying the cosine rule:

$$e^2 = d^2 + f^2 - 2df \cos E$$

i.e. $\quad e^2 = (35.0)^2 + (25.0)^2 - [2(35.0)(25.0) \cos 64°]$

$$= 1225 + 625 - 767.1 = 1083$$

from which, $\quad e = \sqrt{1083} = \mathbf{32.91\,mm}$

Applying the sine rule:

$$\frac{32.91}{\sin 64°} = \frac{25.0}{\sin F}$$

from which, $\quad \sin F = \dfrac{25.0 \sin 64°}{32.91} = 0.6828$

Thus $\quad \angle F = \sin^{-1} 0.6828 = 43.06°$ or $136.94°$

by symmetry in Fig. 10.18. $F = 136.94°$ is not possible in this case since $136.94° + 64°$ is greater than $180°$. Thus only $F = \mathbf{43.06°}$ is valid.

$$\angle D = 180° - 64° - 43.06° = \mathbf{72.94°}$$

Area of triangle $DEF = \frac{1}{2} df \sin E$

$$= \tfrac{1}{2}(35.0)(25.0) \sin 64° = \mathbf{393.2\,mm^2}$$

> *Problem 20.* A triangle *ABC* has sides $a = 9.0$ cm, $b = 7.5$ cm and $c = 6.5$ cm. Determine its three angles and its area.

Triangle *ABC* is shown in Fig. 10.23. It is usual to first calculate the largest angle to determine whether the triangle is acute or obtuse. In this case the largest angle is A (i.e. opposite the longest side).

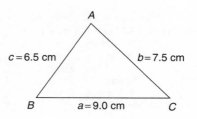

Fig. 10.23

Applying the cosine rule:

$$a^2 = b^2 + c^2 - 2bc \cos A$$

from which, $\quad 2bc \cos A = b^2 + c^2 - a^2$

and $\quad \cos A = \dfrac{b^2 + c^2 - a^2}{2bc}$

$$= \frac{7.5^2 + 6.5^2 - 9.0^2}{2(7.5)(6.5)} = 0.1795$$

Hence $A = \cos^{-1} 0.1795 = \mathbf{79.66°}$ (or $280.34°$ by symmetry in Fig. 10.18(b); however, since there are only $180°$ in a triangle this latter value is impossible).

Applying the sine rule:

$$\frac{9.0}{\sin 79.66°} = \frac{7.5}{\sin B}$$

from which, $\quad \sin B = \dfrac{7.5 \sin 79.66°}{9.0} = 0.8198$

Hence $\quad B = \sin^{-1} 0.8198 = \mathbf{55.06°}$

and $\quad C = 180° - 79.66° - 55.06° = \mathbf{45.28°}$

$$\text{Area} = \sqrt{[s(s-a)(s-b)(s-c)]}$$

where $\quad s = \dfrac{a+b+c}{2} = \dfrac{9.0 + 7.5 + 6.5}{2} = 11.5 \text{ cm}$

Hence **area** $= \sqrt{[11.5(11.5 - 9.0)(11.5 - 7.5)(11.5 - 6.5)]}$

$$= \sqrt{[11.5(2.5)(4.0)(5.0)]} = \mathbf{23.98\,cm^2}$$

Alternatively,

$$\text{area} = \tfrac{1}{2} ab \sin C = \tfrac{1}{2}(9.0)(7.5) \sin 45.28° = \mathbf{23.98\,cm^2}$$

Now try the following exercise

> **Exercise 37 Further problems on the solution of triangles and their areas (Answers on page 302)**
>
> In Problems 1 and 2, use the cosine and sine rules to solve the triangles *PQR* and find their areas.
>
> 1. $q = 12$ cm, $r = 16$ cm, $P = 54°$
> 2. $q = 3.25$ m, $r = 4.42$ m, $P = 105°$

In Problems 3 and 4, use the cosine and sine rules to solve the triangles XYZ and find their areas.

3. $x = 10.0$ cm, $y = 8.0$ cm, $z = 7.0$ cm.

4. $x = 21$ mm, $y = 34$ mm, $z = 42$ mm.

10.11 Practical situations involving trigonometry

There are a number of **practical situations** where the use of trigonometry is needed to find unknown sides and angles of triangles. This is demonstrated in the following worked problems.

Problem 21. A room 8.0 m wide has a span roof which slopes at 33° on one side and 40° on the other. Find the length of the roof slopes, correct to the nearest centimetre.

A section of the roof is shown in Fig. 10.24.

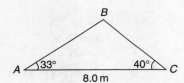

Fig. 10.24

Angle at ridge, $B = 180° - 33° - 40° = 107°$

From the sine rule:

$$\frac{8.0}{\sin 107°} = \frac{a}{\sin 33°}$$

from which, $\quad a = \dfrac{8.0 \sin 33°}{\sin 107°} = 4.556$ m

Also from the sine rule:

$$\frac{8.0}{\sin 107°} = \frac{c}{\sin 40°}$$

from which, $\quad c = \dfrac{8.0 \sin 40°}{\sin 107°} = 5.377$ m

Hence the roof slopes are 4.56 m and 5.38 m, correct to the nearest centimetre.

Problem 22. Two voltage phasors are shown in Fig. 10.25. If $V_1 = 40$ V and $V_2 = 100$ V determine the value of their resultant (i.e. length OA) and the angle the resultant makes with V_1

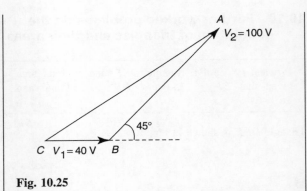

Fig. 10.25

Angle $OBA = 180° - 45° = 135°$

Applying the cosine rule:

$$OA^2 = V_1^2 + V_2^2 - 2V_1V_2 \cos OBA$$
$$= 40^2 + 100^2 - \{2(40)(100) \cos 135°\}$$
$$= 1600 + 10\,000 - \{-5657\}$$
$$= 1600 + 10\,000 + 5657 = 17\,257$$

The resultant $\quad OA = \sqrt{17\,257} = 131.4$ V

Applying the sine rule:

$$\frac{131.4}{\sin 135°} = \frac{100}{\sin AOB}$$

from which, $\quad \sin AOB = \dfrac{100 \sin 135°}{131.4} = 0.5381$

Hence angle $AOB = \sin^{-1} 0.5381 = 32.55°$ or $147.45°$, which is impossible in this case.

Hence the resultant voltage is 131.4 volts at 32.55° to V_1

Problem 23. In Fig. 10.26, PR represents the inclined jib of a crane and is 10.0 m long. PQ is 4.0 m long. Determine the inclination of the jib to the vertical and the length of tie QR.

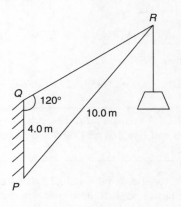

Fig. 10.26

Applying the sine rule:

$$\frac{PR}{\sin 120°} = \frac{PQ}{\sin R}$$

from which, $\quad \sin R = \dfrac{PQ \sin 120°}{PR}$

$$= \frac{(4.0) \sin 120°}{10.0} = 0.3464$$

Hence $\angle R = \sin^{-1} 0.3464 = 20.27°$ (or $159.73°$, which is impossible in this case).

$$\angle P = 180° - 120° - 20.27° = \mathbf{39.73°}$$

which is the inclination of the jib to the vertical.

Applying the sine rule:

$$\frac{10.0}{\sin 120°} = \frac{QR}{\sin 39.73°}$$

from which, $\quad$ **length of tie, $QR = \dfrac{10.0 \sin 39.73°}{\sin 120°}$**

$$= \mathbf{7.38\,m}$$

Problem 24. The area of a field is in the form of a quadrilateral $ABCD$ as shown in Fig. 10.27. Determine its area.

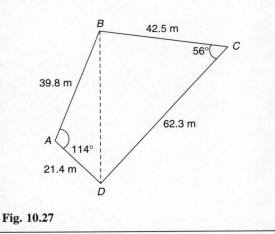

Fig. 10.27

A diagonal drawn from B to D divides the quadrilateral into two triangles.

Area of quadrilateral $ABCD$

$$= \text{area of triangle } ABD + \text{area of triangle } BCD$$

$$= \tfrac{1}{2}(39.8)(21.4) \sin 114° + \tfrac{1}{2}(42.5)(62.3) \sin 56°$$

$$= 389.04 + 1097.5 = \mathbf{1487\,m^2}$$

Now try the following exercise

Exercise 38 **Further problems on practical situations involving trigonometry (Answers on page 302)**

1. A ship P sails at a steady speed of 45 km/h in a direction of W 32° N (i.e. a bearing of 302°) from a port. At the same time another ship Q leaves the port at a steady speed of 35 km/h in a direction N 15° E (i.e. a bearing of 015°). Determine their distance apart after 4 hours.

2. A jib crane is shown in Fig. 10.28. If the tie rod PR is 8.0 long and PQ is 4.5 m long determine (a) the length of jib RQ and (b) the angle between the jib and the tie rod.

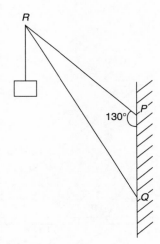

Fig. 10.28

3. A building site is in the form of a quadrilateral as shown in Fig. 10.29, and its area is $1510\,m^2$. Determine the length of the perimeter of the site.

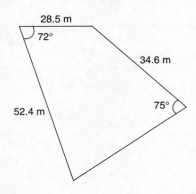

Fig. 10.29

4. Determine the length of members *BF* and *EB* in the roof truss shown in Fig. 10.30.

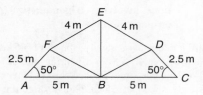

Fig. 10.30

5. Calculate, correct to 3 significant figures, the co-ordinates *x* and *y* to locate the hole centre at *P* shown in Fig. 10.31.

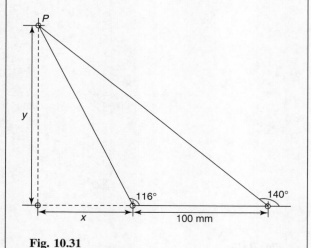

Fig. 10.31

Assignment 3

This assignment covers the material contained in Chapters 8 to 10. The marks for each question are shown in brackets at the end of each question.

1. Solve the following pairs of simultaneous equations:

   (a) $7x - 3y = 23$ (b) $3a - 8 + \dfrac{b}{8} = 0$

   $2x + 4y = -8$ $b + \dfrac{a}{2} = \dfrac{21}{4}$ (12)

2. In an engineering process two variables *x* and *y* are related by the equation $y = ax + \dfrac{b}{x}$ where *a* and *b* are constants. Evaluate *a* and *b* if $y = 15$ when $x = 1$ and $y = 13$ when $x = 3$ (5)

3. Determine the gradient and intercept on the *y*-axis for the equation:

   $$y = -5x + 2$$ (2)

4. The equation of a line is $2y = 4x + 7$. A table of corresponding values is produced and is as shown below. Complete the table and plot a graph of *y* against *x*. Determine the gradient of the graph.   (6)

x	−3	−2	−1	0	1	2	3
y	−2.5					7.5	

5. Plot the graphs $y = 3x + 2$ and $\dfrac{y}{2} + x = 6$ on the same axes and determine the co-ordinates of their point of intersection.   (7)

6. Figure A3.1 shows a plan view of a kite design. Calculate the lengths of the dimensions shown as *a* and *b*.   (4)

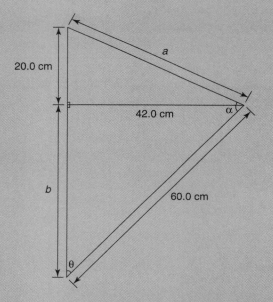

Fig. A3.1

7. In Fig. A3.1, evaluate (a) angle θ (b) angle α.   (5)

8. Determine the area of the plan view of a kite shown in Fig. A3.1.   (4)

9. Evaluate, each correct to 4 significant figures:

   (a) $\sin 231.78°$ (b) $\cos 151°16'$ (c) $\tan \dfrac{3\pi}{8}$ (3)

10. Determine the acute angle $\cos^{-1} 0.4117$ (i) in degrees and minutes, and (ii) in radians (correct to 2 decimal places).   (2)

Multi-choice questions on applied mathematics

Exercise 39 (Answers on page 303)

1. The relationship between the temperature in degrees Fahrenheit (F) and the temperature in degrees Celsius (C) is given by: $F = \frac{9}{5}C + 32$

   135°F is equivalent to:

   (a) 43°C (b) 57.2° (c) 185.4°C (d) 184°C

2. When $a = 2$, $b = 1$ and $c = -3$, the engineering expression $2ab - bc + abc$ is equal to:

   (a) 13 (b) −5 (c) 7 (d) 1

3. 11 mm expressed as a percentage of 41 mm is:

   (a) 2.68, correct to 3 significant figures

   (b) 26.83, correct to 2 decimal places

   (c) 2.6, correct to 2 significant figures

   (d) 0.2682, correct to 4 decimal places

4. When two resistors R_1 and R_2 are connected in parallel the formula $\frac{1}{R_T} = \frac{1}{R_1} + \frac{1}{R_2}$ is used to determine the total resistance R_T. If $R_1 = 470\,\Omega$ and $R_2 = 2.7\,k\Omega$, R_T (correct to 3 significant figures) is equal to:

   (a) $2.68\,\Omega$ (b) $400\,\Omega$ (c) $473\,\Omega$ (d) $3170\,\Omega$

5. $1\frac{1}{3} + 1\frac{2}{3} \div 2\frac{2}{3} - \frac{1}{3}$ is equal to:

   (a) $1\frac{5}{8}$ (b) $\frac{19}{24}$ (c) $2\frac{1}{21}$ (d) $1\frac{2}{7}$

6. Four engineers can complete a task in 5 hours. Assuming the rate of work remains constant, six engineers will complete the task in:

   (a) 126 h   (b) 4 h 48 min

   (c) 3 h 20 min   (d) 7 h 30 min

7. In an engineering equation $\frac{3^4}{3^r} = \frac{1}{9}$. The value of r is:

   (a) −6 (b) 2 (c) 6 (d) −2

8. A graph of resistance against voltage for an electrical circuit is shown in Fig. M1. The equation relating resistance R and voltage V is:

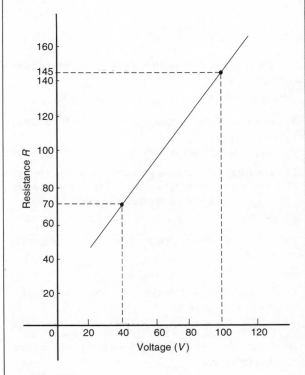

Fig. M1

(a) $R = 1.45\,V + 40$ (b) $R = 0.8\,V + 20$

(c) $R = 1.45\,V + 20$ (d) $R = 1.25\,V + 20$

9. $\dfrac{1}{9.83} - \dfrac{1}{16.91}$ is equal to:

(a) 0.0425, correct to 4 significant figures

(b) 4.26×10^{-2}, correct to 2 decimal places

(c) −0.141, correct to 3 decimal places

(d) 0.042, correct to 2 significant figures

10. A formula for the focal length f of a convex lens is $\dfrac{1}{f} = \dfrac{1}{u} + \dfrac{1}{v}$. When $f = 4$ and $u = 6$, v is:

(a) −2   (b) $\dfrac{1}{12}$ (c) 12   (d) $-\dfrac{1}{2}$

11. The engineering expression $\left(\dfrac{p+2p}{p}\right) \times 3p - 2p$ simplifies to:

(a) $7p$ (b) $6p$ (c) $9p^2 - 2p$ (d) $3p$

12. $(9.2 \times 10^2 + 1.1 \times 10^3)$ cm in standard form is equal to:

(a) 10.3×10^2 cm   (b) 20.2×10^2 cm

(c) 10.3×10^3 cm   (d) 2.02×10^3 cm

13. The current I in an a.c. circuit is given by:

$$I = \dfrac{V}{\sqrt{R^2 + X^2}}$$

When $R = 4.8$, $X = 10.5$ and $I = 15$, the value of voltage V is:

(a) 173.18   (b) 1.30   (c) 0.98   (d) 229.50

14. In the engineering equation: $2^4 \times 2^t = 64$, the value of t is:

(a) 1   (b) 2   (c) 3   (d) 4

15. The height s of a mass projected vertically upwards at time t is given by: $s = ut - \frac{1}{2}gt$. When $g = 10$, $t = 1.5$ and $s = 3.75$, the value of u is:

(a) 10   (b) −5   (c) +5   (d) −10

16. In the right-angled triangle ABC shown in Fig. M2, sine A is given by:

(a) b/a   (b) c/b   (c) b/c   (d) a/b

17. In the right-angled triangle ABC shown in Fig. M2, cosine C is given by:

(a) a/b   (b) c/b   (c) a/c   (d) b/a

18. In the right-angled triangle shown in Fig. M2, tangent A is given by:

(a) b/c   (b) a/c   (c) a/b   (d) c/a

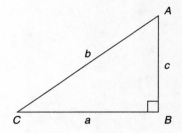

Fig. M2

19. Which of the straight lines shown in Fig. M3 has the equation $y + 4 = 2x$?

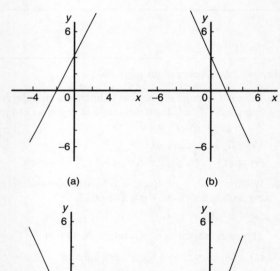

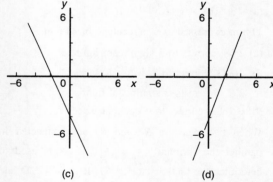

Fig. M3

20. The quantity of heat Q is given by the formula $Q = mc(t_2 - t_1)$. When $m = 5$, $t_1 = 20$, $c = 8$ and $Q = 1200$, the value of t_2 is:

(a) 10   (b) 1.5   (c) 21.5   (d) 50

21. When $p = 3$, $q = -\frac{1}{2}$ and $r = -2$, the engineering expressions $2p^2q^3r^4$ is equal to:

(a) −36   (b) 1296   (c) 36   (d) 18

22. Transposing $v = f\lambda$ to make wavelength λ the subject gives:

(a) $\dfrac{v}{f}$ (b) $v + f$ (c) $f - v$ (d) $\dfrac{f}{v}$

23. The lowest common multiple (LCM) of the numbers 15, 105, 420 and 1155 is:

(a) 4620   (b) 1155   (c) 15   (d) 1695

24. $\dfrac{3}{4} \div 1\dfrac{3}{4}$ is equal to:

(a) $\dfrac{3}{7}$ (b) $1\dfrac{9}{16}$ (c) $1\dfrac{5}{16}$ (d) $2\dfrac{1}{2}$

25. The equation of a straight line graph is $2y = 5 - 4x$. The gradient of the straight line is:

(a) -2 (b) $\dfrac{5}{2}$ (c) -4 (d) 5

26. If $(4^x)^2 = 16$, the value of x is:

(a) 4   (b) 2   (c) 3   (d) 1

27. Tranposing $I = \dfrac{V}{R}$ for resistance R gives:

(a) $I - V$ (b) $\dfrac{V}{I}$ (c) $\dfrac{I}{V}$ (d) VI

28. 23 mm expressed as a percentage of 3.25 cm is:

(a) 0.708, correct to 3 decimal places

(b) 7.08, correct to 2 decimal places

(c) 70.77, correct to 4 significant figures

(d) 14.13, correct to 2 decimal places

29. $16^{-3/4}$ is equal to:

(a) 8   (b) $-\dfrac{1}{2^3}$ (c) 4   (d) $\dfrac{1}{8}$

30. In the engineering equation $\dfrac{(3^2)^3}{(3)(9)^2} = 3^{2x}$, the value of x is:

(a) 0   (b) $\dfrac{1}{2}$ (c) 1   (d) $1\dfrac{1}{2}$

31. If $x = \dfrac{57.06 \times 0.0711}{\sqrt{0.0635}}$ cm, which of the following statements is correct?

(a) $x = 16$ cm, correct to 2 significant figures

(b) $x = 16.09$ cm, correct to 4 significant figures

(c) $x = 1.61 \times 10^1$ cm, correct to 3 decimal places

(d) $x = 16.099$ cm, correct to 3 decimal places

32. Volume $= \dfrac{\text{mass}}{\text{density}}$. The density (in kg/m³) when the mass is 2.532 kg and the volume is 162 cm³ is:

(a) 0.01563 kg/m³   (b) 410.2 kg/m³

(c) 15 630 kg/m³   (d) 64.0 kg/m³

33. $(5.5 \times 10^2)(2 \times 10^3)$ cm in standard form is equal to:

(a) 11×10^6 cm   (b) 1.1×10^6 cm

(c) 11×10^5 cm   (d) 1.1×10^5 cm

34. In the triangular template ABC shown in Fig. M4, the length AC is:

(a) 6.17 cm   (b) 11.17 cm

(c) 9.22 cm   (d) 12.40 cm

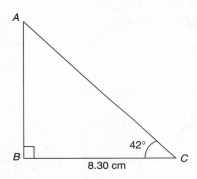

Fig. M4

35. $3k + 2k \times 4k + k \div (6k - 2k)$ simplifies to:

(a) $\dfrac{25k}{4}$ (b) $1 + 2k$

(c) $\dfrac{1}{4} + k(3 + 8k)$ (d) $20k^2 + \dfrac{1}{4}$

36. Correct to 3 decimal places, $\sin(-2.6 \text{ rad})$ is:

(a) 0.516   (b) -0.045 (c) -0.516 (d) 0.045

37. The engineering expression $a^3b^2c \times ab^{-4}c^3 \div a^2bc^{-2}$ is equivalent to:

(a) $a^2b^5c^2$ (b) $\dfrac{a^2c^6}{b^3}$ (c) a^6bc^2 (d) $a^{3/2}b^{-7}c^{-3/2}$

38. The highest common factor (HCF) of the numbers 12, 30 and 42 is:

(a) 42   (b) 12   (c) 420   (d) 6

39. Resistance R ohms varies with temperature t according to the formula $R = R_0(1 + \alpha t)$. Given $R = 21\,\Omega$, $\alpha = 0.004$ and $t = 100$, R_0 has a value of:

(a) $21.4\,\Omega$ (b) $29.4\,\Omega$ (c) $15\,\Omega$ (d) $0.067\,\Omega$

40. The solution of the simultaneous equations

$3x - 2y = 13$ and $2x + 5y = -4$ is:

(a) $x = -2, y = 3$ (b) $x = 1, y = -5$

(c) $x = 3, y = -2$ (d) $x = -7, y = 2$

41. An engineering equation is $y = 2a^2b^3c^4$. When $a = 2$, $b = -\frac{1}{2}$ and $c = 2$, the value of y is:

    (a) -16 (b) -12 (c) 16 (d) 12

42. Electrical resistance $R = \dfrac{\rho l}{a}$; transposing this equation for l gives:

    (a) $\dfrac{Ra}{\rho}$ (b) $\dfrac{R}{a\rho}$ (c) $\dfrac{a}{R\rho}$ (d) $\dfrac{\rho a}{R}$

43. For the right-angled triangle PQR shown in Fig. M5, angle R is equal to:

    (a) $41.41°$ (b) $48.59°$
    (c) $36.87°$ (d) $53.13°$

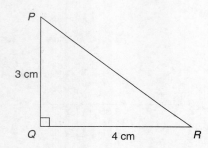

3 cm

Q   4 cm   R

Fig. M5

44. The value of $\dfrac{0.01372}{7.5}$ m is equal to:

    (a) 1.829 m, correct to 4 significant figures

    (b) 0.001829 m, correct to 6 decimal places

    (c) 0.0182 m, correct to 4 significant figures

    (d) 1.829×10^{-2} m, correct to 3 decimal places

45. A graph of y against x, two engineering quantities, produces a straight line. A table of values is shown below:

x	2	-1	p
y	9	3	5

    The value of p is:

    (a) $-\frac{1}{2}$ (b) -2 (c) 3 (d) 0

46. The engineering expression $\dfrac{(16 \times 4)^2}{(8 \times 2)^4}$ is equal to:

    (a) 4 (b) 2^{-4} (c) $\dfrac{1}{2^2}$ (d) 1

47. The area A of a triangular piece of land of sides a, b and c may be calculated using:

$$A = \sqrt{[s(s-a)(s-b)(s-c)]}$$

where $s = \dfrac{a+b+c}{2}$. When $a = 15$ m, $b = 11$ m and $c = 8$ m, the area, correct to the nearest square metre, is:

    (a) $1836\,\text{m}^2$ (b) $648\,\text{m}^2$ (c) $445\,\text{m}^2$ (d) $43\,\text{m}^2$

48. 0.001010 cm expressed in standard form is:

    (a) 1010×10^{-6} cm   (b) 1.010×10^{-3} cm

    (c) 0.101×10^{-2} cm   (d) 1.010×10^2 cm

49. In a system of pylleys, the effort P required to raise a load W is given by $P = aW + b$, where a and b are constants. If $W = 40$ when $P = 12$ and $W = 90$ when $P = 22$, the values of a and b are:

    (a) $a = 5, b = \frac{1}{4}$ (b) $a = 1, b = -28$

    (c) $a = \frac{1}{3}, b = -8$ (d) $a = \frac{1}{5}, b = 4$

50. $(16^{-1/4} - 27^{-2/3})$ is equal to:

    (a) $\dfrac{7}{18}$ (b) -7 (c) $1\dfrac{8}{9}$ (d) $-8\dfrac{1}{2}$

51. If $\cos A = \dfrac{12}{13}$, then $\sin A$ is equal to:

    (a) $\dfrac{5}{13}$ (b) $\dfrac{13}{12}$ (c) $\dfrac{5}{12}$ (d) $\dfrac{12}{5}$

52. The area of triangle XYZ in Fig. M6 is:

    (a) $24.22\,\text{cm}^2$ (b) $19.35\,\text{cm}^2$
    (c) $38.72\,\text{cm}^2$ (d) $32.16\,\text{cm}^2$

53. 3.810×10^{-3} is equal to:

    (a) 0.003810 (b) 0.0381 (c) 3810 (d) 0.3810

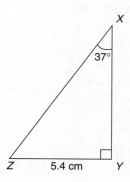

X

37°

Z   5.4 cm   Y

Fig. M6

54. A graph relating effort E (plotted vertically) against load L (plotted horizontally) for a set of pulleys is given by $L + 30 = 6E$. The gradient of the graph is:

    (a) $\frac{1}{6}$ (b) 5 (c) 6 (d) $\frac{1}{5}$

55. The value of $42 \div 3 - 2 \times 3 + 15 \div (1+4)$ is:

    (a) 39   (b) 131   (c) $43\frac{1}{5}$ (d) 11

56. The value, correct to 3 decimal places, of $\cos\left(\dfrac{-3\pi}{4}\right)$ is:

    (a) 0.999   (b) 0.707   (c) -0.999 (d) -0.707

57. $\dfrac{5^2 \times 5^{-3}}{5^{-4}}$ is equivalent to:

    (a) 5^{-5} (b) 5^{24} (c) 5^3 (d) 5

58. The value of $24 \div \frac{8}{3} - 16[2 + (-3 \times 4) - 10]$ is:

    (a) -311 (b) 329   (c) -256 (d) 384

59. A triangle has sides $a = 9.0$ cm, $b = 8.0$ cm and $c = 6.0$ cm. Angle A is equal to:

    (a) 82.42°   (b) 56.49°   (c) 78.58°   (d) 79.87°

60. The value of $\dfrac{2}{5}$ of $\left(4\dfrac{1}{2} - 3\dfrac{1}{4}\right) + 5 \div \dfrac{5}{16} - \dfrac{1}{4}$ is:

    (a) $17\dfrac{7}{20}$ (b) $80\dfrac{1}{2}$ (c) $16\dfrac{1}{4}$ (d) 88

Section 2

Energy Applications

11

SI units and density

At the end of this chapter you should be able to:

- state the seven basic SI units and their symbols
- understand prefixes to make units larger or smaller
- solve simple problems involving length, area, volume and mass
- define density in terms of mass and volume
- define relative density
- appreciate typical values of densities and relative densities for common materials
- perform calculations involving density, mass, volume and relative density

11.1 SI units

The system of units used in engineering and science is the **Systeme Internationale d'Unites** (International system of units), usually abbreviated to SI units, and is based on the metric system. This was introduced in 1960 and is now adopted by the majority of countries as the official system of measurement.

The basic units in the SI system are listed below with their symbols:

Quantity	Unit and symbol
length	metre, m
mass	kilogram, kg
time	second, s
electric current	ampere, A
thermodynamic temperature	kelvin, K
luminous intensity	candela, cd
amount of substance	mole, mol

SI units may be made larger or smaller by using **prefixes** which denote multiplication or division by a particular amount. The eight most common multiples, with their meaning, are listed below:

Prefix	Name	Meaning
T	tera	multiply by 1 000 000 000 000 (i.e. $\times 10^{12}$)
G	giga	multiply by 1 000 000 000 (i.e. $\times 10^{9}$)
M	mega	multiply by 1 000 000 (i.e. $\times 10^{6}$)
k	kilo	multiply by 1000 (i.e. $\times 10^{3}$)
m	milli	divide by 1000 (i.e. $\times 10^{-3}$)
μ	micro	divide by 1 000 000 (i.e. $\times 10^{-6}$)
n	nano	divide by 1 000 000 000 (i.e. $\times 10^{-9}$)
p	pico	divide by 1 000 000 000 000 (i.e. $\times 10^{-12}$)

Length is the distance between two points. The standard unit of length is the **metre**, although the **centimetre, cm, millimetre, mm and kilometre, km**, are often used.

$$1\,cm = 10\,mm, \ 1\,m = 100\,cm = 1000\,mm \text{ and}$$

$$1\,km = 1000\,m$$

Area is a measure of the size or extent of a plane surface and is measured by multiplying a length by a length. If the lengths are in metres then the unit of area is the **square metre, m^2**.

$$1\,m^2 = 1\,m \times 1\,m = 100\,cm \times 100\,cm$$

$$= 10\,000\,cm^2 \text{ or } 10^4\,cm^2$$

$$= 1000\,mm \times 1000\,mm = 1\,000\,000\,mm^2 \text{ or } 10^6\,mm^2$$

Conversely, $1\,cm^2 = 10^{-4}\,m^2$ and $1\,mm^2 = 10^{-6}\,m^2$

Volume is a measure of the space occupied by a solid and is measured by multiplying a length by a length by a length. If the lengths are in metres then the unit of volume is in **cubic metres, m^3**

$$1\,m^3 = 1\,m \times 1\,m \times 1\,m$$
$$= 100\,cm \times 100\,cm \times 100\,cm = 10^6\,cm^3$$
$$= 1000\,mm \times 1000\,mm \times 1000\,mm = 10^9\,mm^3$$

Conversely, $1\,cm^3 = 10^{-6}\,m^3$ and $1\,mm^3 = 10^{-9}\,m^3$

Another unit used to measure volume, particularly with liquids, is the **litre, l**, where $1\,l = 1000\,cm^3$

Mass is the amount of matter in a body and is measured in **kilograms, kg**.

$$1\,kg = 1000\,g \text{ (or conversely, } 1\,g = 10^{-3}\,kg)$$

and 1 tonne (t) $= 1000\,kg$

Problem 1. Express (a) a length of 36 mm in metres, (b) $32\,400\,mm^2$ in square metres, and (c) $8\,540\,000\,mm^3$ in cubic metres.

(a) $1\,m = 10^3\,mm$ or $1\,mm = 10^{-3}\,m$. Hence,
$$36\,mm = 36 \times 10^{-3}\,m = \frac{36}{10^3}\,m = \frac{36}{1000}\,m = \mathbf{0.036\,m}$$

(b) $1\,m^2 = 10^6\,mm^2$ or $1\,mm^2 = 10^{-6}\,m^2$. Hence,
$$32\,400\,mm^2 = 32\,400 \times 10^{-6}\,m^2 = \frac{32\,400}{1\,000\,000}\,m^2$$
$$= \mathbf{0.0324\,m^2}$$

(c) $1\,m^3 = 10^9\,mm^3$ or $1\,mm^3 = 10^{-9}\,m^2$. Hence,
$$8\,540\,000\,mm^3 = 8\,540\,000 \times 10^{-9}\,m^3 = \frac{8\,540\,000}{10^9}\,m^3$$
$$= \mathbf{8.54 \times 10^{-3}\,m^3} \text{ or } \mathbf{0.00854\,m^3}$$

Problem 2. Determine the area of a room 15 m long by 8 m wide in (a) m^2 (b) cm^2 (c) mm^2

(a) Area of room $= 15\,m \times 8\,m = \mathbf{120\,m^2}$

(b) $120\,m^2 = 120 \times 10^4\,cm^2$, since $1\,m^2 = 10^4\,cm^2$
$$= \mathbf{1\,200\,000\,cm^2} \text{ or } \mathbf{1.2 \times 10^6\,cm^2}$$

(c) $120\,m^2 = 120 \times 10^6\,mm^2$, since $1\,m^2 = 10^6\,mm^2$
$$= \mathbf{120\,000\,000\,mm^2} \text{ or } \mathbf{0.12 \times 10^9\,mm^2}$$

(Note, it is usual to express the power of 10 as a multiple of 3, i.e. $\times 10^3$ or $\times 10^6$ or $\times 10^{-9}$, and so on.)

Problem 3. A cube has sides each of length 50 mm. Determine the volume of the cube in cubic metres.

Volume of cube $= 50\,mm \times 50\,mm \times 50\,mm = 125\,000\,mm^3$
$1\,mm^3 = 10^{-9}\,m^3$, thus
volume $= 125\,000 \times 10^{-9}\,m^3 = \mathbf{0.125 \times 10^{-3}\,m^3}$

Problem 4. A container has a capacity of 2.5 litres. Calculate its volume in (a) m^3 (b) mm^3

Since 1 litre $= 1000\,cm^3$, 2.5 litres $= 2.5 \times 1000\,cm^3$
$$= 2500\,cm^3$$

(a) $2500\,cm^3 = 2500 \times 10^{-6}\,m^3$
$$= \mathbf{2.5 \times 10^{-3}\,m^3} \text{ or } \mathbf{0.0025\,m^3}$$

(b) $2500\,cm^3 = 2500 \times 10^3\,mm^3$
$$= \mathbf{2\,500\,000\,mm^3} \text{ or } \mathbf{2.5 \times 10^6\,mm^3}$$

Now try the following exercise

Exercise 40   Further problems on SI units (Answers on page 303)

1. Express
   (a) a length of 52 mm in metres
   (b) $20\,000\,mm^2$ in square metres
   (c) $10\,000\,000\,mm^3$ in cubic metres
2. A garage measures 5 m by 2.5 m. Determine the area in
   (a) m^2 (b) mm^2
3. The height of the garage in Question 2 is 3 m. Determine the volume in
   (a) m^3 (b) mm^3
4. A bottle contains 6.3 litres of liquid. Determine the volume in
   (a) m^3 (b) cm^3 (c) mm^3

11.2 Density

Density is the mass per unit volume of a substance. The symbol used for density is ρ (Greek letter rho) and its units are kg/m^3.

Density $= \dfrac{mass}{volume}$, i.e.

$$\boxed{\rho = \frac{m}{V}} \quad \text{or} \quad \boxed{m = \rho V} \quad \text{or} \quad \boxed{V = \frac{m}{\rho}}$$

where m is the mass in kg, V is the volume in m^3 and ρ is the density in kg/m^3.

Some **typical values of densities** include:

Aluminium	2700 kg/m^3	Steel	7800 kg/m^3
Cast iron	7000 kg/m^3	Petrol	700 kg/m^3
Cork	250 kg/m^3	Lead	11 400 kg/m^3
Copper	8900 kg/m^3	Water	1000 kg/m^3

The relative density of a substance is the ratio of the density of the substance to the density of water, i.e.

$$\text{relative density} = \frac{\text{density of substance}}{\text{density of water}}$$

Relative density has no units, since it is the ratio of two similar quantities. Typical values of relative densities can be determined from above (since water has a density of 1000 kg/m^3), and include:

Aluminium	2.7	Steel	7.8
Cast iron	7.0	Petrol	0.7
Cork	0.25	Lead	11.4
Copper	8.9		

The relative density of a liquid may be measured using a **hydrometer**.

Problem 5. Determine the density of 50 cm^3 of copper if its mass is 445 g.

Volume = 50 cm^3 = 50×10^{-6} m^3 and
mass = 445 g = 445×10^{-3} kg

$$\textbf{Density} = \frac{\text{mass}}{\text{volume}} = \frac{445 \times 10^{-3}\,\text{kg}}{50 \times 10^{-6}\,\text{m}^3} = \frac{445}{50} \times 10^3$$

$$= \textbf{8.9} \times \textbf{10}^3\,\textbf{kg/m}^3 \text{ or } \textbf{8900 kg/m}^3$$

Problem 6. The density of aluminium is 2700 kg/m^3. Calculate the mass of a block of aluminium which has a volume of 100 cm^3

Density, $\rho = 2700$ kg/m^3 and
volume $V = 100$ cm^3 = 100×10^{-6} m^3

Since density = mass/volume, then

$$\text{mass} = \text{density} \times \text{volume}$$

Hence

$$\textbf{mass} = \rho V = 2700\,\text{kg/m}^3 \times 100 \times 10^{-6}\,\text{m}^3$$

$$= \frac{2700 \times 100}{10^6}\,\text{kg} = \textbf{0.270 kg} \text{ or } \textbf{270 g}$$

Problem 7. Determine the volume, in litres, of 20 kg of paraffin oil of density 800 kg/m^3

$$\text{Density} = \frac{\text{mass}}{\text{volume}} \text{ hence volume} = \frac{\text{mass}}{\text{density}}$$

$$\text{Thus volume} = \frac{m}{\rho} = \frac{20\,\text{kg}}{800\,\text{kg/m}^3} = \frac{1}{40}\,\text{m}^3$$

$$= \frac{1}{40} \times 10^6\,\text{cm}^3 = 25\,000\,\text{cm}^3$$

$$1\,\text{litre} = 1000\,\text{cm}^3 \text{ hence } 25\,000\,\text{cm}^3 = \frac{25\,000}{1000} = \textbf{25 litres}$$

Problem 8. Determine the relative density of a piece of steel of density 7850 kg/m^3. Take the density of water as 1000 kg/m^3

$$\textbf{Relative density} = \frac{\text{density of steel}}{\text{density of water}} = \frac{7850}{1000} = \textbf{7.85}$$

Problem 9. A piece of metal 200 mm long, 150 mm wide and 10 mm thick has a mass of 2700 g. What is the density of the metal?

$$\text{Volume of metal} = 200\,\text{mm} \times 150\,\text{mm} \times 10\,\text{mm}$$

$$= 300\,000\,\text{mm}^3 = 3 \times 10^5\,\text{mm}^3$$

$$= \frac{3 \times 10^5}{10^9}\,\text{m}^3 = 3 \times 10^{-4}\,\text{m}^3$$

$$\text{Mass} = 2700\,\text{g} = 2.7\,\text{kg}$$

$$\textbf{Density} = \frac{\text{mass}}{\text{volume}} = \frac{2.7\,\text{kg}}{3 \times 10^{-4}\,\text{m}^3} = 0.9 \times 10^4\,\text{kg/m}^3$$

$$= \textbf{9000 kg/m}^3$$

Problem 10. Cork has a relative density of 0.25. Calculate (a) the density of cork and (b) the volume in cubic centimetres of 50 g of cork. Take the density of water to be 1000 kg/m^3

(a) Relative density $= \dfrac{\text{density of cork}}{\text{density of water,}}$ from which

density of cork = relative density $\times$ density of water,
i.e. **density of cork**, $\rho = 0.25 \times 1000 = \textbf{250 kg/m}^3$

(b) Density $= \dfrac{\text{mass}}{\text{volume}}$, from which

volume $= \dfrac{\text{mass}}{\text{density}}$

Mass, m = 50 g = 50×10^{-3} kg

Hence **volume**, $V = \dfrac{m}{\rho} = \dfrac{50 \times 10^{-3}\,\text{kg}}{250\,\text{kg/m}^3} = \dfrac{0.05}{250}\,\text{m}^3$

$\qquad\qquad = \dfrac{0.05}{250} \times 10^6\,\text{cm}^3 = \textbf{200 cm}^3$

Now try the following exercises

Exercise 41   Further problems on density (Answers on page 303)

1. Determine the density of 200 cm^3 of lead which has a mass of 2280 g

2. The density of iron is 7500 kg/m^3. If the volume of a piece of iron is 200 cm^3, determine its mass

3. Determine the volume, in litres, of 14 kg of petrol of density 700 kg/m^3

4. The density of water is 1000 kg/m^3. Determine the relative density of a piece of copper of density 8900 kg/m^3

5. A piece of metal 100 mm long, 80 mm wide and 20 mm thick has a mass of 1280 g. Determine the density of the metal.

6. Some oil has a relative density of 0.80. Determine

    (a) the density of the oil, and

    (b) the volume of 2 kg of oil.

    Take the density of water as 1000 kg/m^3

Exercise 42   Short answer questions on SI units and density

1. State the SI units for length, mass and time.

2. State the SI units for electric current and thermodynamic temperature.

3. What is the meaning of the following prefixes?

    (a) M   (b) m   (c) μ   (d) k

In Questions 4 to 8, complete the statements

4. 1 m = ... mm and 1 km = ... m

5. 1 m^2 = ... cm^2 and 1 cm^2 = ... mm^2

6. 1 l = ... cm^3 and 1 m^3 = ... mm^3

7. 1 kg = ... g and 1 t = ... kg

8. 1 mm^2 = ... m^2 and 1 cm^3 = ... m^3

9. Define density

10. What is meant by 'relative density'?

11. Relative density of liquids may be measured using a
    . . .

Exercise 43   Multi-choice questions on SI units and density (Answers on page 303)

1. Which of the following statements is true?

    1000 mm^3 is equivalent to:

    (a) 1 m^3 (b) 10^{-3} m^3 (c) 10^{-6} m^3 (d) 10^{-9} m^3

2. Which of the following statements is true?

    (a) 1 mm^2 = 10^{-4} m^2 (b) 1 cm^3 = 10^{-3} m^3
    (c) 1 mm^3 = 10^{-6} m^3 (d) 1 km^2 = 10^{10} cm^2

3. Which of the following statements is false?

    1000 litres is equivalent to:

    (a) 10^3 m^3 (b) 10^6 cm^3
    (c) 10^9 mm^3 (d) 10^3 cm^3

4. Let mass = A, volume = B and density = C. Which of the following statements is false ?

    (a) $C = A - B$ (b) $C = \dfrac{A}{B}$
    (c) $B = \dfrac{C}{A}$ (d) $A = BC$

5. The density of 100 cm^3 of a material having a mass of 700 g is:

    (a) 70 000 kg/m^3 (b) 7000 kg/m^3
    (c) 7 kg/m^3 (d) 70 kg/m^3

6. An alloy has a relative density of 10. If the density of water is 1000 kg/m^3, the density of the alloy is:

    (a) 100 kg/m^3 (b) 0.01 kg/m^3
    (c) 10 000 kg/m^3 (d) 1010 kg/m^3

7. Three fundamental physical quantities in the SI system are:
    (a)   mass, velocity, length
    (b)   time, length, mass
    (c)   energy, time, length
    (d)   velocity, length, mass

8. 60 μs is equivalent to:

    (a) 0.06 s   (b) 0.00006 s
    (c) 1000 minutes   (d) 0.6 s

9. The density (in kg/m^3) when the mass is 2.532 kg and the volume is 162 cm^3 is:

    (a) 0.01563 kg/m^3 (b) 410.2 kg/m^3
    (c) 15 630 kg/m^3 (d) 64.0 kg/m^3

10. Which of the following statements is true?
    100 cm^3 is equal to:

    (a) 10^{-4} m^3 (b) 10^{-2} m^3
    (c) 0.1 m^3 (d) 10^{-6} m^3

12

Work, energy and power

12.1 Work

If a body moves as a result of a force being applied to it, the force is said to do work on the body. The amount of work done is the product of the applied force and the distance, i.e.

**work done = force × distance moved in the
direction of the force**

The unit of work is the **joule, J**, which is defined as the amount of work done when a force of 1 newton acts for a distance of 1 m in the direction of the force. Thus,

$1 J = 1 N m$.

If a graph is plotted of experimental values of force (on the vertical axis) against distance moved (on the horizontal axis) a force/distance graph or work diagram is produced. **The area under the graph represents the work done**.

For example, a constant force of 20 N used to raise a load a height of 8 m may be represented on a force/distance graph as shown in Fig. 12.1. The area under the graph shown shaded represents the work done. Hence

work done $= 20\,\text{N} \times 8\,\text{m} = \textbf{160 J}$

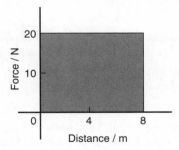

Fig. 12.1

Similarly, a spring extended by 20 mm by a force of 500 N may be represented by the work diagram shown in Fig. 12.2, where

work done = shaded area

$= \frac{1}{2} \times \text{base} \times \text{height}$

$= \frac{1}{2} \times (20 \times 10^{-3})\,\text{m} \times 500\,\text{N} = \textbf{5 J}$

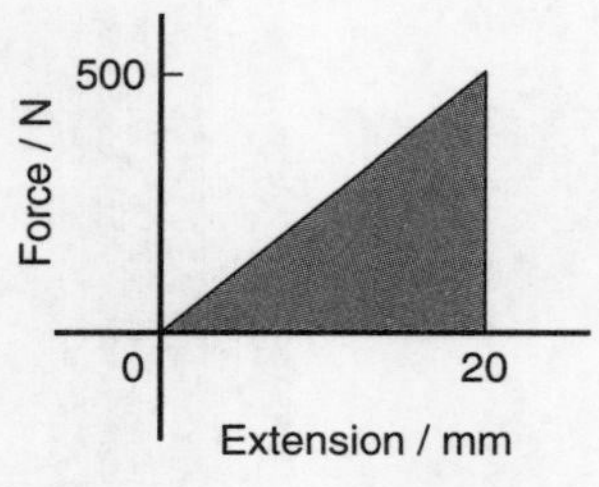

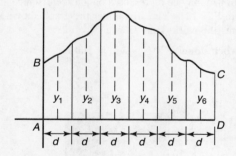

Fig. 12.2

It is shown in Chapter 26 that force = mass × acceleration, and that if an object is dropped from a height it has a constant acceleration of around $9.81 \, \text{m/s}^2$. Thus if a mass of 8 kg is lifted vertically 4 m, the work done is given by:

$$\text{work done} = \text{force} \times \text{distance}$$

$$= (\text{mass} \times \text{acceleration}) \times \text{distance}$$

$$= (8 \times 9.81) \times 4 = 313.92 \, \text{N}$$

The work done by a variable force may be found by determining the area enclosed by the force/distance graph using an approximate method such as the **mid-ordinate rule**.

To determine the area ABCD of Fig. 12.3 using the mid-ordinate rule:

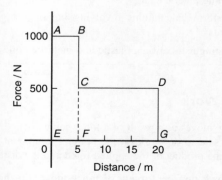

Fig. 12.3

(i)   Divide base AD into any number of equal intervals, each of width d (the greater the number of intervals, the greater the accuracy).

(ii) Erect ordinates in the middle of each interval (shown by broken lines in Fig. 12.3).

(iii) Accurately measure ordinates y_1, y_2, y_3, etc.

(iv) Area ABCD $= d(y_1 + y_2 + y_3 + y_4 + y_5 + y_6)$.

In general, the mid-ordinate rule states:

Area = (width of interval) (sum of mid-ordinates)

Problem 1.   Calculate the work done when a force of 40 N pushes an object a distance of 500 m in the same direction as the force.

$$\text{Work done} = \text{force} \times \text{distance moved in the direction}$$
$$\text{of the force}$$

$$= 40 \, \text{N} \times 500 \, \text{m} = 20\,000 \, \text{J}$$

$$\text{(since } 1 \, \text{J} = 1 \, \text{Nm)}$$

i.e. **work done = 20 kJ**

Problem 2.   Calculate the work done when a mass is lifted vertically by a crane to a height of 5 m, the force required to lift the mass being 98 N.

When work is done in lifting then:

$$\text{work done} = (\text{weight of the body})$$
$$\times (\text{vertical distance moved})$$

Weight is the downward force due to the mass of an object. Hence

work done $= 98 \, \text{N} \times 5 \, \text{m} = \textbf{490 J}$

Problem 3.   A motor supplies a constant force of 1 kN which is used to move a load a distance of 5 m. The force is then changed to a constant 500 N and the load is moved a further 15 m. Draw the force/distance graph for the operation and from the graph determine the work done by the motor.

The force/distance graph or work diagram is shown in Fig. 12.4. Between points A and B a constant force of 1000 N moves the load 5 m; between points C and D a constant force of 500 N moves the load from 5 m to 20 m

Fig. 12.4

Total work done = area under the force/distance graph

$$= \text{area ABFE} + \text{area CDGF}$$

$$= (1000\,\text{N} \times 5\,\text{m}) + (500\,\text{N} \times 15\,\text{m})$$

$$= 5000\,\text{J} + 7500\,\text{J} = 12\,500\,\text{J}$$

$$= \textbf{12.5\,kJ}$$

Problem 4. A spring, initially in a relaxed state, is extended by 100 mm. Determine the work done by using a work diagram if the spring requires a force of 0.6 N per mm of stretch.

Force required for a 100 mm extension = 100 mm × 0.6 N/mm = 60 N. Figure 12.5 shows the force/extension graph or work diagram representing the increase in extension in proportion to the force, as the force is increased from 0 to 60 N. The work done is the area under the graph, hence

$$\text{work done} = \tfrac{1}{2} \times \text{base} \times \text{height}$$

$$= \tfrac{1}{2} \times 100\,\text{mm} \times 60\,\text{N}$$

$$= \tfrac{1}{2} \times 100 \times 10^{-3}\,\text{m} \times 60\,\text{N} = \textbf{3\,J}$$

(Alternatively, average force during extension $= \dfrac{(60-0)}{2}$ = 30 N and total extension = 100 mm = 0.1 m, hence

$$\text{work done} = \text{average force} \times \text{extension}$$

$$= 30\,\text{N} \times 0.1\,\text{m} = 3\,\text{J})$$

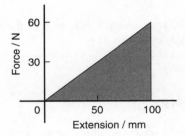

Fig. 12.5

Problem 5. A spring requires a force of 10 N to cause an extension of 50 mm. Determine the work done in extending the spring (a) from zero to 30 mm, and (b) from 30 mm to 50 mm.

Figure 12.6 shows the force/extension graph for the spring.

(a) Work done in extending the spring from zero to 30 mm is given by area ABO of Fig. 12.6, i.e.

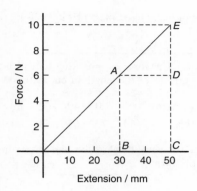

Fig. 12.6

$$\text{work done} = \tfrac{1}{2} \times \text{base} \times \text{height}$$

$$= \tfrac{1}{2} \times 30 \times 10^{-3}\,\text{m} \times 6\,\text{N}$$

$$= 90 \times 10^{-3}\,\text{J} = \textbf{0.09\,J}$$

(b) Work done in extending the spring from 30 mm to 50 mm is given by area ABCE of Fig. 12.6, i.e.

$$\text{work done} = \text{area ABCD} + \text{area ADE}$$

$$= (20 \times 10^{-3}\,\text{m} \times 6\,\text{N})$$

$$+ \tfrac{1}{2} \times (20 \times 10^{-3}\,\text{m})(4\,\text{N})$$

$$= 0.12\,\text{J} + 0.04\,\text{J} = \textbf{0.16\,J}$$

Problem 6. Calculate the work done when a mass of 20 kg is lifted vertically through a distance of 5.0 m. Assume that the acceleration due to gravity is 9.81 m/s².

The force to be overcome when lifting a mass of 20 kg vertically upwards is mg, i.e. 20 × 9.81 = 196.2 N (see Chapter's 25 and 26)

Work done = force × distance = 196.2 × 5.0 = **981 J**

Problem 7. Water is pumped vertically upwards through a distance of 50.0 m and the work done is 294.3 kJ. Determine the number of litres of water pumped. (1 litre of water has a mass of 1 kg).

Work done = force × distance, i.e.

$$2\,94\,300 = \text{force} \times 50.0, \text{ from which}$$

$$\text{force} = \frac{2\,94\,300}{50.0} = 5886\,\text{N}$$

The force to be overcome when lifting a mass m kg vertically upwards is mg, i.e. (m × 9.81) N (see Chapters 25 and 26).

Thus 5886 = m × 9.81, from which mass,

$$m = \frac{5886}{9.81} = 600\,\text{kg}$$

Since 1 litre of water has a mass of 1 kg, **600 litres of water are pumped**.

Problem 8. The force on a cutting tool of a shaping machine varies over the length of cut as follows:

Distance (mm)	0	20	40	60	80	100
Force (kN)	60	72	65	53	44	50

Determine the work done as the tool moves through a distance of 100 mm.

The force/distance graph for the given data is shown in Fig. 12.7. The work done is given by the area under the graph; the area may be determined by an approximate method. Using the mid-ordinate rule, with each strip of width 20 mm, mid-ordinates y_1, y_2, y_3, y_4 and y_5 are erected as shown, and each is measured.

$$\text{Area under curve} = (\text{width of each strip})$$
$$\times (\text{sum of mid-ordinate values})$$
$$= (20)(69 + 69.5 + 59 + 48 + 45.5)$$
$$= (20)(291)$$
$$= 5820 \text{ kN mm} = 5820 \text{ Nm} = 5820 \text{ J}$$

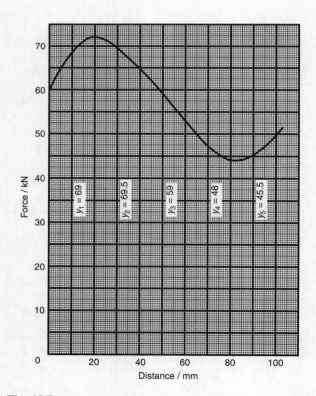

Fig. 12.7

Hence the work done as the tool moves through 100 mm is **5.82 kJ**.

Now try the following exercise

Exercise 44   Further problems on work (Answers on page 303)

1. Determine the work done when a force of 50 N pushes an object 1.5 km in the same direction as the force.

2. Calculate the work done when a mass of weight 200 N is lifted vertically by a crane to a height of 100 m.

3. A motor supplies a constant force of 2 kN to move a load 10 m. The force is then changed to a constant 1.5 kN and the load is moved a further 20 m.

   Draw the force/distance graph for the complete operation, and, from the graph, determine the total work done by the motor.

4. A spring, initially relaxed, is extended 80 mm. Draw a work diagram and hence determine the work done if the spring requires a force of 0.5 N/mm of stretch.

5. A spring requires a force of 50 N to cause an extension of 100 mm. Determine the work done in extending the spring (a) from 0 to 100 mm, and (b) from 40 mm to 100 mm.

6. The resistance to a cutting tool varies during the cutting stroke of 800 mm as follows: (i) the resistance increases uniformly from an initial 5000 N to 10 000 N as the tool moves 500 mm, and (ii) the resistance falls uniformly from 10 000 N to 6000 N as the tool moves 300 mm.

   Draw the work diagram and calculate the work done in one cutting stroke.

12.2 Energy

Energy is the capacity, or ability, to do work. The unit of energy is the joule, the same as for work. Energy is expended when work is done. There are several forms of energy and these include:

(i)   Mechanical energy
(ii) Heat or thermal energy
(iii) Electrical energy
(iv) Chemical energy
(v)   Nuclear energy
(vi) Light energy
(vii) Sound energy

Energy may be converted from one form to another. **The principle of conservation of energy** states that the total amount of energy remains the same in such conversions, i.e. energy cannot be created or destroyed.

Some examples of energy conversions include:

(i) Mechanical energy is converted to electrical energy by a generator.

(ii) Electrical energy is converted to mechanical energy by a motor.

(iii) Heat energy is converted to mechanical energy by a steam engine.

(iv) Mechanical energy is converted to heat energy by friction.

(v) Heat energy is converted to electrical energy by a solar cell.

(vi) Electrical energy is converted to heat energy by an electric fire.

(vii) Heat energy is converted to chemical energy by living plants.

(viii) Chemical energy is converted to heat energy by burning fuels.

(ix) Heat energy is converted to electrical energy by a thermocouple.

(x) Chemical energy is converted to electrical energy by batteries.

(xi) Electrical energy is converted to light energy by a light bulb.

(xii) Sound energy is converted to electrical energy by a microphone.

(xiii) Electrical energy is converted to chemical energy by electrolysis.

Efficiency is defined as the ratio of the useful output energy to the input energy. The symbol for efficiency is η (Greek letter eta). Hence

$$\text{efficiency, } \eta = \frac{\text{useful output energy}}{\text{input energy}}$$

Efficiency has no units and is often stated as a percentage. A perfect machine would have an efficiency of 100%. However, all machines have an efficiency lower than this due to friction and other losses. Thus, if the input energy to a motor is 1000 J and the output energy is 800 J then the efficiency is

$$\frac{800}{1000} \times 100\% = \mathbf{80\%}$$

Problem 9. A machine exerts a force of 200 N in lifting a mass through a height of 6 m. If 2 kJ of energy are supplied to it, what is the efficiency of the machine?

Work done in lifting mass = force × distance moved

$$= \text{weight of body} \times \text{distance moved}$$

$$= 200\,\text{N} \times 6\,\text{m} = 1200\,\text{J}$$

$$= \text{useful energy output}$$

Energy input = 2 kJ = 2000 J

$$\textbf{Efficiency, } \eta = \frac{\text{useful output energy}}{\text{input energy}}$$

$$= \frac{1200}{2000} = \mathbf{0.6} \text{ or } \mathbf{60\%}$$

Problem 10. Calculate the useful output energy of an electric motor which is 70% efficient if it uses 600 J of electrical energy.

$$\text{Efficiency, } \eta = \frac{\text{useful output energy}}{\text{input energy}}$$

thus

$$\frac{70}{100} = \frac{\text{output energy}}{600\,\text{J}}$$

from which, **output energy** $= \dfrac{70}{100} \times 600 = \mathbf{420\,J}$

Problem 11. 4 kJ of energy are supplied to a machine used for lifting a mass. The force required is 800 N. If the machine has an efficiency of 50%, to what height will it lift the mass?

$$\text{Efficiency, } \eta = \frac{\text{useful output energy}}{\text{input energy}}$$

i.e.

$$\frac{50}{100} = \frac{\text{output energy}}{4000\,\text{J}}$$

from which, output energy $= \dfrac{50}{100} \times 4000 = 2000\,\text{J}$

Work done = force × distance moved, hence

$$2000\,\text{J} = 800\,\text{N} \times \text{height, from which,}$$

$$\textbf{height} = \frac{2000\,\text{J}}{800\,\text{N}} = \mathbf{2.5\,m}$$

Problem 12. A hoist exerts a force of 500 N in raising a load through a height of 20 m. The efficiency of the hoist gears is 75% and the efficiency of the motor is 80%. Calculate the input energy to the hoist.

The hoist system is shown diagrammatically in Fig. 12.8.

$$\text{Output energy} = \text{work done} = \text{force} \times \text{distance}$$

$$= 500\,\text{N} \times 20\,\text{m} = 10\,000\,\text{J}$$

Fig. 12.8

For the gearing,

$$\text{efficiency} = \frac{\text{output energy}}{\text{input energy}}$$

i.e. $\dfrac{75}{100} = \dfrac{10\,000}{\text{input energy}}$ from which,

the input energy to the gears $= 10\,000 \times \dfrac{100}{75} = 13\,333\,\text{J}$

The input energy to the gears is the same as the output energy of the motor. Thus, for the motor,

$$\text{efficiency} = \frac{\text{output energy}}{\text{input energy}}$$

i.e. $\dfrac{80}{100} = \dfrac{13\,333}{\text{input energy}}$

Hence **input energy to the hoist** $= 13\,333 \times \dfrac{100}{80} = 16\,670\,\text{J}$

$$= \mathbf{16.67\,kJ}$$

Now try the following exercise

Exercise 45   Further problems on energy (Answers on page 303)

1. A machine lifts a mass of weight 490.5 N through a height of 12 m when 7.85 kJ of energy is supplied to it. Determine the efficiency of the machine.

2. Determine the output energy of an electric motor which is 60% efficient if it uses 2 kJ of electrical energy.

3. A machine which is used for lifting a particular mass is supplied with 5 kJ of energy. If the machine has an efficiency of 65% and exerts a force of 812.5 N to what height will it lift the mass?

4. A load is hoisted 42 m and requires a force of 100 N. The efficiency of the hoist gear is 60% and that of the motor is 70%. Determine the input energy to the hoist.

12.3   Power

Power is a measure of the rate at which work is done or at which energy is converted from one form to another.

$$\boxed{\text{Power } P = \frac{\text{energy used}}{\text{time taken}}} \quad \text{or} \quad \boxed{P = \frac{\text{work done}}{\text{time taken}}}$$

The unit of power is the **watt, W**, where 1 watt is equal to 1 joule per second. The watt is a small unit for many purposes and a larger unit called the kilowatt, kW, is used, where $1\,\text{kW} = 1000\,\text{W}$.

The power output of a motor which does 120 kJ of work in 30 s is thus given by

$$P = \frac{120\,\text{kJ}}{30\,\text{s}} = 4\,\text{kW}$$

(For electrical power, see Chapter 13).

Since work done = force × distance, then

$$\text{Power} = \frac{\text{work done}}{\text{time taken}} = \frac{\text{force} \times \text{distance}}{\text{time taken}}$$

$$= \text{force} \times \frac{\text{distance}}{\text{time taken}}$$

However, $\dfrac{\text{distance}}{\text{time taken}} = \text{velocity}$.

Hence $\boxed{\textbf{power} = \textbf{force} \times \textbf{velocity}}$

Problem 13.   The output power of a motor is 8 kW. How much work does it do in 30 s?

$$\text{Power} = \frac{\text{work done}}{\text{time taken}}, \text{from which,}$$

$$\textbf{work done} = \text{power} \times \text{time} = 8000\,\text{W} \times 30\,\text{s}$$

$$= 240\,000\,\text{J} = \mathbf{240\,kJ}$$

Problem 14.   Calculate the power required to lift a mass through a height of 10 m in 20 s if the force required is 3924 N

Work done = force × distance moved

$$= 3924\,\text{N} \times 10\,\text{m} = 39\,240\,\text{J}$$

$$\textbf{Power} = \frac{\text{work done}}{\text{time taken}} = \frac{39\,240\,\text{J}}{20\,\text{s}}$$

$$= \mathbf{1962\,W} \text{ or } \mathbf{1.962\,kW}$$

Problem 15.   10 kJ of work is done by a force in moving a body uniformly through 125 m in 50 s. Determine (a) the value of the force and (b) the power.

(a)   Work done = force × distance, hence

$$10\,000\,\text{J} = \text{force} \times 125\,\text{m}, \text{from which,}$$

$$\textbf{force} = \frac{10\,000\,\text{J}}{125\,\text{m}} = \mathbf{80\,N}$$

(b)   **Power** $= \dfrac{\text{work done}}{\text{time taken}} = \dfrac{10\,000\,\text{J}}{50\,\text{s}} = \mathbf{200\,W}$

Problem 16. A car hauls a trailer at 90 km/h when exerting a steady pull of 600 N. Calculate (a) the work done in 30 minutes and (b) the power required.

(a) Work done = force × distance moved.
The distance moved in 30 min, i.e. $\frac{1}{2}$ h, at 90 km/h = 45 km. Hence

work done = 600 N × 45 000 m = **27 000 kJ** or **27 MJ**

(b) **Power required** = $\dfrac{\text{work done}}{\text{time taken}} = \dfrac{27 \times 10^6\,\text{J}}{30 \times 60\,\text{s}}$

= **15 000 W** or **15 kW**

Problem 17. To what height will a mass of weight 981 N be raised in 40 s by a machine using a power of 2 kW?

Work done = force × distance. Hence,
work done = 981 N × height.

$$\text{Power} = \frac{\text{work done}}{\text{time taken}}, \text{ from which,}$$

work done = power × time taken

= 2000 W × 40 s = 80 000 J

Hence 80 000 = 981 N × height, from which,

$$\textbf{height} = \frac{80\,000\,\text{J}}{981\,\text{N}} = \textbf{81.55 m}$$

Problem 18. A planing machine has a cutting stroke of 2 m and the stroke takes 4 seconds. If the constant resistance to the cutting tool is 900 N calculate for each cutting stroke (a) the power consumed at the tool point, and (b) the power input to the system if the efficiency of the system is 75%

(a) Work done in each cutting stroke = force × distance

= 900 N × 2 m

= 1800 J

Power consumed at tool point = $\dfrac{\text{work done}}{\text{time taken}}$

= $\dfrac{1800\,\text{J}}{4\,\text{s}}$ = **450 W**

(b) Efficiency = $\dfrac{\text{output energy}}{\text{input energy}} = \dfrac{\text{output power}}{\text{input power}}$

Hence $\dfrac{75}{100} = \dfrac{450}{\text{input power}}$ from which,

input power = 450 × $\dfrac{100}{75}$ = **600 W**

Problem 19. An electric motor provides power to a winding machine. The input power to the motor is 2.5 kW and the overall efficiency is 60%. Calculate (a) the output power of the machine, (b) the rate at which it can raise a 300 kg load vertically upwards.

(a) Efficiency, $\eta = \dfrac{\text{power output}}{\text{power input}}$

i.e. $\dfrac{60}{100} = \dfrac{\text{power output}}{2500}$ from which,

power output = $\dfrac{60}{100} \times 2500$

= **1500 W** or **1.5 kW**

(b) Power output = force × velocity, from which,

velocity = $\dfrac{\text{power output}}{\text{force}}$

Force acting on the 300 kg load due to gravity

= 300 kg × 9.81 m/s^2

= 2943 N

Hence, **velocity** = $\dfrac{1500}{2943}$ = **0.510 m/s** or **510 mm/s**

Problem 20. A lorry is travelling at a constant velocity of 72 km/h. The force resisting motion is 800 N. Calculate the tractive power necessary to keep the lorry moving at this speed.

Power = force × velocity
The force necessary to keep the lorry moving at constant speed is equal and opposite to the force resisting motion, i.e. 800 N.

Velocity = 72 km/h = $\dfrac{72 \times 60 \times 60}{1000}$ m/s = 20 m/s

Hence, power = 800 N × 20 m/s = 16 000 Nm/s

= 16 000 J/s = 16 000 W or 16 kW

Thus the tractive power needed to keep the lorry moving at a constant speed of 72 km/h is 16 kW

Problem 21. The variation of tractive force with distance for a vehicle which is accelerating from rest is:

force (kN)	8.0	7.4	5.8	4.5	3.7	3.0
distance (m)	0	10	20	30	40	50

Determine the average power necessary if the time taken to travel the 50 m from rest is 25 s

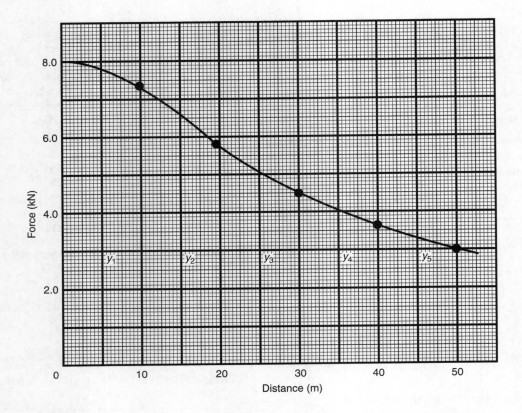

Fig. 12.9

The force/distance diagram is shown in Fig. 12.9. The work done is determined from the area under the curve. Using the mid-ordinate rule with five intervals gives:

$$\text{area} = (\text{width of interval})\,(\text{sum of mid-ordinate})$$

$$= (10)[y_1 + y_2 + y_3 + y_4 + y_5]$$

$$= (10)[7.8 + 6.6 + 5.1 + 4.0 + 3.3]$$

$$= (10)[26.8] = 268\,\text{kNm},$$

i.e. work done $= 268\,\text{kJ}$

Average power $= \dfrac{\text{work done}}{\text{time taken}}$

$$= \dfrac{2\,68\,000\,\text{J}}{25\,\text{s}}$$

$$= \mathbf{10\,720\,W}\ \text{or}\ \mathbf{10.72\,kW}$$

Now try the following exercise

Exercise 46   Further problems on power (Answers on page 303)

1. The output power of a motor is 10 kW. How much work does it do in 1 minute?

2. Determine the power required to lift a load through a height of 20 m in 12.5 s if the force required is 2.5 kN.

3. 25 kJ of work is done by a force in moving an object uniformly through 50 m in 40 s. Calculate (a) the value of the force, and (b) the power.

4. A car towing another at 54 km/h exerts a steady pull of 800 N. Determine (a) the work done in $\frac{1}{4}$ hr, and (b) the power required.

5. To what height will a mass of weight 500 N be raised in 20 s by a motor using 4 kW of power?

6. The output power of a motor is 10 kW. Determine (a) the work done by the motor in 2 hours, and (b) the energy used by the motor if it is 72% efficient.

7. A car is travelling at a constant speed of 81 km/h. The frictional resistance to motion is 0.60 kN. Determine the power required to keep the car moving at this speed.

8. A constant force of 2.0 kN is required to move the table of a shaping machine when a cut is being made. Determine the power required if the stroke of 1.2 m is completed in 5.0 s

9. The variation of force with distance for a vehicle which is decelerating is as follows:

Distance (m)	600	500	400	300	200	100	0
Force (kN)	24	20	16	12	8	4	0

   If the vehicle covers the 600 m in 1.2 minutes, find the power needed to bring the vehicle to rest.

10. A cylindrical bar of steel is turned in a lathe. The tangential cutting force on the tool is 0.5 kN and the cutting speed is 180 mm/s. Determine the power absorbed in cutting the steel.

12.4 Potential and kinetic energy

Mechanical engineering is concerned principally with two kinds of energy, potential energy and kinetic energy.

Potential energy is energy due to the position of the body. The force exerted on a mass of m kg is mg N (where $g = 9.81$ m/s^2, the acceleration due to gravity). When the mass is lifted vertically through a height h m above some datum level, the work done is given by: force × distance = $(mg)(h)$ J. This work done is stored as potential energy in the mass.

Hence,

$$\text{potential energy} = mgh \text{ joules}$$

(the potential energy at the datum level being taken as zero)

Kinetic energy is the energy due to the motion of a body. Suppose a force F acts on an object of mass m originally at rest (i.e. $u = 0$) and accelerates it to a velocity v in a distance s:

$$\text{work done} = \text{force} \times \text{distance}$$
$$= Fs = (ma)(s) \quad \text{(if no energy is lost)}$$

where a is the acceleration

Since $v^2 = u^2 + 2as$ (see Chapter 29) and $u = 0$,

$$v^2 = 2as, \text{ from which } a = \frac{v^2}{2s}$$

hence, work done $= (ma)(s) = (m)\left(\dfrac{v^2}{2s}\right)(s) = \dfrac{1}{2}mv^2$

This energy is called the kinetic energy of the mass m, i.e.

$$\text{kinetic energy} = \frac{1}{2}mv^2 \text{ joules}$$

As stated in Section 12.2, energy may be converted from one form to another. The **principle of conservation of energy** states that the total amount of energy remains the same in such conversions, i.e. energy cannot be created or destroyed.

In mechanics, the potential energy possessed by a body is frequently converted into kinetic energy, and vice versa. When a mass is falling freely, its potential energy decreases as it loses height, and its kinetic energy increases as its velocity increases. Ignoring air frictional losses, at all times:

Potential energy + kinetic energy = a constant

If friction is present, then work is done overcoming the resistance due to friction and this is dissipated as heat. Then,

Initial energy = final energy + work done

overcoming frictional resistance

Kinetic energy is not always conserved in collisions. Collisions in which kinetic energy is conserved (i.e. stays the same) are called **elastic collisions**, and those in which it is not conserved are termed **inelastic collisions**.

Problem 22. A car of mass 800 kg is climbing an incline at 10° to the horizontal. Determine the increase in potential energy of the car as it moves a distance of 50 m up the incline.

With reference to Fig. 12.10,

$$\sin 10° = \frac{\text{opposite}}{\text{hypotenuse}} = \frac{h}{50}, \text{ from which,}$$

$$h = 50 \sin 10° = 8.682 \text{ m}$$

(see Chapter 10 for more on trigonometry)

Hence, increase in potential energy

$$= mgh = 800 \text{ kg} \times 9.81 \text{ m/s}^2 \times 8.682 \text{ m}$$

$$= \textbf{68 140 J} \text{ or } \textbf{68.14 kJ}$$

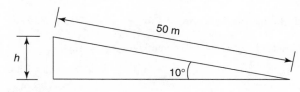

Fig. 12.10

Problem 23. At the instant of striking, a hammer of mass 30 kg has a velocity of 15 m/s. Determine the kinetic energy in the hammer.

Kinetic energy $= \frac{1}{2}mv^2 = \frac{1}{2}(30 \text{ kg})(15 \text{ m/s})^2$

i.e. **kinetic energy in hammer = 3375 J or 3.375 kJ**

Problem 24. A lorry having a mass of 1.5 t is travelling along a level road at 72 km/h. When the brakes are applied, the speed decreases to 18 km/h. Determine how much the kinetic energy of the lorry is reduced.

Initial velocity of lorry, $v_1 = 72$ km/h

$$= 72\frac{km}{h} \times 1000\frac{m}{km} \times \frac{1h}{3600\,s}$$

$$= \frac{72}{3.6} = 20 \text{ m/s},$$

final velocity of lorry, $v_2 = \dfrac{18}{3.6} = 5$ m/s and

mass of lorry, $m = 1.5\,t = 1500$ kg

Initial kinetic energy of the lorry $= \dfrac{1}{2}\,mv_1^2 = \dfrac{1}{2}(1500)(20)^2$

$$= 300 \text{ kJ}$$

Final kinetic energy of the lorry $= \dfrac{1}{2}\,mv_2^2 = \dfrac{1}{2}(1500)(5)^2$

$$= 18.75 \text{ kJ}$$

Hence, **the change in kinetic energy** $= 300 - 18.75$

$$= \textbf{281.25 kJ}$$

(Part of this reduction in kinetic energy is converted into heat energy in the brakes of the lorry and is hence dissipated in overcoming frictional forces and air friction).

Problem 25. A canister containing a meteorology balloon of mass 4 kg is fired vertically upwards from a gun with an initial velocity of 400 m/s. Neglecting the air resistance, calculate (a) its initial kinetic energy, (b) its velocity at a height of 1 km, (c) the maximum height reached.

(a) Initial kinetic energy $= \frac{1}{2}\,mv^2 = \frac{1}{2}(4)(400)^2 = \textbf{320 kJ}$

(b) At a height of 1 km, potential energy $= mgh$

$$= 4 \times 9.81 \times 1000$$

$$= 39.24 \text{ kJ}$$

By the principle of conservation of energy: potential energy + kinetic energy at 1 km = initial kinetic energy.

Hence $39\,240 + \dfrac{1}{2}\,mv^2 = 320\,000$

from which, $\dfrac{1}{2}(4)v^2 = 320\,000 - 39\,240 = 280\,760$

Hence $v = \sqrt{\left(\dfrac{2 \times 280\,760}{4}\right)} = 374.7$ m/s

i.e. **the velocity of the canister at a height of 1 km is 374.7 m/s**

(c) At the maximum height, the velocity of the canister is zero and all the kinetic energy has been converted into potential energy. Hence potential energy = initial kinetic energy = 320 000 J (from part (a)). Then

$$320\,000 = mgh = (4)(9.81)(h), \text{ from which,}$$

height $h = \dfrac{320\,000}{(4)(9.81)} = 8155$ m

i.e. **the maximum height reached is 8155 m**

Problem 26. A pile-driver of mass 500 kg falls freely through a height of 1.5 m on to a pile of mass 200 kg. Determine the velocity with which the driver hits the pile. If, at impact, 3 kJ of energy are lost due to heat and sound, the remaining energy being possessed by the pile and driver as they are driven together into the ground a distance of 200 mm, determine (a) the common velocity immediately after impact, (b) the average resistance of the ground.

The potential energy of the pile-driver is converted into kinetic energy. Thus

$$\text{potential energy} = \text{kinetic energy,}$$

i.e. $mgh = \dfrac{1}{2}\,mv^2$, from which,

velocity $v = \sqrt{2\,gh} = \sqrt{(2)(9.81)(1.5)}$

$$= 5.42 \text{ m/s}$$

Hence, **the pile-driver hits the pile at a velocity of 5.42 m/s**

(a) Before impact, kinetic energy of pile driver

$$= \frac{1}{2}\,mv^2 = \frac{1}{2}(500)(5.42)^2 = 7.34 \text{ kJ}$$

Kinetic energy after impact $= 7.34 - 3 = 4.34$ kJ. Thus the pile-driver and pile together have a mass of $500 + 200 = 700$ kg and possess kinetic energy of 4.34 kJ.

Hence $4.34 \times 10^3 = \dfrac{1}{2}\,mv^2 = \dfrac{1}{2}(700)v^2$ from which,

velocity $v = \sqrt{\left(\dfrac{2 \times 4.34 \times 10^3}{700}\right)} = 3.52$ m/s

Thus, **the common velocity after impact is 3.52 m/s**

(b) The kinetic energy after impact is absorbed in overcoming the resistance of the ground, in a distance of 200 mm.

Kinetic energy = work done = resistance × distance

i.e. 4.34×10^3 = resistance × 0.200, from which,

resistance $= \dfrac{4.34 \times 10^3}{0.200} = 21\,700$ N

Hence, **the average resistance of the ground is 21.7 kN**

Problem 27. A car of mass 600 kg reduces speed from 90 km/h to 54 km/h in 15 s. Determine the braking power required to give this change of speed.

Change in kinetic energy of car $= \frac{1}{2}mv_1^2 - \frac{1}{2}mv_2^2$

where m = mass of car = 600 kg,

$$v_1 = \text{initial velocity} = 90\,\text{km/h} = \frac{90}{3.6}\,\text{m/s} = 25\,\text{m/s}$$

and v_2 = final velocity $= 54\,\text{km/h} = \frac{54}{3.6}\,\text{m/s} = 15\,\text{m/s}$

Hence, change in kinetic energy $= \frac{1}{2}m(v_1^2 - v_2^2)$

$$= \frac{1}{2}(600)(25^2 - 15^2)$$

$$= 120\,000\,\text{J}$$

Braking power $= \dfrac{\text{change in energy}}{\text{time taken}}$

$$= \frac{120\,000\,\text{J}}{15\text{s}} = \textbf{8000 W or 8 kW}$$

Now try the following exercises

Exercise 47 Further problems on potential and kinetic energy (Answers on page 303)

(Assume the acceleration due to gravity, $g = 9.81\,\text{m/s}^2$)

1. An object of mass 400 g is thrown vertically upwards and its maximum increase in potential energy is 32.6 J. Determine the maximum height reached, neglecting air resistance.

2. A ball bearing of mass 100 g rolls down from the top of a chute of length 400 m inclined at an angle of 30° to the horizontal. Determine the decrease in potential energy of the ball bearing as it reaches the bottom of the chute.

3. A vehicle of mass 800 kg is travelling at 54 km/h when its brakes are applied. Find the kinetic energy lost when the car comes to rest.

4. Supplies of mass 300 kg are dropped from a helicopter flying at an altitude of 60 m. Determine the potential energy of the supplies relative to the ground at the instant of release, and its kinetic energy as it strikes the ground.

5. A shell of mass 10 kg is fired vertically upwards with an initial velocity of 200 m/s. Determine its initial kinetic energy and the maximum height reached,

correct to the nearest metre, neglecting air resistance.

6. The potential energy of a mass is increased by 20.0 kJ when it is lifted vertically through a height of 25.0 m. It is now released and allowed to fall freely. Neglecting air resistance, find its kinetic energy and its velocity after it has fallen 10.0 m.

7. A pile-driver of mass 400 kg falls freely through a height of 1.2 m on to a pile of mass 150 kg. Determine the velocity with which the driver hits the pile. If, at impact, 2.5 kJ of energy are lost due to heat and sound, the remaining energy being possessed by the pile and driver as they are driven together into the ground a distance of 150 mm, determine (a) the common velocity after impact, (b) the average resistance of the ground.

Exercise 48 Short answer questions on work, energy and power

1. Define work in terms of force applied and distance moved
2. Define energy, and state its unit
3. Define the joule
4. The area under a force/distance graph represents ....
5. Name five forms of energy
6. State the principle of conservation of energy
7. Give two examples of conversion of heat energy to other forms of energy
8. Give two examples of conversion of electrical energy to other forms of energy
9. Give two examples of conversion of chemical energy to other forms of energy
10. Give two examples of conversion of mechanical energy to other forms of energy
11. (a) Define efficiency in terms of energy input and energy output
    (b) State the symbol used for efficiency
12. Define power and state its unit
13. Define potential energy
14. The change in potential energy of a body of mass m kg when lifted vertically upwards to a height h m is given by.......
15. What is kinetic energy?
16. The kinetic energy of a body of mass m kg and moving at a velocity of v m/s is given by.......
17. Distinguish between elastic and inelastic collisions

Exercise 49   Multi-choice questions on work, energy and power (Answers on page 303)

1. State which of the following is incorrect:

    (a) $1\,W = 1\,J/s$

    (b) $1\,J = 1\,N/m$

    (c) $\eta = \dfrac{\text{output energy}}{\text{input energy}}$

    (d) energy = power × time

2. An object is lifted 2000 mm by a crane. If the force required is 100 N, the work done is:

    (a) $\dfrac{1}{20}\,Nm$ (b) 200 kNm   (c) 200 Nm   (d) 20 J

3. A motor having an efficiency of 0.8 uses 800 J of electrical energy. The output energy of the motor is:

    (a) 800 J   (b) 1000 J   (c) 640 J   (d) 6.4 J

4. 6 kJ of work is done by a force in moving an object uniformly through 120 m in 1 minute. The force applied is:

    (a) 50 N   (b) 20 N   (c) 720 N   (d) 12 N

5. For the object in question 4, the power developed is:

    (a) 6 kW   (b) 12 kW   (c) 5/6 W   (d) 0.1 kW

6. Which of the following statements is false?

    (a) The unit of energy and work is the same

    (b) The area under a force/distance graph gives the work done

    (c) Electrical energy is converted to mechanical energy by a generator

    (d) Efficiency is the ratio of the useful output energy to the input energy

7. A machine using a power of 1 kW requires a force of 100 N to raise a mass in 10 s. The height the mass is raised in this time is:

    (a) 100 m   (b) 1 km   (c) 10 m   (d) 1 m

8. A force/extension graph for a spring is shown in Fig. 12.11 Which of the following statements is false?
    The work done in extending the spring:

    (a) from 0 to 100 mm is 5 J

    (b) from 0 to 50 mm is 1.25 J

    (c) from 20 mm to 60 mm is 1.6 J

    (d) from 60 mm to 100 mm is 3.75 J

9. A vehicle of mass 1 tonne climbs an incline of 30° to the horizontal. Taking the acceleration due to gravity as $10\,m/s^2$, the increase in potential energy of the vehicle as it moves a distance of 200 m up the incline is:

    (a) 1 kJ   (b) 2 MJ   (c) 1 MJ   (d) 2 kJ

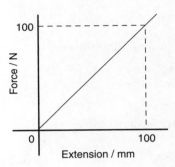

Fig. 12.11

10. A bullet of mass 100 g is fired from a gun with an initial velocity of 360 km/h. Neglecting air resistance, the initial kinetic energy possessed by the bullet is:

    (a) 6.48 kJ   (b) 500 J   (c) 500 kJ   (d) 6.48 MJ

11. A small motor requires 50 W of electrical power in order to produce 40 W of mechanical energy output. The efficiency of the motor is:

    (a) 10%   (b) 80%   (c) 40%   (d) 90%

12. A load is lifted 4000 mm by a crane. If the force required to lift the mass is 100 N, the work done is:

    (a) 400 J   (b) 40 Nm   (c) 25 J   (d) 400 kJ

13. A machine exerts a force of 100 N in lifting a mass through a height of 5 m. If 1 kJ of energy is supplied, the efficiency of the machine is:

    (a) 10%   (b) 20%   (c) 100%   (d) 50%

14. At the instant of striking, a hammer of mass 40 kg has a velocity of 10 m/s. The kinetic energy in the hammer is:

    (a) 2 kJ   (b) 1 kJ   (c) 400 J   (d) 8 kJ

15. A machine which has an efficiency of 80% raises a load of 50 N through a vertical height of 10 m. The work input to the machine is:

    (a) 400 J   (b) 500 J   (c) 800 J   (d) 625 J

Assignment 4

This assignment covers the material contained in Chapters 11 and 12. The marks for each question are shown in brackets at the end of each question.

1. Express (a) a length of 125 mm in metres, (b) 25 000 mm^2 in square metres, and (c) 7 500 000 mm^3 in cubic metres. (6)

2. Determine the area of a room 12 m long by 9 m wide in (a) cm^2 (b) mm^2 (4)

3. A container has a capacity of 10 litres. Calculate its volume in (a) m^3 (b) mm^3 (4)

4. The density of cast iron is 7000 kg/m^3. Calculate the mass of a block of aluminium which has a volume of 200 cm^3 (4)

5. Determine the volume, in litres, of 14 kg of petrol of density 700 kg/m^3 (4)

6. A spring, initially in a relaxed state, is extended by 80 mm. Determine the work done by using a work diagram if the spring requires a force of 0.7 N per mm of stretch. (4)

7. Water is pumped vertically upwards through a distance of 40.0 m and the work done is 176.58 kJ. Determine the number of litres of water pumped. (1 litre of water has a mass of 1 kg). (4)

8. 3 kJ of energy are supplied to a machine used for lifting a mass. The force required is 1 kN. If the machine has an efficiency of 60%, to what height will it lift the mass? (4)

9. When exerting a steady pull of 450 N, a lorry travels at 80 km/h. Calculate (a) the work done in 15 minutes and (b) the power required. (4)

10. An electric motor provides power to a winding machine. The input power to the motor is 4.0 kW and the overall efficiency is 75%. Calculate (a) the output power of the machine, (b) the rate at which it can raise a 509.7 kg load vertically upwards. (4)

11. A tank of mass 4800 kg is climbing an incline at 12° to the horizontal. Determine the increase in potential energy of the tank as it moves a distance of 40 m up the incline. (4)

12. A car of mass 500 kg reduces speed from 108 km/h to 36 km/h in 20 s. Determine the braking power required to give this change of speed. (4)

13

An introduction to electric circuits

At the end of this chapter you should be able to:

- recognize common electrical circuit diagram symbols
- understand that electric current is the rate of movement of charge and is measured in amperes
- appreciate that the unit of charge is the coulomb
- calculate charge or quantity of electricity Q from $Q = It$
- understand that a potential difference between two points in a circuit is required for current to flow
- appreciate that the unit of p.d. is the volt
- understand that resistance opposes current flow and is measured in ohms
- appreciate what an ammeter, a voltmeter, an ohmmeter, a multimeter and a C.R.O. measure
- state Ohm's law as

$$V = IR \quad \text{or} \quad I = \frac{V}{R} \quad \text{or} \quad R = \frac{V}{I}$$

- use Ohm's law in calculations, including multiples and sub-multiples of units
- describe a conductor and an insulator, giving examples of each
- appreciate that electrical power P is given by

$$P = VI = I^2R = \frac{V^2}{R} \text{ watts}$$

- calculate electrical power
- define electrical energy and state its unit
- calculate electrical energy
- state the three main effects of an electric current, giving practical examples of each
- explain the importance of fuses in electrical circuits

13.1 Standard symbols for electrical components

Symbols are used for components in electrical circuit diagrams and some of the more common ones are shown in Fig. 13.1.

13.2 Electric current and quantity of electricity

All **atoms** consist of **protons, neutrons** and **electrons**. The protons, which have positive electrical charges, and the neutrons, which have no electrical charge, are contained within the **nucleus**. Removed from the nucleus are minute negatively charged particles called electrons. Atoms of different materials differ from one another by having different numbers of protons, neutrons and electrons. An equal number of protons and electrons exist within an atom and it is said to be electrically balanced, as the positive and negative charges cancel each other out. When there are more than two electrons in an atom the electrons are arranged into **shells** at various distances from the nucleus.

All atoms are bound together by powerful forces of attraction existing between the nucleus and its electrons. Electrons in the outer shell of an atom, however, are attracted to their nucleus less powerfully than are electrons whose shells are nearer the nucleus.

It is possible for an atom to lose an electron; the atom, which is now called an **ion**, is not now electrically balanced, but is positively charged and is thus able to attract an electron to itself from another atom. Electrons that move from one atom to another are called free electrons and such random motion can continue indefinitely. However, if an electric pressure or **voltage** is applied across any material there is a tendency for electrons to move in a particular direction.

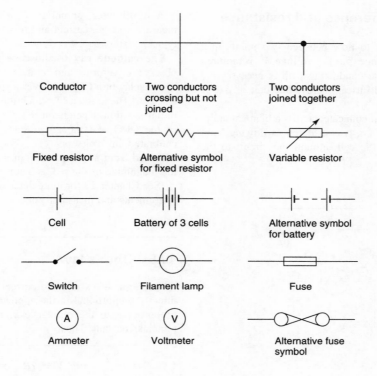

Fig. 13.1

This movement of free electrons, known as **drift**, constitutes an electric current flow.

Thus current is the rate of movement of charge.

Conductors are materials that contain electrons that are loosely connected to the nucleus and can easily move through the material from one atom to another. **Insulators** are materials whose electrons are held firmly to their nucleus.

The unit used to measure the **quantity of electrical charge Q** is called the **coulomb C** (where 1 coulomb = 6.24×10^{18} electrons).

If the drift of electrons in a conductor takes place at the rate of one coulomb per second the resulting current is said to be a current of one ampere.

Thus   1 ampere = 1 coulomb per second or $1\,A = 1\,C/s$

Hence 1 coulomb = 1 ampere second or $1\,C = 1\,As$

Generally, if I is the current in amperes and t the time in seconds during which the current flows, then $I \times t$ represents the quantity of electrical charge in coulombs, i.e. quantity of electrical charge transferred,

$$Q = I \times t \text{ coulombs}$$

Problem 1.   What current must flow if 0.24 coulombs is to be transferred in 15 ms?

Since the quantity of electricity, $Q = It$, then

$$\textbf{current, } I = \frac{Q}{t} = \frac{0.24}{15 \times 10^{-3}} = \frac{0.24 \times 10^3}{15}$$

$$= \frac{240}{15} = \textbf{16 A}$$

Problem 2.   If a current of 10 A flows for four minutes, find the quantity of electricity transferred.

Quantity of electricity, $Q = It$ coulombs

$I = 10\,A$ and $t = 4 \times 60 = 240\,s$

Hence, **charge, $Q = 10 \times 240 = $ 2400 C**

Now try the following exercise

Exercise 50   Further problems on charge (Answers on page 303)

1. In what time would a current of 10 A transfer a charge of 50 C?

2. A current of 6 A flows for 10 minutes. What charge is transferred?

3. How long must a current of 100 mA flow so as to transfer a charge of 80 C?

13.3 Potential difference and resistance

For a continuous current to flow between two points in a circuit a **potential difference (p.d.)** or **voltage**, V, is required between them; a complete conducting path is necessary to and from the source of electrical energy. The unit of p.d. is the **volt, V**.

Figure 13.2 shows a cell connected across a filament lamp. Current flow, by convention, is considered as flowing from the positive terminal of the cell, around the circuit to the negative terminal.

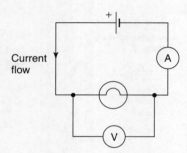

Fig. 13.2

The flow of electric current is subject to friction. This friction, or opposition, is called **resistance R** and is the property of a conductor that limits current. The unit of resistance is the **ohm**; 1 ohm is defined as the resistance which will have a current of 1 ampere flowing through it when 1 volt is connected across it,

i.e. $\boxed{\text{resistance } R = \dfrac{\textbf{Potential difference}}{\textbf{current}}}$

13.4 Basic electrical measuring instruments

An **ammeter** is an instrument used to measure current and must be connected **in series** with the circuit. Figure 13.2 shows an ammeter connected in series with the lamp to measure the current flowing through it. Since all the current in the circuit passes through the ammeter it must have a very **low resistance**.

A **voltmeter** is an instrument used to measure p.d. and must be connected **in parallel** with the part of the circuit whose p.d. is required. In Fig. 13.2, a voltmeter is connected in parallel with the lamp to measure the p.d. across it. To avoid a significant current flowing through it a voltmeter must have a very **high resistance**.

An **ohmmeter** is an instrument for measuring resistance.

A **multimeter**, or universal instrument, may be used to measure voltage, current and resistance. An 'Avometer' is a typical example.

The **cathode ray oscilloscope (CRO)** may be used to observe waveforms and to measure voltages and currents. The display of a CRO involves a spot of light moving across a screen. The amount by which the spot is deflected from its initial position depends on the p.d. applied to the terminals of the CRO and the range selected. The displacement is calibrated in 'volts per cm'. For example, if the spot is deflected 3 cm and the volts/cm switch is on 10 V/cm then the magnitude of the p.d. is 3 cm × 10 V/cm, i.e. 30 V.

(See Chapter 23 for more detail about electrical measuring instruments and measurements.)

13.5 Ohm's law

Ohm's law states that the current I flowing in a circuit is directly proportional to the applied voltage V and inversely proportional to the resistance R, provided the temperature remains constant. Thus,

$$\boxed{I = \dfrac{V}{R} \quad \text{or} \quad V = IR \quad \text{or} \quad R = \dfrac{V}{I}}$$

Problem 3. The current flowing through a resistor is 0.8 A when a p.d. of 20 V is applied. Determine the value of the resistance.

From Ohm's law, **resistance**,

$$R = \frac{V}{I} = \frac{20}{0.8} = \frac{200}{8} = 25\,\Omega$$

13.6 Multiples and sub-multiples

Currents, voltages and resistance's can often be very large or very small. Thus **multiples and sub-multiples** of units are often used, as stated in chapter 11. The most common ones, with an example of each, are listed in Table 13.1.

Problem 4. Determine the p.d. which must be applied to a 2 kΩ resistor in order that a current of 10 mA may flow.

Resistance $R = 2\,\text{k}\Omega = 2 \times 10^3 = 2000\,\Omega$, and

current $I = 10\,\text{mA} = 10 \times 10^{-3}\,\text{A}$ or $\dfrac{10}{10^3}\,\text{A}$ or

$\dfrac{10}{1000}\,\text{A} = 0.01\,\text{A}$

Table 13.1

Prefix	Name	Meaning	Example
M	mega	multiply by 1 000 000 (i.e. $\times 10^6$)	$2\,M\Omega = 2\,000\,000$ ohms
k	kilo	multiply by 1000 (i.e. $\times 10^3$)	$10\,kV = 10\,000$ volts
m	milli	divide by 1000 (i.e. $\times 10^{-3}$)	$25\,mA = \dfrac{25}{1000}\,A$ $= 0.025$ amperes
μ	micro	divide by 1 000 000 (i.e. $\times 10^{-6}$)	$50\,\mu V = \dfrac{50}{1\,000\,000}\,V$ $= 0.00005$ volts

From Ohm's law, **potential difference**,

$$V = IR = (0.01)(2000) = \mathbf{20\,V}$$

Problem 5. A coil has a current of 50 mA flowing through it when the applied voltage is 12 V. What is the resistance of the coil?

Resistance,

$$R = \frac{V}{I} = \frac{12}{50 \times 10^{-3}} = \frac{12 \times 10^3}{50} = \frac{12\,000}{50} = \mathbf{240\,\Omega}$$

Problem 6. The current/voltage relationship for two resistors A and B is as shown in Fig. 13.3. Determine the value of the resistance of each resistor.

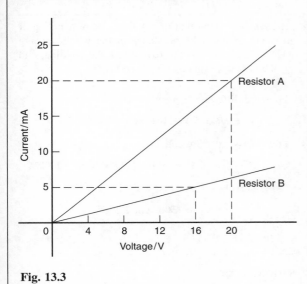

Fig. 13.3

For resistor A, $\quad R = \dfrac{V}{I} = \dfrac{20\,V}{20\,mA} = \dfrac{20}{0.02} = \dfrac{2000}{2}$

$$= \mathbf{1000\,\Omega} \quad \text{or} \quad \mathbf{1\,k\Omega}$$

For resistor B, $\quad R = \dfrac{V}{I} = \dfrac{16\,V}{5\,mA} = \dfrac{16}{0.005} = \dfrac{16\,000}{5}$

$$= \mathbf{3200\,\Omega} \quad \text{or} \quad \mathbf{3.2\,k\Omega}$$

Now try the following exercise

Exercise 51 Further problems on Ohm' law (Answers on page 303)

1. The current flowing through a heating element is 5 A when a p.d. of 35 V is applied across it. Find the resistance of the element.

2. An electric light bulb of resistance 960 Ω is connected to a 240 V supply. Determine the current flowing in the bulb.

3. Graphs of current against voltage for two resistors P and Q are shown in Fig. 13.4. Determine the value of each resistor.

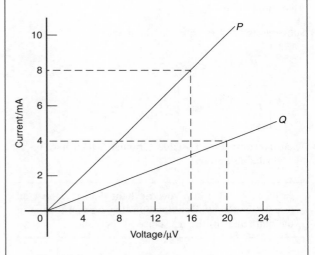

Fig. 13.4

4. Determine the p.d. which must be applied to a $5\,k\Omega$ resistor such that a current of $6\,mA$ may flow.

13.7   Conductors and insulators

A **conductor** is a material having a low resistance which allows electric current to flow in it. All metals are conductors and some examples include copper, aluminium, brass, platinum, silver, gold and carbon.

An **insulator** is a material having a high resistance which does not allow electric current to flow in it. Some examples of insulators include plastic, rubber, glass, porcelain, air, paper, cork, mica, ceramics and certain oils.

13.8   Electrical power and energy

Electrical Power

Power P in an electrical circuit is given by the product of potential difference V and current I. The unit of power is the **watt, W**.

Hence $\boxed{P = V \times I \text{ watts}}$ (1)

From Ohm's law, $V = IR$

Substituting for V in equation (1) gives:

$$P = (IR) \times I$$

i.e. $\boxed{P = I^2R \text{ watts}}$

Also, from Ohm's law, $I = \dfrac{V}{R}$

Substituting for I in equation (1) gives:

$$P = V \times \frac{V}{R}$$

i.e. $\boxed{P = \dfrac{V^2}{R} \text{ watts}}$

There are thus three possible formulae which may be used for calculating power.

Problem 7.   A $100\,W$ electric light bulb is connected to a $250\,V$ supply. Determine (a) the current flowing in the bulb, and (b) the resistance of the bulb.

Power, $P = V \times I$, from which, current $I = \dfrac{P}{V}$

(a) **Current**, $I = \dfrac{100}{250} = \dfrac{10}{25} = \dfrac{2}{5} = \mathbf{0.4\,A}$

(b) **Resistance**, $R = \dfrac{V}{I} = \dfrac{250}{0.4} = \dfrac{2500}{4} = \mathbf{625\,\Omega}$

Problem 8.   Calculate the power dissipated when a current of $4\,mA$ flows through a resistance of $5\,k\Omega$.

Power, $P = I^2R = (4 \times 10^{-3})^2(5 \times 10^3)$

$$= 16 \times 10^{-6} \times 5 \times 10^3 = 80 \times 10^{-3}$$

$$= \mathbf{0.08\,W} \quad \text{or} \quad \mathbf{80\,mW}$$

Alternatively, since

$$I = 4 \times 10^{-3} \quad \text{and} \quad R = 5 \times 10^3$$

then from Ohm's law, voltage

$$V = IR = 4 \times 10^{-3} \times 5 \times 10^3 = 20\,V$$

Hence, **power** $P = V \times I = 20 \times 4 \times 10^{-3} = \mathbf{80\,mW}$

Problem 9.   An electric kettle has a resistance of $30\,\Omega$. What current will flow when it is connected to a $240\,V$ supply? Find also the power rating of the kettle.

Current, $I = \dfrac{V}{I} = \dfrac{240}{30} = \mathbf{8\,A}$

Power, $P = VI = 240 \times 8 = 1920\,W = \mathbf{1.92\,kW}$

$$= \textbf{power rating of kettle}$$

Problem 10.   A current of $5\,A$ flows in the winding of an electric motor, the resistance of the winding being $100\,\Omega$. Determine (a) the p.d. across the winding, and (b) the power dissipated by the coil.

(a) Potential difference across winding,

$$V = IR = 5 \times 100 = \mathbf{500\,V}$$

(b) Power dissipated by coil,

$$P = I^2R = 5^2 \times 100 = \mathbf{2500\,W} \quad \text{or} \quad \mathbf{2.5\,kW}$$

(Alternatively, $P = V \times I = 500 \times 5$

$$= \mathbf{2500\,W} \text{or} \mathbf{2.5\,kW})$$

Electrical Energy

$$\boxed{\textbf{Electrical energy} = \textbf{power} \times \textbf{time}}$$

If the power is measured in watts and the time in seconds then the unit of energy is watt-seconds or **joules**. If the power is measured in kilowatts and the time in hours then the unit of energy is **kilowatt-hours**, often called the '**unit of electricity**'. The 'electricity meter' in the home records the number of kilowatt-hours used and is thus an energy meter.

Problem 11. A 12 V battery is connected across a load having a resistance of 40 Ω. Determine the current flowing in the load, the power consumed and the energy dissipated in 2 minutes.

Current, $I = \dfrac{V}{R} = \dfrac{12}{40} = \mathbf{0.3\,A}$

Power consumed, $P = VI = (12)(0.3) = \mathbf{3.6\,W}$

Energy dissipated = power × time

$$= (3.6\,\text{W})(2 \times 60\,\text{s}) = \mathbf{432\,J}$$

(since $1\,\text{J} = 1\,\text{Ws}$)

Problem 12. A source of e.m.f. of 15 V supplies a current of 2 A for 6 minutes. How much energy is provided in this time?

Energy = power × time, and power = voltage × current

Hence, **energy** $= VIt = 15 \times 2 \times (6 \times 60)$

$$= 10\,800\,\text{Ws or J} = \mathbf{10.8\,kJ}$$

Problem 13. Electrical equipment in an office takes a current of 13 A from a 240 V supply. Estimate the cost per week of electricity if the equipment is used for 30 hours each week and 1 kWh of energy costs 7p.

Power $= VI$ watts $= 240 \times 13 = 3120\,\text{W} = 3.12\,\text{kW}$

Energy used per week = power × time

$$= (3.12\,\text{kW}) \times (30\,\text{h}) = 93.6\,\text{kWh}$$

Cost at 7p per kWh $= 93.6 \times 7 = 655.2\text{p}$

Hence **weekly cost of electricity = £6.55**

Problem 14. An electric heater consumes 3.6 MJ when connected to a 250 V supply for 40 minutes. Find the power rating of the heater and the current taken from the supply.

Power $= \dfrac{\text{energy}}{\text{time}} = \dfrac{3.6 \times 10^6\,\text{J}}{40 \times 60\,\text{s}}$ (or W) $= 1500\,\text{W}$

i.e. **power rating of heater = 1.5 kW**

Power, $P = VI$, thus $I = \dfrac{P}{V} = \dfrac{1500}{250} = 6\,\text{A}$

Hence the current taken from the supply is **6 A**

Problem 15. A business uses two 3 kW fires for an average of 20 hours each per week, and six 150 W lights for 30 hours each per week. If the cost of electricity is 6.4p per unit, determine the weekly cost of electricity to the business.

Energy = power × time

Energy used by one 3 kW fire in 20 hours

$$= 3\,\text{kW} \times 20\,\text{h} = 60\,\text{kWh}$$

Hence weekly energy used by two 3 kW fires

$$= 2 \times 60 = 120\,\text{kWh}$$

Energy used by one 150 W light for 30 hours

$$= 150\,\text{W} \times 30\,\text{h} = 4500\,\text{Wh} = 4.5\,\text{kWh}$$

Hence weekly energy used by six 150 W lamps

$$= 6 \times 4.5 = 27\,\text{kWh}$$

Total energy used per week $= 120 + 27 = 147\,\text{kWh}$

1 unit of electricity $= 1\,\text{kWh}$ of energy

Thus, **weekly cost of energy** at 6.4p per kWh

$$= 6.4 \times 147 = 940.8\text{p} = \mathbf{£9.41}$$

Now try the following exercise

Exercise 52   Further problems on power and energy (Answers on page 303)

1. The hot resistance of a 250 V filament lamp is 625 Ω. Determine the current taken by the lamp and its power rating.

2. Determine the resistance of an electric fire which takes a current of 12 A from a 240 V supply. Find also the power rating of the fire and the energy used in 20 h.

3. Determine the power dissipated when a current of 10 mA flows through an appliance having a resistance of 8 kΩ.

4. 85.5 J of energy are converted into heat in 9 s. What power is dissipated?

5. A current of 4 A flows through a conductor and 10 W is dissipated. What p.d. exists across the ends of the conductor?

6. Find the power dissipated when:
   (a) a current of 5 mA flows through a resistance of 20 kΩ
   (b) a voltage of 400 V is applied across a 120 kΩ resistor
   (c) a voltage applied to a resistor is 10 kV and the current flow is 4 mA

7. A d.c. electric motor consumes 72 MJ when connected to 400 V supply for 2 h 30 min. Find the power rating of the motor and the current taken from the supply.

8. A p.d. of 500 V is applied across the winding of an electric motor and the resistance of the winding is 50 Ω. Determine the power dissipated by the coil.

9. In a household during a particular week three 2 kW fires are used on average 25 h each and eight 100 W light bulbs are used on average 35 h each. Determine the cost of electricity for the week if 1 unit of electricity costs 7p.

10. Calculate the power dissipated by the element of an electric fire of resistance 30 Ω when a current of 10 A flows in it. If the fire is on for 30 hours in a week determine the energy used. Determine also the weekly cost of energy if electricity costs 6.5p per unit.

13.9 Main effects of electric current

The three main effects of an electric current are:

(a) magnetic effect   (b) chemical effect   (c) heating effect

Some practical applications of the effects of an electric current include:

Magnetic effect: bells, relays, motors, generators, transformers, telephones, car-ignition and lifting magnets (see chapter 20)

Chemical effect: primary and secondary cells and electroplating (see chapter 14)

Heating effect: cookers, water heaters, electric fires, irons, furnaces, kettles and soldering irons

13.10 Fuses

A **fuse** is used to prevent overloading of electrical circuits. The fuse, which is made of material having a low melting point, utilizes the heating effect of an electric current. A fuse is placed in an electrical circuit and if the current becomes too large the fuse wire melts and so breaks the circuit. A circuit diagram symbol for a fuse is shown in Fig. 13.1, on page 91.

Problem 16. If 5 A, 10 A and 13 A fuses are available, state which is most appropriate for the following appliances which are both connected to a 240 V supply:

(a) Electric toaster having a power rating of 1 kW.

(b) Electric fire having a power rating of 3 kW.

Power, $P = VI$, from which, current $I = \dfrac{P}{V}$

(a) For the toaster, current
$$I = \frac{P}{V} = \frac{1000}{240} = \frac{100}{24} = 4.17 \text{ A}$$
Hence a **5 A fuse** is most appropriate

(b) For the fire, current
$$I = \frac{P}{V} = \frac{3000}{240} = \frac{300}{24} = 12.5 \text{ A}$$
Hence a **13 A fuse** is most appropriate

Now try the following exercises

Exercise 53 Further problem on fuses (Answer on page 303)

1. A television set having a power rating of 120 W and an electric lawnmower of power rating 1 kW are both connected to a 250 V supply. If 3 A, 5 A and 13 A fuses are available state which is the most appropriate for each appliance.

Exercise 54 Short answer questions on the introduction to electric circuits

1. Draw the preferred symbols for the following components used when drawing electrical circuit diagrams:
   (a) fixed resistor   (b) cell
   (c) filament lamp   (d) fuse
   (e) voltmeter

2. State the unit of (a) current (b) potential difference (c) resistance

3. State an instrument used to measure (a) current (b) potential difference (c) resistance

4. What is a multimeter?

5. State Ohm's law

6. State the meaning of the following abbreviations of prefixes used with electrical units:
   (a) k   (b) μ   (c) m   (d) M

7. What is a conductor? Give four examples

8. What is an insulator? Give four examples

9. Complete the following statement:
   'An ammeter has a... resistance and must be connected... with the load'

10. Complete the following statement:
    'A voltmeter has a... resistance and must be connected... with the load'

11. State the unit of electrical power. State three formulae used to calculate power

12. State two units used for electrical energy

13. State the three main effects of an electric current and give two examples of each

14. What is the function of a fuse in an electrical circuit?

Exercise 55 Multi-choice problems on the introduction to electric circuits (Answers on page 304)

1. The ohm is the unit of:
   (a) charge (b) electric potential
   (c) current (d) resistance

2. The current which flows when 0.1 coulomb is transferred in 10 ms is:
   (a) 10 A (b) 1 A (c) 10 mA (d) 100 mA

3. The p.d. applied to a 1 kΩ resistance in order that a current of 100 μA may flow is:
   (a) 1 V (b) 100 V (c) 0.1 V (d) 10 V

4. Which of the following formulae for electrical power is incorrect?
   (a) VI (b) $\dfrac{V}{I}$ (c) I^2R (d) $\dfrac{V^2}{R}$

5. The power dissipated by a resistor of 4 Ω when a current of 5 A passes through it is:
   (a) 6.25 W (b) 20 W (c) 80 W (d) 100 W

6. Which of the following statements is true?
   (a) Electric current is measured in volts
   (b) 200 kΩ resistance is equivalent to 2 MΩ
   (c) An ammeter has a low resistance and must be connected in parallel with a circuit
   (d) An electrical insulator has a high resistance

7. A current of 3 A flows for 50 h through a 6 Ω resistor. The energy consumed by the resistor is:

(a) 0.9 kWh (b) 2.7 kWh
(c) 9 kWh (d) 27 kWh

8. What must be known in order to calculate the energy used by an electrical appliance?
   (a) voltage and current
   (b) current and time of operation
   (c) power and time of operation
   (d) current and resistance

9. Voltage drop is the:
   (a) difference in potential between two points
   (b) maximum potential
   (c) voltage produced by a source
   (d) voltage at the end of a circuit

10. The energy used by a 3 kW heater in 1 minute is:
    (a) 180 000 J (b) 3000 J (c) 180 J (d) 50 J

11. Electromotive force is provided by:
    (a) resistance's
    (b) a conducting path
    (c) an electrical supply source
    (d) an electric current

12. A 240 V, 60 W lamp has a working resistance of:
    (a) 1400 ohm (b) 60 ohm
    (c) 960 ohm (d) 325 ohm

13. When an atom loses an electron, the atom:
    (a) experiences no effect at all
    (b) becomes positively charged
    (c) disintegrates
    (d) becomes negatively charged

14. The largest number of 60 W electric light bulbs which can be operated from a 240 V supply fitted with 5 A fuse is:
    (a) 20 (b) 48 (c) 12 (d) 4

15. The unit of current is the:
    (a) joule (b) coulomb (c) ampere (d) volt

16. State which of the following is incorrect.
    (a) $1\,W = 1\,J\,s^{-1}$
    (b) $1\,J = 1\,N/m$
    (c) $\eta = \dfrac{\text{output energy}}{\text{input energy}}$
    (d) energy = power × time

17. The coulomb is a unit of:
    (a) voltage
    (b) power
    (c) energy
    (d) quantity of electricity

18. If in the circuit shown in Fig. 13.5, the reading on
    the voltmeter is 4 V and the reading on the ammeter
    is 20 mA, the resistance of resistor *R* is:

    (a) 0.005 Ω   (b) 5 Ω   (c) 80 Ω   (d) 200 Ω

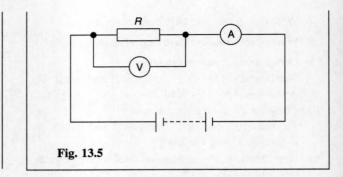

Fig. 13.5

14

Chemical effects of electricity

At the end of this chapter you should be able to:

- understand electrolysis and its applications, including electroplating
- appreciate the purpose and construction of a simple cell
- explain polarisation and local action
- explain corrosion and its effects
- define the terms e.m.f., E, and internal resistance, r, of a cell
- perform calculations using $V = E - Ir$
- determine the total e.m.f. and total internal resistance for cells connected in series and in parallel
- distinguish between primary and secondary cells
- explain the construction and practical applications of the Leclanché, mercury, lead-acid and alkaline cells
- list the advantages and disadvantages of alkaline cells over lead-acid cells
- understand the term 'cell capacity' and state its unit

14.1 Introduction

A material must contain **charged particles** to be able to conduct electric current. In **solids**, the current is carried by **electrons**. Copper, lead, aluminium, iron and carbon are some examples of solid conductors. In **liquids and gases**, the current is carried by the part of a molecule which has acquired an electric charge, called **ions**. These can possess a positive or negative charge, and examples include hydrogen ion H^+, copper ion Cu^{++} and hydroxyl ion OH^-. Distilled water contains no ions and is a poor conductor of

electricity, whereas salt water contains ions and is a fairly good conductor of electricity.

14.2 Electrolysis

Electrolysis is the decomposition of a liquid compound by the passage of electric current through it. Practical applications of electrolysis include the electroplating of metals (see Section 14.3), the refining of copper and the extraction of aluminium from its ore.

An **electrolyte** is a compound which will undergo electrolysis. Examples include salt water, copper sulphate and sulphuric acid.

The **electrodes** are the two conductors carrying current to the electrolyte. The positive-connected electrode is called the **anode** and the negative-connected electrode the **cathode**.

When two copper wires connected to a battery are placed in a beaker containing a salt water solution, current will flow through the solution. Air bubbles appear around the wires as the water is changed into hydrogen and oxygen by electrolysis.

14.3 Electroplating

Electroplating uses the principle of electrolysis to apply a thin coat of one metal to another metal. Some practical applications include the tin-plating of steel, silver-plating of nickel alloys and chromium-plating of steel. If two copper electrodes connected to a battery are placed in a beaker containing copper sulphate as the electrolyte it is found that the cathode (i.e. the electrode connected to the negative terminal of the battery) gains copper whilst the anode loses copper.

14.4   The simple cell

The purpose of an **electric cell** is to convert chemical energy into electrical energy.

A **simple cell** comprises two dissimilar conductors (electrodes) in an electrolyte. Such a cell is shown in Fig. 14.1, comprising copper and zinc electrodes. An electric current is found to flow between the electrodes. Other possible electrode pairs exist, including zinc-lead and zinc-iron. The electrode potential (i.e. the p.d. measured between the electrodes) varies for each pair of metals. By knowing the e.m.f. of each metal with respect to some standard electrode, the e.m.f. of any pair of metals may be determined. The standard used is the hydrogen electrode. The **electrochemical series** is a way of listing elements in order of electrical potential, and Table 14.1 shows a number of elements in such a series.

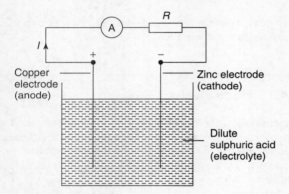

Fig. 14.1

Table 14.1   Part of the electrochemical series

Potassium
sodium
aluminium
zinc
iron
lead
hydrogen
copper
silver
carbon

In a simple cell two faults exist – those due to **polarisation** and **local action**.

Polarisation

If the simple cell shown in Fig. 14.1 is left connected for some time, the current I decreases fairly rapidly. This is because of the formation of a film of hydrogen bubbles on the copper anode. This effect is known as the polarisation of the cell. The hydrogen prevents full contact between the copper electrode and the electrolyte and this increases the internal resistance of the cell. The effect can be overcome by using a chemical depolarising agent or depolariser, such as potassium dichromate which removes the hydrogen bubbles as they form. This allows the cell to deliver a steady current.

Local action

When commercial zinc is placed in dilute sulphuric acid, hydrogen gas is liberated from it and the zinc dissolves. The reason for this is that impurities, such as traces of iron, are present in the zinc which set up small primary cells with the zinc. These small cells are short-circuited by the electrolyte, with the result that localised currents flow causing corrosion. This action is known as local action of the cell. This may be prevented by rubbing a small amount of mercury on the zinc surface, which forms a protective layer on the surface of the electrode.

When two metals are used in a simple cell the electrochemical series may be used to predict the behaviour of the cell:

(i) The metal that is higher in the series acts as the negative electrode, and vice-versa. For example, the zinc electrode in the cell shown in Fig. 14.1 is negative and the copper electrode is positive.

(ii) The greater the separation in the series between the two metals the greater is the e.m.f. produced by the cell.

The electrochemical series is representative of the order of reactivity of the metals and their compounds:

(i) The higher metals in the series react more readily with oxygen and vice-versa.

(ii) When two metal electrodes are used in a simple cell the one that is higher in the series tends to dissolve in the electrolyte.

14.5   Corrosion

Corrosion is the gradual destruction of a metal in a damp atmosphere by means of simple cell action. In addition to the presence of moisture and air required for rusting, an electrolyte, an anode and a cathode are required for corrosion. Thus, if metals widely spaced in the electrochemical series are used in contact with each other in the presence of an electrolyte, corrosion will occur. For example, if a brass valve is fitted to a heating system made of steel, corrosion will occur.

The **effects of corrosion** include the weakening of structures, the reduction of the life of components and materials, the wastage of materials and the expense of replacement.

Corrosion may be **prevented** by coating with paint, grease, plastic coatings and enamels, or by plating with tin or chromium. Also, iron may be galvanised, i.e. plated with zinc, the layer of zinc helping to prevent the iron from corroding.

14.6 E.m.f. and internal resistance of a cell

The **electromotive force (e.m.f.)**, E, of a cell is the p.d. between its terminals when it is not connected to a load (i.e. the cell is on 'no load'). The e.m.f. of a cell is measured by using a **high resistance voltmeter** connected in parallel with the cell. The voltmeter must have a high resistance otherwise it will pass current and the cell will not be on 'no-load'. For example, if the resistance of a cell is $1\,\Omega$ and that of a voltmeter $1\,\text{M}\Omega$ then the equivalent resistance of the circuit is $1\,\text{M}\Omega+1\,\Omega$, i.e. approximately $1\,\text{M}\Omega$, hence no current flows and the cell is not loaded. The voltage available at the terminals of a cell falls when a load is connected. This is caused by the **internal resistance** of the cell which is the opposition of the material of the cell to the flow of current. The internal resistance acts in series with other resistance's in the circuit. Figure 14.2 shows a cell of e.m.f. E volts and internal resistance, r, and XY represents the terminals of the cell.

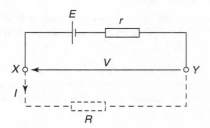

Fig. 14.2

When a load (shown as resistance R) is not connected, no current flows and the terminal p.d., $V = E$. When R is connected a current I flows which causes a voltage drop in the cell, given by Ir. The p.d. available at the cell terminals is less than the e.m.f. of the cell and is given by:

$$V = E - Ir$$

Thus if a battery of e.m.f. 12 volts and internal resistance $0.01\,\Omega$ delivers a current of 100 A, the terminal p.d.,

$$V = 12 - (100)\,(0.01) = 12 - 1 = 11\,\text{V}$$

When different values of potential difference V across a cell or power supply are measured for different values of current

I, a graph may be plotted as shown in Fig. 14.3. Since the e.m.f. E of the cell or power supply is the p.d. across its terminals on no load (i.e. when $I = 0$), then E is as shown by the broken line.

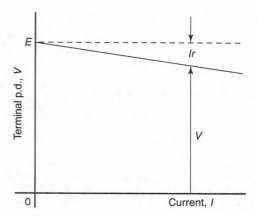

Fig. 14.3

Since $V = E - Ir$ then the internal resistance may be calculated from

$$r = \frac{E - V}{I}$$

When a current is flowing in the direction shown in Fig. 14.2 the cell is said to be **discharging** ($E > V$).

When a current flows in the opposite direction to that shown in Fig. 14.2 the cell is said to be **charging** ($V > E$).

A **battery** is a combination of more than one cell. The cells in a battery may be connected in series or in parallel.

(i) **For cells connected in series:**

Total e.m.f. = sum of cell's e.m.f.'s.

Total internal resistance = sum of cell's internal resistances.

(ii) **For cells connected in parallel:**

If each cell has the same e.m.f. and internal resistance:
Total e.m.f. = e.m.f. of one cell.

Total internal resistance of n cells = $\frac{1}{n} \times$ internal resistance of one cell.

Problem 1. Eight cells, each with an internal resistance of $0.2\,\Omega$ and an e.m.f. of 2.2 V are connected (a) in series, (b) in parallel. Determine the e.m.f. and the internal resistance of the batteries so formed.

(a) When connected in series,

total e.m.f. = sum of cell's e.m.f.

$$= 2.2 \times 8 = \mathbf{17.6\,V}$$

Total internal resistance

   = sum of cell's internal resistance

   $= 0.2 \times 8 = \mathbf{1.6\,\Omega}$

(b) When connected in parallel,

   total e.m.f. = e.m.f. of one cell

   $= \mathbf{2.2\,V}$

Total internal resistance of 8 cells

   $= \frac{1}{8} \times$ internal resistance of one cell

   $= \frac{1}{8} \times 0.2 = \mathbf{0.025\,\Omega}$

Problem 2. A cell has an internal resistance of $0.02\,\Omega$ and an e.m.f. of $2.0\,V$. Calculate its terminal p.d. if it delivers (a) $5\,A$ (b) $50\,A$.

(a) Terminal p.d. $V = E - Ir$

where E = e.m.f. of cell,

   I = current flowing

and r = internal resistance of cell

   $E = 2.0\,V, I = 5\,A$ and $r = 0.02\,\Omega$

Hence **terminal p.d.**

   $V = 2.0 - (5)(0.02) = 2.0 - 0.1$

   $= \mathbf{1.9\,V}$

(b) When the current is $50\,A$, terminal p.d.,

   $V = E - Ir = 2.0 - 50(0.02)$

i.e. $V = 2.0 - 1.0 = \mathbf{1.0\,V}$

Thus the terminal p.d. decreases as the current drawn increases.

Problem 3. Ten $1.5\,V$ cells, each having an internal resistance of $0.2\,\Omega$, are connected in series to a load of $58\,\Omega$. Determine (a) the current flowing in the circuit and (b) the p.d. at the battery terminals.

(a) For ten cells, battery emf, $E = 10 \times 1.5 = 15\,V$, and the total internal resistance, $r = 10 \times 0.2 = 2\,\Omega$.

   When connected to a $58\,\Omega$ load the circuit is as shown in Fig. 14.4.

   $\text{Current } I = \dfrac{\text{e.m.f.}}{\text{total resistance}} = \dfrac{15}{58 + 2} = \dfrac{15}{60}$

   $= \mathbf{0.25\,A}$

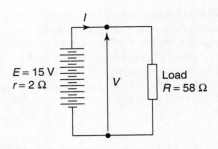

Fig. 14.4

(b) P.d. at battery terminals,

   $V = E - Ir$

i.e. $V = 15 - (0.25)(2) = \mathbf{14.5\,V}$

Now try the following exercise

Exercise 56   Further problems on e.m.f. and internal resistance of a cell (Answers on page 304)

1. Twelve cells, each with an internal resistance of $0.24\,\Omega$ and an e.m.f. of $1.5\,V$ are connected
   (a) in series   (b) in parallel.

   Determine the e.m.f. and internal resistance of the batteries so formed

2. A cell has an internal resistance of $0.03\,\Omega$ and an e.m.f. of $2.2\,V$. Calculate its terminal p.d. if it delivers (a) $1\,A$, (b) $20\,A$, (c) $50\,A$.

3. The p.d. at the terminals of a battery is $16\,V$ when no load is connected and $14\,V$ when a load taking $8\,A$ is connected. Determine the internal resistance of the battery.

4. A battery of e.m.f. $20\,V$ and internal resistance $0.2\,\Omega$ supplies a load taking $10\,A$. Determine the p.d. at the battery terminals and the resistance of the load.

5. Ten $2.2\,V$ cells, each having an internal resistance of $0.1\,\Omega$ are connected in series to a load of $21\,\Omega$. Determine

   (a) the current flowing in the circuit, and

   (b) the p.d. at the battery terminals

6. For the circuits shown in Fig. 14.5 the resistors represent the internal resistance of the batteries. Find, in each case:

   (i) the total e.m.f. across PQ

   (ii) the total equivalent internal resistances of the batteries

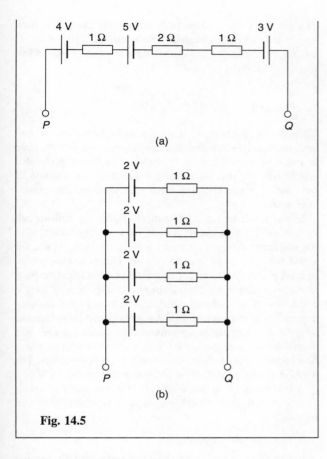

Fig. 14.5

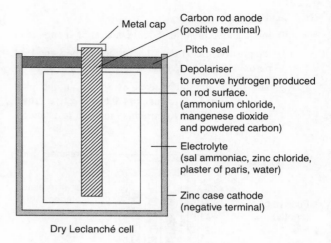

Dry Leclanché cell

Fig. 14.6

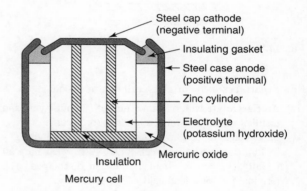

Mercury cell

Fig. 14.7

14.7 Primary cells

Primary cells cannot be recharged, that is, the conversion of chemical energy to electrical energy is irreversible and the cell cannot be used once the chemicals are exhausted. Examples of primary cells include the Leclanché cell and the mercury cell.

Lechlanché cell

A typical dry Lechlanché cell is shown in Fig. 14.6. Such a cell has an e.m.f. of about 1.5 V when new, but this falls rapidly if in continuous use due to polarisation. The hydrogen film on the carbon electrode forms faster than can be dissipated by the depolariser. The Lechlanché cell is suitable only for intermittent use, applications including torches, transistor radios, bells, indicator circuits, gas lighters, controlling switch-gear, and so on. The cell is the most commonly used of primary cells, is cheap, requires little maintenance and has a shelf life of about 2 years.

Mercury cell

A typical mercury cell is shown in Fig. 14.7. Such a cell has an e.m.f. of about 1.3 V which remains constant for a rel- atively long time. Its main advantages over the Lechlanché

cell is its smaller size and its long shelf life. Typical prac- tical applications include hearing aids, medical electronics, cameras and for guided missiles.

14.8 Secondary cells

Secondary cells can be recharged after use, that is, the con- version of chemical energy to electrical energy is reversible and the cell may be used many times. Examples of secondary cells include the lead-acid cell and the alkaline cell. Practical applications of such cells include car batteries, telephone cir- cuits and for traction purposes – such as milk delivery vans and fork lift trucks.

Lead-acid cell

A typical lead-acid cell is constructed of:

(i) A container made of glass, ebonite or plastic.

(ii) **Lead plates**

   (a) the negative plate (cathode) consists of spongy lead

   (b) the positive plate (anode) is formed by pressing lead peroxide into the lead grid.

   The plates are interleaved as shown in the plan view of Fig. 14.8 to increase their effective cross-sectional area and to minimize internal resistance.

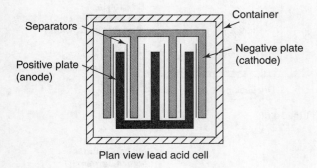

Plan view lead acid cell

Fig. 14.8

(iii) **Separators** made of glass, celluloid or wood.

(iv) An **electrolyte** which is a mixture of sulphuric acid and distilled water. The relative density (or specific gravity) of a lead-acid cell, which may be measured using a hydrometer, varies between about 1.26 when the cell is fully charged to about 1.19 when discharged. The terminal p.d. of a lead-acid cell is about 2 V.

When a cell supplies current to a load it is said to be **discharging**. During discharge:

(i) the lead peroxide (positive plate) and the spongy lead (negative plate) are converted into lead sulphate, and

(ii) the oxygen in the lead peroxide combines with hydrogen in the electrolyte to form water. The electrolyte is therefore weakened and the relative density falls.

The terminal p.d. of a lead-acid cell when fully discharged is about 1.8 V. A cell is **charged** by connecting a d.c. supply to its terminals, the positive terminal of the cell being connected to the positive terminal of the supply. The charging current flows in the reverse direction to the discharge current and the chemical action is reversed. During charging:

(i) the lead sulphate on the positive and negative plates is converted back to lead peroxide and lead respectively, and

(ii) the water content of the electrolyte decreases as the oxygen released from the electrolyte combines with the lead of the positive plate. The relative density of the electrolyte thus increases.

The colour of the positive plate when fully charged is dark brown and when discharged is light brown. The colour of the negative plate when fully charged is grey and when discharged is light grey.

Alkaline cell

There are two main types of alkaline cell – the nickel-iron cell and the nickel-cadmium cell. In both types the positive plate is made of nickel hydroxide enclosed in finely perforated steel tubes, the resistance being reduced by the addition of pure nickel or graphite. The tubes are assembled into nickel-steel plates.

In the nickel-iron cell, (sometimes called the **Edison cell** or **nife cell**), the negative plate is made of iron oxide, with the resistance being reduced by a little mercuric oxide, the whole being enclosed in perforated steel tubes and assembled in steel plates. In the nickel-cadmium cell the negative plate is made of cadmium. The electrolyte in each type of cell is a solution of potassium hydroxide which does not undergo any chemical change and thus the quantity can be reduced to a minimum. The plates are separated by insulating rods and assembled in steel containers which are then enclosed in a non-metallic crate to insulate the cells from one another. The average discharge p.d. of an alkaline cell is about 1.2 V.

Advantages of an alkaline cell (for example, a nickel-cadmium cell or a nickel-iron cell) over a lead-acid cell include:

(i) More robust construction.

(ii) Capable of withstanding heavy charging and discharging currents without damage.

(iii) Has a longer life.

(iv) For a given capacity is lighter in weight.

(v) Can be left indefinitely in any state of charge or discharge without damage.

(vi) Is not self-discharging.

Disadvantages of an alkaline cell over a lead-acid cell include:

(i) Is relatively more expensive.

(ii) Requires more cells for a given e.m.f.

(iii) Has a higher internal resistance.

(iv) Must be kept sealed.

(v) Has a lower efficiency.

Alkaline cells may be used in extremes of temperature, in conditions where vibration is experienced or where duties require long idle periods or heavy discharge currents. Practical examples include traction and marine work, lighting in railway carriages, military portable radios and for starting diesel and petrol engines.

However, the lead-acid cell is the most common one in practical use.

14.9 Cell capacity

The **capacity** of a cell is measured in ampere-hours (Ah). A fully charged 50 Ah battery rated for 10 h discharge can be discharged at a steady current of 5 A for 10 h, but if the load current is increased to 10 A then the battery is discharged in 3–4 h, since the higher the discharge current, the lower is the effective capacity of the battery. Typical discharge characteristics for a lead-acid cell are shown in Fig. 14.9.

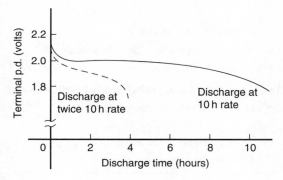

Fig. 14.9

Now try the following exercises

> **Exercise 57   Short answer questions on the chemical effects of electricity**
>
> 1. What is electrolysis?
>
> 2. What is an electrolyte?
>
> 3. Conduction in electrolytes is due to....
>
> 4. A positive-connected electrode is called the... and the negative-connected electrode the....
>
> 5. State two practical applications of electrolysis.
>
> 6. The purpose of an electric cell is to convert... to....
>
> 7. Make a labelled sketch of a simple cell.
>
> 8. What is the electrochemical series?
>
> 9. With reference to a simple cell, explain briefly what is meant by
>
>   (a) polarisation   (b) local action
>
> 10. What is corrosion? Name two effects of corrosion and state how they may be prevented.

11. What is meant by the e.m.f. of a cell? How may the e.m.f. of a cell be measured?

12. Define internal resistance.

13. If a cell has an e.m.f. of E volts, an internal resistance of r ohms and supplies a current I amperes to a load, the terminal p.d. V volts is given by: $V = ...$

14. Name the two main types of cells.

15. Explain briefly the difference between primary and secondary cells.

16. Name two types of primary cells.

17. Name two types of secondary cells.

18. State three typical applications of primary cells.

19. State three typical applications of secondary cells.

20. In what unit is the capacity of a cell measured?

> **Exercise 58   Multi-choice questions on the chemical effects of electricity (Answers on page 304)**
>
> 1. A battery consists of:
>
>   (a) a cell   (b) a circuit
>
>   (c) a generator   (d) a number of cells
>
> 2. The terminal p.d. of a cell of e.m.f. 2 V and internal resistance $0.1\,\Omega$ when supplying a current of 5 A will be:
>
>   (a) 1.5 V   (b) 2 V   (c) 1.9 V   (d) 2.5 V
>
> 3. Five cells, each with an e.m.f. of 2 V and internal resistance $0.5\,\Omega$ are connected in series. The resulting battery will have:
>
>   (a) an e.m.f. of 2 V and an internal resistance of $0.5\,\Omega$
>
>   (b) an e.m.f. of 10 V and an internal resistance of $2.5\,\Omega$
>
>   (c) an e.m.f. of 2 V and an internal resistance of $0.1\,\Omega$
>
>   (d) an e.m.f. of 10 V and an internal resistance of $0.1\,\Omega$
>
> 4. If the five cells of question 3 are connected in parallel the resulting battery will have:
>
>   (a) an e.m.f. of 2 V and an internal resistance of $0.5\,\Omega$
>
>   (b) an e.m.f. of 10 V and an internal resistance of $2.5\,\Omega$
>
>   (c) an e.m.f. of 2 V and an internal resistance of $0.1\,\Omega$

(d) an e.m.f. of 10 V and an internal resistance of 0.1 Ω

5. Which of the following statements is false?

   (a) A Leclanché cell is suitable for use in torches

   (b) A nickel-cadnium cell is an example of a primary cell

   (c) When a cell is being charged its terminal p.d. exceeds the cell e.m.f.

   (d) A secondary cell may be recharged after use

6. Which of the following statements is false?

   When two metal electrodes are used in a simple cell, the one that is higher in the electrochemical series:

   (a) tends to dissolve in the electrolyte

   (b) is always the negative electrode

   (c) reacts most readily with oxygen

   (d) acts an an anode

7. Five 2 V cells, each having an internal resistance of 0.2 Ω are connected in series to a load of resistance 14 Ω. The current flowing in the circuit is:

   (a) 10 A   (b) 1.4 A   (c) 1.5 A   (d) $\frac{2}{3}$ A

8. For the circuit of question 7, the p.d. at the battery terminals is:

   (a) 10 V   (b) $9\frac{1}{3}$ V   (c) 0 V   (d) $10\frac{2}{3}$ V

9. Which of the following statements is true?

   (a) The capacity of a cell is measured in volts

   (b) A primary cell converts electrical energy into chemical energy

   (c) Galvanising iron helps to prevent corrosion

   (d) A positive electrode is termed the cathode

10. The greater the internal resistance of a cell:

    (a) the greater the terminal p.d.

    (b) the less the e.m.f.

    (c) the greater the e.m.f.

    (d) the less the terminal p.d.

11. The negative pole of a dry cell is made of:

    (a) carbon   (b) copper   (c) zinc   (d) mercury

12. The energy of a secondary cell is usually renewed:

    (a) by passing a current through it

    (b) it cannot be renewed at all

    (c) by renewing its chemicals

    (d) by heating it

15

Capacitors and inductors

At the end of this chapter you should be able to:

- explain how a capacitor stores energy

- Appreciate $Q = It$, $E = \dfrac{V}{d}$, $D = \dfrac{Q}{A}$, $Q = CV$ and $\dfrac{D}{E} = \varepsilon_0 \varepsilon_r$ and perform calculations using these formulae

- state the unit of capacitance

- define a dielectric

- state the value of ε_0 and appreciate typical values of ε_r

- appreciate that the capacitance C of a parallel plate capacitor is given by $C = \dfrac{\varepsilon_0 \varepsilon_r A(n-1)}{d}$

- perform calculations involving the parallel plate capacitor

- define dielectric strength

- determine the energy stored in a capacitor and state its unit

- describes practical types of capacitor

- define inductance and state its unit

- state the factors which affect the inductance of an inductor

- describe examples of practical inductors and their circuit diagrams

- calculate the energy stored in the magnetic field of an inductor and state its unit

15.1 Capacitors and capacitance

A **capacitor** is a device capable of storing electrical energy. Figure 15.1 shows a capacitor consisting of a pair of parallel

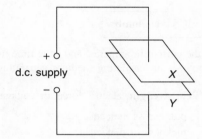

Fig. 15.1

metal plates X and Y separated by an insulator, which could be air. Since the plates are electrical conductors each will contain a large number of mobile electrons. Because the plates are connected to a d.c. supply the electrons on plate X, which have a small negative charge, will be attracted to the positive pole of the supply and will be repelled from the negative pole of the supply on to plate Y. X will become positively charged due to its shortage of electrons whereas Y will have a negative charge due to its surplus of electrons.

The difference in charge between the plates results in a p.d. existing between them, the flow of electrons dying away and ceasing when the p.d. between the plates equals the supply voltage. The plates are then said to be **charged** and there exists an **electric field** between them. Figure 15.2 shows a side view of the plates with the field represented by 'lines of electrical flux'. If the plates are disconnected from the supply and connected together through a resistor the surplus of electrons on the negative plate will flow through the resistor to the positive plate. This is called **discharging**. The current flow decreases to zero as the charges on the plates reduce. The current flowing in the resistor causes it to liberate heat showing that **energy is stored in the electric field**.

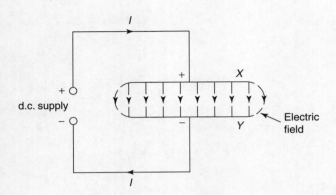

Fig. 15.2

Summary of important formulae and definitions

From Chapter 13, charge Q is given by:

$$Q = I \times t \text{ coulombs}$$

where I is the current in amperes, and t the time in seconds.

A **dielectric** is an insulating medium separating charged surfaces.

Electric field strength, electric force, or voltage gradient,

$$E = \frac{\text{p.d. across dielectric}}{\text{thickness of dielectric}}$$

i.e. $$E = \frac{V}{d} \text{ volts/m}$$

Electric flux density,

$$D = \frac{Q}{A} \text{ C/m}^2$$

Charge Q on a capacitor is proportional the applied voltage, V, i.e. $Q \propto V$

$$Q = CV$$

where the constant of proportionality, C, is the capacitance

Capacitance, $$C = \frac{Q}{V}$$

The unit of capacitance is the **farad, F** (or more usually $\mu F = 10^{-6} F$ or $pF = 10^{-12} F$), which is defined as the capacitance of a capacitor when a p.d. of one volt appears across the plates when charged with one coulomb.

Every system of electrical conductors possesses capacitance. For example, there is capacitance between the conductors of overhead transmission lines and also between the wires of a telephone cable. In these examples the capacitance

is undesirable but has to be accepted, minimised or compensated for. There are other situations, such as in capacitors, where capacitance is a desirable property.

The ratio of electric flux density, D, to electric field strength, E, is called **absolute permittivity, ε,** of a dielectric.

Thus $$\frac{D}{E} = \varepsilon$$

Permittivity of free space is a constant, given by

$$\varepsilon_0 = 8.85 \times 10^{-12} \text{ F/m}.$$

Relative permittivity,

$$\varepsilon_r = \frac{\text{flux density of the field in a dielectric}}{\text{flux density of the field in a vacuum}}$$

(ε_r has no units). Examples of the values of ε_r include: air $= 1$, polythene $= 2.3$, mica $= 3–7$, glass $= 5–10$, ceramics $= 6–1000$.

Absolute permittivity, $\varepsilon = \varepsilon_0 \varepsilon_r$, thus

$$\frac{D}{E} = \varepsilon_0 \varepsilon_r$$

Problem 1.   (a) Determine the p.d. across a $4 \mu F$ capacitor when charged with 5 mC (b) Find the charge on a 50 pF capacitor when the voltage applied to it is 2 kV

(a) $C = 4 \mu F = 4 \times 10^{-6} F$ and $Q = 5 \text{ mC} = 5 \times 10^{-3} C$

Since $C = \dfrac{Q}{V}$

then $V = \dfrac{Q}{C} = \dfrac{5 \times 10^{-3}}{4 \times 10^{-6}} = \dfrac{5 \times 10^{6}}{4 \times 10^{3}} = \dfrac{5000}{4}$

Hence p.d. $V = 1250 \text{ V}$ or   1.25 kV

(b) $C = 50 \text{ pF} = 50 \times 10^{-12} F$ and $V = 2 \text{ kV} = 2000 \text{ V}$

$$Q = CV = 50 \times 10^{-12} \times 2000$$

$$= \frac{5 \times 2}{10^8} = 0.1 \times 10^{-6}$$

Hence, charge $Q = 0.1 \mu C$

Problem 2.   A direct current of 4 A flows into a previously uncharged $20 \mu F$ capacitor for 3 ms. Determine the p.d. between the plates.

$I = 4 \text{ A}$, $C = 20 \mu F = 20 \times 10^{-6} F$ and $t = 3 \text{ ms} = 3 \times 10^{-3} \text{ s}$.

$$Q = It = 4 \times 3 \times 10^{-3} C$$

$$V = \frac{Q}{C} = \frac{4 \times 3 \times 10^{-3}}{20 \times 10^{-6}} = \frac{12 \times 10^{6}}{20 \times 10^{3}} = 0.6 \times 10^{3}$$

$$= 600 \text{ V}$$

Hence, the p.d. between the plates is 600 V

Problem 3. A $5\,\mu F$ capacitor is charged so that the p.d. between its plates is $800\,V$. Calculate how long the capacitor can provide an average discharge current of $2\,mA$.

$C = 5\,\mu F = 5 \times 10^{-6}\,F$, $V = 800\,V$ and

$$I = 2\,mA = 2 \times 10^{-3}\,A.$$

$$Q = CV = 5 \times 10^{-6} \times 800 = 4 \times 10^{-3}\,C$$

Also, $Q = It$.

Thus, $t = \dfrac{Q}{I} = \dfrac{4 \times 10^{-3}}{2 \times 10^{-3}} = 2\,s$

Hence, the capacitor can provide an average discharge current of $2\,mA$ for $2\,s$

Problem 4. Two parallel rectangular plates measuring $20\,cm$ by $40\,cm$ carry an electric charge of $0.2\,\mu C$. Calculate the electric flux density. If the plates are spaced $5\,mm$ apart and the voltage between them is $0.25\,kV$ determine the electric field strength.

Area $A = 20\,cm \times 40\,cm = 800\,cm^2 = 800 \times 10^{-4}\,m^2$ and charge $Q = 0.2\,\mu C = 0.2 \times 10^{-6}\,C$.

Electric flux density,

$$D = \frac{Q}{A} = \frac{0.2 \times 10^{-6}}{800 \times 10^{-4}} = \frac{0.2 \times 10^4}{800 \times 10^6}$$

$$= \frac{2000}{800} \times 10^{-6} = 2.5\,\mu C/m^2$$

Voltage $V = 0.25\,kV = 250\,V$ and plate spacing,

$$d = 5\,mm = 5 \times 10^{-3}\,m.$$

Electric field strength,

$$E = \frac{V}{d} = \frac{250}{5 \times 10^{-3}} = 50\,kV/m$$

Now try the following exercise

Exercise 59   Further problems on capacitors and capacitance (Answers on page 304)

(Where appropriate take ε_0 as 8.85×10^{-12} F/m.)

1. Find the charge on a $10\,\mu F$ capacitor when the applied voltage is $250\,V$.

2. Determine the voltage across a $1000\,pF$ capacitor to charge it with $2\,\mu C$.

3. The charge on the plates of a capacitor is $6\,mC$ when the potential between them is $2.4\,kV$. Determine the capacitance of the capacitor.

4. For how long must a charging current of $2\,A$ be fed to a $5\,\mu F$ capacitor to raise the p.d. between its plates by $500\,V$.

5. A direct current of $10\,A$ flows into a previously uncharged $5\,\mu F$ capacitor for $1\,ms$. Determine the p.d. between the plates.

6. A capacitor uses a dielectric $0.04\,mm$ thick and operates at $30\,V$. What is the electric field strength across the dielectric at this voltage?

7. A charge of $1.5\,\mu C$ is carried on two parallel rectangular plates each measuring $60\,mm$ by $80\,mm$. Calculate the electric flux density. If the plates are spaced $10\,mm$ apart and the voltage between them is $0.5\,kV$ determine the electric field strength.

15.2   The parallel plate capacitor

For a parallel-plate capacitor, experiments show that capacitance C is proportional to the area A of a plate, inversely proportional to the plate spacing d (i.e. the dielectric thickness) and depends on the nature of the dielectric:

$$\text{Capacitance, } C = \frac{\varepsilon_0 \varepsilon_r A(n-1)}{d} \text{ farads}$$

where $\varepsilon_0 = 8.85 \times 10^{-12}$ F/m (constant),
  ε_r = relative permittivity,
  A = area of one of the plates, in m^2,
  d = thickness of dielectric in m
and n = number of plates

Problem 5. (a) A ceramic capacitor has an effective plate area of $4\,cm^2$ separated by $0.1\,mm$ of ceramic of relative permittivity 100. Calculate the capacitance of the capacitor in picofarads. (b) If the capacitor in part (a) is given a charge of $1.2\,\mu C$ what will be the p.d. between the plates?

(a) Area $A = 4\,cm^2 = 4 \times 10^{-4}\,m^2$,

  $d = 0.1\,mm = 0.1 \times 10^{-3}\,m$, $\varepsilon_0 = 8.85 \times 10^{-12}$ F/m,
  $\varepsilon_r = 100$ and $n = 2$

$$\text{Capacitance, } C = \frac{\varepsilon_0 \varepsilon_r A}{d} \text{ farads}$$

$$= \frac{8.85 \times 10^{-12} \times 100 \times 4 \times 10^{-4}}{0.1 \times 10^{-3}}\,F$$

$$= \frac{8.85 \times 4}{10^{10}}\,F = \frac{8.85 \times 4 \times 10^{12}}{10^{10}}\,pF$$

$$= 3540\,pF$$

(b) $Q = CV$ thus $V = \dfrac{Q}{C} = \dfrac{1.2 \times 10^{-6}}{3540 \times 10^{-12}}\,V = 339\,V$

Problem 6. A waxed paper capacitor has two parallel plates, each of effective area $800 \, cm^2$. If the capacitance of the capacitor is $4425 \, pF$ determine the effective thickness of the paper if its relative permittivity is 2.5

$A = 800 \, cm^2 = 800 \times 10^{-4} \, m^2 = 0.08 \, m^2$,
$C = 4425 \, pF = 4425 \times 10^{-12} \, F$, $\varepsilon_0 = 8.85 \times 10^{-12} \, F/m$ and
$\varepsilon_r = 2.5$

Since $C = \dfrac{\varepsilon_0 \varepsilon_r A}{d}$ then $d = \dfrac{\varepsilon_0 \varepsilon_r A}{C}$

$$= \frac{8.85 \times 10^{-12} \times 2.5 \times 0.08}{4425 \times 10^{-12}} = 0.0004 \, m$$

Hence, the thickness of the paper is 0.4 mm

Problem 7. A parallel plate capacitor has nineteen interleaved plates each $75 \, mm$ by $75 \, mm$ separated by mica sheets $0.2 \, mm$ thick. Assuming the relative permittivity of the mica is 5, calculate the capacitance of the capacitor.

$n = 19$ thus $n - 1 = 18$, $A = 75 \times 75 = 5625 \, mm^2 = 5625 \times 10^{-6} \, m^2$, $\varepsilon_r = 5$, $\varepsilon_0 = 8.85 \times 10^{-12} \, F/m$ and $d = 0.2 \, mm = 0.2 \times 10^{-3} \, m$.

Capacitance, $C = \dfrac{\varepsilon_0 \varepsilon_r A(n-1)}{d}$

$$= \frac{8.85 \times 10^{-12} \times 5 \times 5625 \times 10^{-6} \times 18}{0.2 \times 10^{-3}} \, F$$

$$= \mathbf{0.0224 \, \mu F} \quad \text{or} \quad \mathbf{22.4 \, nF}$$

Now try the following exercise

Exercise 60   Further problems on parallel plate capacitors (Answers on page 304)

(Where appropriate take ε_0 as $8.85 \times 10^{-12} \, F/m$)

1. A capacitor consists of two parallel plates each of area $0.01 \, m^2$, spaced $0.1 \, mm$ in air. Calculate the capacitance in picofarads.

2. A waxed paper capacitor has two parallel plates, each of effective area $0.2 \, m^2$. If the capacitance is $4000 \, pF$ determine the effective thickness of the paper if its relative permittivity is 2.

3. How many plates has a parallel plate capacitor having a capacitance of $5 \, nF$, if each plate is $40 \, mm$ by $40 \, mm$ and each dielectric is $0.102 \, mm$ thick with a relative permittivity of 6.

4. A parallel plate capacitor is made from 25 plates, each $70 \, mm$ by $120 \, mm$ interleaved with mica of relative permittivity 5. If the capacitance of the capacitor is $3000 \, pF$ determine the thickness of the mica sheet.

5. A capacitor is constructed with parallel plates and has a value of $50 \, pF$. What would be the capacitance of the capacitor if the plate area is doubled and the plate spacing is halved?

15.3   Capacitors connected in parallel and series

For n **parallel-connected capacitors**, the equivalent capacitance, C, is given by:

$$C = C_1 + C_2 + C_3 \cdots + C_n$$

(similar to **resistors** connected in **series**.)
   Also, total charge, Q, is given by:

$$Q = Q_1 + Q_2 + Q_3$$

For n **series-connected capacitors** the equivalent capacitance, C, is given by:

$$\frac{1}{C} = \frac{1}{C_1} + \frac{1}{C_2} + \frac{1}{C_3} + \cdots + \frac{1}{C_n}$$

(similar to **resistors** connected in **parallel**.)
   When connected in series the charge on each capacitor is the same.

Problem 8. Calculate the equivalent capacitance of two capacitors of $6 \, \mu F$ and $4 \, \mu F$ connected (a) in parallel and (b) in series.

(a) In parallel, equivalent capacitance

$$C = C_1 + C_2 = 6 \, \mu F + 4 \, \mu F = \mathbf{10 \, \mu F}$$

(b) For the special case of **two capacitors in series:**

$$\frac{1}{C} = \frac{1}{C_1} + \frac{1}{C_2} = \frac{C_2 + C_1}{C_1 + C_2}$$

Hence $C = \dfrac{C_1 C_2}{C_1 + C_2} \quad \left(\text{i.e.} \dfrac{\text{product}}{\text{sum}} \right)$

Thus $C = \dfrac{6 \times 4}{6 + 4} = \dfrac{24}{10} = \mathbf{2.4 \, \mu F}$

Problem 9. What capacitance must be connected in series with a $30 \, \mu F$ capacitor for the equivalent capacitance to be $12 \, \mu F$?

Let $C = 12\,\mu\text{F}$ (the equivalent capacitance), $C_1 = 30\,\mu\text{F}$ and C_2 be the unknown capacitance.

For two capacitors in series

$$\frac{1}{C} = \frac{1}{C_1} + \frac{1}{C_2}$$

Hence $\quad \dfrac{1}{C_2} = \dfrac{1}{C} - \dfrac{1}{C_1} = \dfrac{C_1 - C}{CC_1}$

and $\quad C_2 = \dfrac{CC_1}{C_1 - C} = \dfrac{12 \times 30}{30 - 12} = \dfrac{360}{18} = \mathbf{20\,\mu F}$

Problem 10. Capacitance's of $3\,\mu\text{F}$, $6\,\mu\text{F}$ and $12\,\mu\text{F}$ are connected in series across a 350 V supply. Calculate (a) the equivalent circuit capacitance, (b) the charge on each capacitor, and (c) the p.d. across each capacitor.

The circuit diagram is shown in Fig. 15.3.

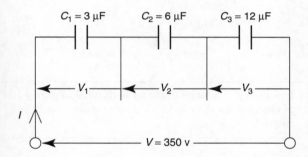

$C_1 = 3\,\mu\text{F}$ $C_2 = 6\,\mu\text{F}$ $C_3 = 12\,\mu\text{F}$

$V = 350\ \text{v}$

Fig. 15.3

(a) The equivalent circuit capacitance C for three capacitors in series is given by:

$$\frac{1}{C} = \frac{1}{C_1} + \frac{1}{C_2} + \frac{1}{C_3}$$

i.e. $\quad \dfrac{1}{C} = \dfrac{1}{3} + \dfrac{1}{6} + \dfrac{1}{12} = \dfrac{4 + 2 + 1}{12} = \dfrac{7}{12}$

Hence the equivalent circuit capacitance,

$$C = \frac{12}{7} = 1\frac{5}{7}\mu\mathbf{F} \quad \text{or} \quad \mathbf{1.714\,\mu F}$$

(b) Total charge $Q_T = CV$,

hence $\quad Q_T = \dfrac{12}{7} \times 10^{-6} \times 350 = 600\,\mu\text{C} \quad$ or $\quad 0.6\,\text{mC}$

Since the capacitors are connected in series 0.6 mC is the charge on each of them.

(c) The voltage across the $3\,\mu\text{F}$ capacitor,

$$V_1 = \frac{Q}{C_1} = \frac{0.6 \times 10^{-3}}{3 \times 10^{-6}} = \mathbf{200\,V}$$

The voltage across the $6\,\mu\text{F}$ capacitor,

$$V_2 = \frac{Q}{C_2} = \frac{0.6 \times 10^{-3}}{6 \times 10^{-6}} = \mathbf{100\,V}$$

The voltage across the $12\,\mu\text{F}$ capacitor,

$$V_3 = \frac{Q}{C_3} = \frac{0.6 \times 10^{-3}}{12 \times 10^{-6}} = \mathbf{50\,V}$$

[Check: In a series circuit $V = V_1 + V_2 + V_3$

$$V_1 + V_2 + V_3 = 200 + 100 + 50$$
$$= 350\,\text{V} = \text{supply voltage}]$$

In practice, capacitors are rarely connected in series unless they are of the same capacitance. The reason for this can be seen from the above problem where the lowest valued capacitor (i.e. $3\,\mu\text{F}$) has the highest p.d. across it (i.e. 200 V) which means that if all the capacitors have an identical construction they must all be rated at the highest voltage.

Now try the following exercise

Exercise 61   Further problems on capacitors in parallel and series (Answers on page 304)

1. Capacitors of $2\,\mu\text{F}$ and $6\,\mu\text{F}$ are connected (a) in parallel and (b) in series. Determine the equivalent capacitance in each case.

2. Find the capacitance to be connected in series with a $10\,\mu\text{F}$ capacitor for the equivalent capacitance to be $6\,\mu\text{F}$.

3. What value of capacitance would be obtained if capacitors of $0.15\,\mu\text{F}$ and $0.10\,\mu\text{F}$ are connected (a) in series and (b) in parallel.

4. Two $6\,\mu\text{F}$ capacitors are connected in series with one having a capacitance of $12\,\mu\text{F}$. Find the total equivalent circuit capacitance. What capacitance must be added in series to obtain a capacitance of $1.2\,\mu\text{F}$?

5. For the arrangement shown in Fig. 15.4 find (a) the equivalent circuit capacitance and (b) the voltage across a $4.5\,\mu\text{F}$ capacitor.

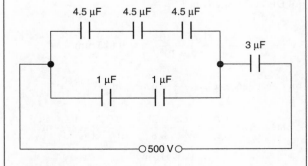

$4.5\,\mu\text{F}$ $4.5\,\mu\text{F}$ $4.5\,\mu\text{F}$ $3\,\mu\text{F}$

$1\,\mu\text{F}$ $1\,\mu\text{F}$

$500\ \text{V}$

Fig. 15.4

6. In Fig. 15.5 capacitors *P*, *Q* and *R* are identical and the total equivalent capacitance of the circuit is $3\,\mu F$. Determine the values of *P*, *Q* and *R*.

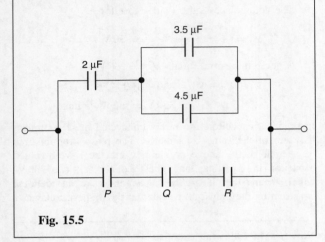

3.5 µF

2 µF

4.5 µF

P   *Q*   *R*

Fig. 15.5

15.4 Dielectric strength

The maximum amount of field strength that a dielectric can withstand is called the dielectric strength of the material.

Dielectric strength, $\boxed{E_m = \dfrac{V_m}{d}}$

Problem 11. A capacitor is to be constructed so that its capacitance is $0.2\,\mu F$ and to take a p.d. of $1.25\,kV$ across its terminals. The dielectric is to be mica which has a dielectric strength of $50\,MV/m$. Find (a) the thickness of the mica needed, and (b) the area of a plate assuming a two-plate construction (Assume ε_r for mica to be 6).

(a) Dielectric strength, $E = \dfrac{V}{d}$

i.e. $d = \dfrac{V}{E} = \dfrac{1.25 \times 10^3}{50 \times 10^6}\,m = \mathbf{0.025\,mm}$

(b) Capacitance, $C = \dfrac{\varepsilon_0 \varepsilon_r A}{d}$ hence,

area $A = \dfrac{Cd}{\varepsilon_0 \varepsilon_r} = \dfrac{0.2 \times 10^{-6} \times 0.025 \times 10^{-3}}{8.85 \times 10^{-12} \times 6}\,m^2$

$= 0.09416\,m^2$

$= \mathbf{941.6\,cm^2}$

15.5 Energy stored in capacitors

The energy, *W*, stored by a capacitor is given by

$$\boxed{W = \frac{1}{2}CV^2 \textbf{ joules}}$$

Problem 12. (a) Determine the energy stored in a $3\,\mu F$ capacitor when charged to $400\,V$ (b) Find also the average power developed if this energy is dissipated in a time of $10\,\mu s$.

(a) **Energy stored,**

$W = \dfrac{1}{2}CV^2$ joules

$= \dfrac{1}{2} \times 3 \times 10^{-6} \times 400^2 = \dfrac{3}{2} \times 16 \times 10^{-2}$

$= \mathbf{0.24\,J}$

(b) **Power** $= \dfrac{\text{energy}}{\text{time}} = \dfrac{0.24}{10 \times 10^{-6}}\,W = \mathbf{24\,kW}$

Problem 13. A $12\,\mu F$ capacitor is required to store $4\,J$ of energy. Find the p.d. to which the capacitor must be charged.

Energy stored $W = \dfrac{1}{2}CV^2$ hence $V^2 = \dfrac{2W}{C}$

and p.d. $V = \sqrt{\dfrac{2W}{C}} = \sqrt{\dfrac{2 \times 4}{12 \times 10^{-6}}}$

$= \sqrt{\dfrac{2 \times 10^6}{3}} = \mathbf{816.5\,V}$

Now try the following exercise

Exercise 62 Further problems on energy stored in capacitors (Answers on page 304)

(Assume $\varepsilon_0 = 8.85 \times 10^{-12}\,F/m$.)

1. When a capacitor is connected across a $200\,V$ supply the charge is $4\,\mu C$. Find (a) the capacitance and (b) the energy stored.

2. Find the energy stored in a $10\,\mu F$ capacitor when charged to $2\,kV$.

3. A $3300\,pF$ capacitor is required to store $0.5\,mJ$ of energy. Find the p.d. to which the capacitor must be charged.

4. A bakelite capacitor is to be constructed to have a capacitance of 0.04 µF and to have a steady working potential of 1 kV maximum. Allowing a safe value of field stress of 25 MV/m find (a) the thickness of bakelite required, (b) the area of plate required if the relative permittivity of bakelite is 5, (c) the maximum energy stored by the capacitor and (d) the average power developed if this energy is dissipated in a time of 20 µs.

15.6 Practical types of capacitor

Practical types of capacitor are characterised by the material used for their dielectric. The main types include: variable air, mica, paper, ceramic, plastic, titanium oxide and electrolytic.

1. **Variable air capacitors.** These usually consist of two sets of metal plates (such as aluminium), one fixed, the other variable. The set of moving plates rotate on a spindle as shown by the end view of Fig. 15.6.

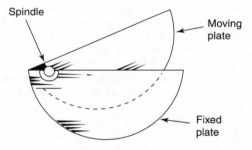

Fig. 15.6

As the moving plates are rotated through half a revolution, the meshing, and therefore the capacitance, varies from a minimum to a maximum value. Variable air capacitors are used in radio and electronic circuits where very low losses are required, or where a variable capacitance is needed. The maximum value of such capacitors is between 500 pF and 1000 pF.

2. **Mica capacitors.** A typical older type construction is shown in Fig. 15.7.
Usually the whole capacitor is impregnated with wax and placed in a bakelite case. Mica is easily obtained in thin sheets and is a good insulator. However, mica is expensive and is not used in capacitors above about 0.2 µF.
A modified form of mica capacitor is the silvered mica type. The mica is coated on both sides with a thin layer of silver which forms the plates. Capacitance is stable and less likely to change with age. Such capacitors have a constant capacitance with change of temperature, a high

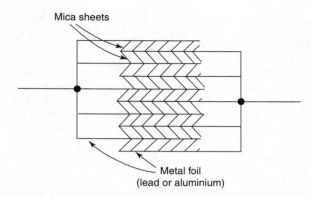

Fig. 15.7

working voltage rating and a long service life and are used in high frequency circuits with fixed values of capacitance up to about 1000 pF.

3. **Paper capacitors.** A typical paper capacitor is shown in Fig. 15.8 where the length of the roll corresponds to the capacitance required.

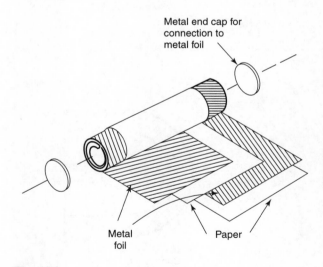

Fig. 15.8

The whole is usually impregnated with oil or wax to exclude moisture, and then placed in a plastic or aluminium container for protection. Paper capacitors are made in various working voltages up to about 150 kV and are used where loss is not very important. The maximum value of this type of capacitor is between 500 pF and 10 µF. Disadvantages of paper capacitors include variation in capacitance with temperature change and a shorter service life than most other types of capacitor.

4. **Ceramic capacitors.** These are made in various forms, each type of construction depending on the value of

capacitance required. For high values, a tube of ceramic material is used as shown in the cross section of Fig. 15.9. For smaller values the cup construction is used as shown in Fig. 15.10, and for still smaller values the disc construction shown in Fig. 15.11 is used. Certain ceramic materials have a very high permittivity and this enables capacitors of high capacitance to be made which are of small physical size with a high working voltage rating. Ceramic capacitors are available in the range 1 pF to 0.1 µF and may be used in high frequency electronic circuits subject to a wide range of temperatures.

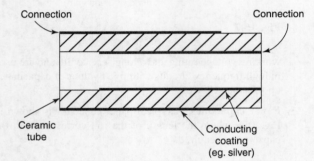

Fig. 15.9

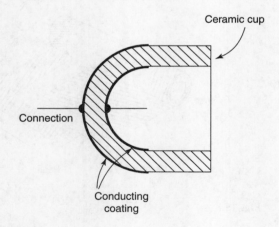

Fig. 15.10

5. **Plastic capacitors.** Some plastic materials such as polystyrene and Teflon can be used as dielectrics. Construction is similar to the paper capacitor but using a plastic film instead of paper. Plastic capacitors operate well under conditions of high temperature, provide a precise value of capacitance, a very long service life and high reliability.

6. **Titanium oxide capacitors** have a very high capacitance with a small physical size when used at a low temperature.

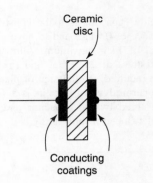

Fig. 15.11

7. **Electrolytic capacitors.** Construction is similar to the paper capacitor with aluminium foil used for the plates and with a thick absorbent material, such as paper, impregnated with an electrolyte (ammonium borate), separating the plates. The finished capacitor is usually assembled in an aluminium container and hermetically sealed. Its operation depends on the formation of a thin aluminium oxide layer on the positive plate by electrolytic action when a suitable direct potential is maintained between the plates. This oxide layer is very thin and forms the dielectric. (The absorbent paper between the plates is a conductor and does not act as a dielectric). Such capacitors **must always be used on d.c.** and must be connected with the correct polarity; if this is not done the capacitor will be destroyed since the oxide layer will be destroyed. Electrolytic capacitors are manufactured with working voltages from 6 V to 600 V, although accuracy is generally not very high. These capacitors possess a much larger capacitance than other types of capacitors of similar dimensions due to the oxide film being only a few microns thick. The fact that they can be used only on d.c. supplies limit their usefulness.

15.7   Discharging capacitors

When a capacitor has been disconnected from the supply it may still be charged and it may retain this charge for some considerable time. Thus precautions must be taken to ensure that the capacitor is automatically discharged after the supply is switched off. This is done by connecting a high value resistor across the capacitor terminals.

15.8   Inductance

Inductance is the property of a circuit whereby there is an e.m.f. induced into the circuit by the change of flux linkages produced by a current change. When the e.m.f. is induced in the same circuit as that in which the current is changing, the property is called **self-inductance** *L*.

The **unit** of inductance is the **henry, *H*:**

> *A circuit has an inductance of one henry when an e.m.f. of one volt is induced in it by a current changing at the rate of one ampere per second.*

Induced e.m.f. in a coil of *N* turns,

$$E = -N\frac{d\Phi}{dt} \text{ volts}$$

where dΦ is the change in flux in Webers, and dt is the time taken for the flux to change in seconds (i.e. $\frac{d\Phi}{dt}$ is the rate of change of flux).

Induced e.m.f. in a coil of inductance *L* henrys,

$$E = -L\frac{dI}{dt} \text{ volts}$$

where d*I* is the change in current in amperes and d*t* is the time taken for the current to change in seconds (i.e. $\frac{dI}{dt}$ is the rate of change of current).

The minus sign in each of the above two equations remind us of its direction (given by Lenz's law).

Problem 14. Determine the e.m.f. induced in a coil of 200 turns when there is a change of flux of 25 mWb linking with it in 50 ms.

Induced e.m.f. $E = -N\dfrac{d\Phi}{dt} = -(200)\left(\dfrac{25 \times 10^{-3}}{50 \times 10^{-3}}\right)$

$$= -100 \text{ volts}$$

Problem 15. A flux of 400 μWb passing through a 150-turn coil is reversed in 40 ms. Find the average e.m.f. induced.

Since the flux reverses, the flux changes from +400 μWb to −400 μWb, a total change of flux of 800 μWb.

Induced e.m.f. $E = -N\dfrac{d\Phi}{dt} = -(150)\left(\dfrac{800 \times 10^{-6}}{40 \times 10^{-3}}\right)$

$$= -\frac{150 \times 800 \times 10^{3}}{40 \times 10^{6}}$$

Hence, the average e.m.f. induced, *E* = −3 volts.

Problem 16. Calculate the e.m.f. induced in a coil of inductance 12 H by a current changing at the rate of 4 A/s.

Induced e.m.f. $E = -L\dfrac{dI}{dt} = -(12)(4) = -48$ **volts**.

Problem 17. An e.m.f. of 1.5 kV is induced in a coil when a current of 4 A collapses uniformly to zero in 8 ms. Determine the inductance of the coil.

Change in current,

$$dI = (4 - 0) = 4 \text{ A}, dt = 8 \text{ ms} = 8 \times 10^{-3}\text{ s},$$

$$\frac{dI}{dt} = \frac{4}{8 \times 10^{-3}} = \frac{4000}{8} = 500 \text{ A/s}$$

and $E = 1.5 \text{ kV} = 1500 \text{ V}$

Since $\qquad |E| = L\dfrac{dI}{dt}$

inductance, $\quad L = \dfrac{|E|}{\dfrac{dI}{dt}} = \dfrac{1500}{500} = \mathbf{3\,H}$

(Note that |*E*| means the 'magnitude of *E*' which disregards the minus sign.)

Now try the following exercise

Exercise 63 Further problems on inductance (Answers on page 304)

1. Find the e.m.f. induced in a coil of 200 turns when there is a change of flux of 30 mWb linking with it in 40 ms.
2. An e.m.f. of 25 V is induced in a coil of 300 turns when the flux linking with it changes by 12 mWb. Find the time, in milliseconds, in which the flux makes the change.
3. An ignition coil having 10 000 turns has an e.m.f. of 8 kV induced in it. What rate of change of flux is required for this to happen?
4. A flux of 0.35 mWb passing through a 125-turn coil is reversed in 25 ms. Find the magnitude of the average e.m.f. induced.
5. Calculate the e.m.f. induced in a coil of inductance 6 H by a current changing at a rate of 15 A/s.

15.9 Inductors

A component called an **inductor** is used when the property of inductance is required in a circuit. The basic form of an inductor is simply a coil of wire. Factors which affect the inductance of an inductor include:

(i) the number of turns of wire – the more turns the higher the inductance.

(ii) the cross-sectional area of the coil of wire – the greater the cross-sectional area the higher the inductance.

(iii) the presence of a magnetic core – when the coil is wound on an iron core the same current sets up a more concentrated magnetic field and the inductance is increased.

(iv) the way the turns are arranged – a short thick coil of wire has a higher inductance than a long thin one.

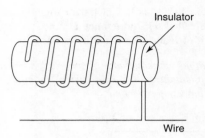

Fig. 15.14

15.10   Practical inductors

Two examples of practical inductors are shown in Fig. 15.12, and the standard electrical circuit diagram symbols for air-cored and iron-cored inductors are shown in Fig. 15.13.

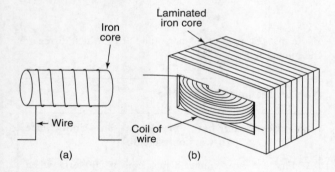

Fig. 15.12

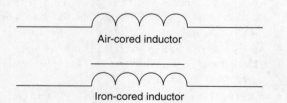

Fig. 15.13

An iron-cored inductor is often called a choke since, when used in a.c. circuits, it has a choking effect, limiting the current flowing through it. Inductance is often undesirable in a circuit. To reduce inductance to a minimum the wire may be bent back on itself, as shown in Fig. 15.14, so that the magnetising effect of one conductor is neutralised by that of the adjacent conductor. The wire may be coiled around an insulator, as shown, without increasing the inductance. Standard resistors may be non-inductively wound in this manner.

15.11   Energy stored by inductors

An inductor possesses an ability to store energy. The energy stored, W, in the magnetic field of an inductor is given by:

$$W = \frac{1}{2}LI^2 \text{ joules}$$

Problem 18.   An 8 H inductor has a current of 3 A flowing through it. How much energy is stored in the magnetic field of the inductor?

Energy stored, $W = \frac{1}{2}LI^2 = \frac{1}{2}(8)(3)^2 = $ **36 joules**

Now try the following exercise

Exercise 64   Further problems on energy stored (Answers on page 304)

1. An inductor of 20 H has a current of 2.5 A flowing in it. Find the energy stored in the magnetic field of the inductor.

2. Calculate the value of the energy stored when a current of 30 mA is flowing in a coil of inductance 400 mH.

3. The energy stored in the magnetic field of an inductor is 80 J when the current flowing in the inductor is 2 A. Calculate the inductance of the coil.

15.12   Inductance of a coil

If a current changing from 0 to I amperes, produces a flux change from 0 to Φ webers, then $dI = I$ and $d\Phi = \Phi$. Then, from Section 15.8,

$$\text{induced e.m.f. } E = \frac{N\Phi}{t} = \frac{LI}{t}$$

from which, **inductance of coil,** $\boxed{L = \frac{N\Phi}{I} \text{ henrys}}$

Problem 19. Calculate the coil inductance when a current of 4 A in a coil of 800 turns produces a flux of 5 mWb linking with the coil.

For a coil, inductance $L = \dfrac{N\Phi}{I} = \dfrac{(800)(5 \times 10^{-3})}{4} = \mathbf{1\,H}$

Problem 20. A flux of 25 mWb links with a 1500 turn coil when a current of 3 A passes through the coil. Calculate (a) the inductance of the coil, (b) the energy stored in the magnetic field, and (c) the average e.m.f. induced if the current falls to zero in 150 ms.

(a) **Inductance,**

$$L = \frac{N\Phi}{I} = \frac{(1500)(25 \times 10^{-3})}{3} = \mathbf{12.5\,H}$$

(b) **Energy stored,**

$$W = \frac{1}{2}LI^2 = \frac{1}{2}(12.5)(3)^2 = \mathbf{56.25\,J}$$

(c) **Induced emf,**

$$E = -L\frac{dI}{dt} = -(12.5)\left(\frac{3-0}{150 \times 10^{-3}}\right) = \mathbf{-250\,V}$$

(Alternatively,

$$E = -N\frac{d\Phi}{dt} = -(1500)\left(\frac{25 \times 10^{-3}}{150 \times 10^{-3}}\right) = \mathbf{-250\,V}$$

since if the current falls to zero so does the flux.)

Problem 21. When a current of 1.5 A flows in a coil the flux linking with the coil is 90 µWb. If the coil inductance is 0.60 H, calculate the number of turns of the coil.

For a coil, $L = \dfrac{N\Phi}{I}$,

thus $N = \dfrac{LI}{\Phi} = \dfrac{(0.6)(1.5)}{90 \times 10^{-6}} = \mathbf{10\,000\ turns}$

Now try the following exercises

Exercise 65 Further problems on the inductance of a coil (Answers on page 304)

1. A flux of 30 mWb links with a 1200 turn coil when a current of 5 A is passing through the coil. Calculate (a) the inductance of the coil, (b) the energy stored in the magnetic field, and (c) the average e.m.f. induced if the current is reduced to zero in 0.20 s.

2. An e.m.f. of 2 kV is induced in a coil when a current of 5 A collapses uniformly to zero in 10 ms. Determine the inductance of the coil.

3. An average e.m.f. of 60 V is induced in a coil of inductance 160 mH when a current of 7.5 A is reversed. Calculate the time taken for the current to reverse.

4. A coil of 2500 turns has a flux of 10 mWb linking with it when carrying a current of 2 A. Calculate the coil inductance and the e.m.f. induced in the coil when the current collapses to zero in 20 ms.

5. When a current of 2 A flows in a coil, the flux linking with the coil is 80 µWb. If the coil inductance is 0.5 H, calculate the number of turns of the coil.

Exercise 66 Short answer questions on capacitors and inductors

1. How can an 'electric field' be established between two parallel metal plates?
2. What is capacitance?
3. State the unit of capacitance.
4. Complete the statement: Capacitance = $\dfrac{\ldots}{\ldots}$
5. Complete the statements:
   (a) 1 µF = ... F   (b) 1 pF = ... F
6. Complete the statement: Electric field strength $E = \dfrac{\ldots}{\ldots}$
7. Complete the statement: Electric flux density $D = \dfrac{\ldots}{\ldots}$
8. Draw the electrical circuit diagram symbol for a capacitor.
9. Name two practical examples where capacitance is present, although undesirable.
10. The insulating material separating the plates of a capacitor is called the ....
11. 10 volts applied to a capacitor results in a charge of 5 coulombs. What is the capacitance of the capacitor?
12. Three 3 µF capacitors are connected in parallel. The equivalent capacitance is ....
13. Three 3 µF capacitors are connected in series. The equivalent capacitance is ....
14. State an advantage of series-connected capacitors.
15. Name three factors upon which capacitance depends.
16. What does 'relative permittivity' mean?
17. Define 'permittivity of free space'.
18. What is meant by the 'dielectric strength' of a material?
19. State the formula used to determine the energy stored by a capacitor.

20. Name five types of capacitor commonly used.
21. Sketch a typical rolled paper capacitor.
22. Explain briefly the construction of a variable air capacitor.
23. State three advantages and one disadvantage of mica capacitors.
24. Name two disadvantages of paper capacitors.
25. Between what values of capacitance are ceramic capacitors normally available.
26. What main advantages do plastic capacitors possess?
27. Explain briefly the construction of an electrolytic capacitor.
28. What is the main disadvantage of electrolytic capacitors?
29. Name an important advantage of electrolytic capacitors.
30. What safety precautions should be taken when a capacitor is disconnected from a supply?
31. Define inductance and name its unit.
32. What factors affect the inductance of an inductor?
33. What is an inductor? Sketch a typical practical inductor.
34. Explain how a standard resistor may be non-inductively wound.
35. Energy W stored in the magnetic field of an inductor is given by: $W = \ldots\ldots$ joules.

Exercise 67   Multi-choice questions on capacitors and inductance (Answers on page 304)

1. The capacitance of a capacitor is the ratio
   (a) charge to p.d. between plates
   (b) p.d. between plates to plate spacing
   (c) p.d. between plates to thickness of dielectric
   (d) p.d. between plates to charge

2. The p.d. across a $10\,\mu F$ capacitor to charge it with $10\,mC$ is
   (a) 10 V   (b) 1 kV   (c) 1 V   (d) 10 V

3. The charge on a $10\,pF$ capacitor when the voltage applied to it is $10\,kV$ is
   (a) $100\,\mu C$ (b) $0.1\,C$ (c) $0.1\,\mu C$ (d) $0.01\,\mu C$

4. Four $2\,\mu F$ capacitors are connected in parallel. The equivalent capacitance is
   (a) $8\,\mu F$ (b) $0.5\,\mu F$ (c) $2\,\mu F$ (d) $6\,\mu F$

5. Four $2\,\mu F$ capacitors are connected in series. The equivalent capacitance is
   (a) $8\,\mu F$ (b) $0.5\,\mu F$ (c) $2\,\mu F$ (d) $6\,\mu F$

6. State which of the following is false.

The capacitance of a capacitor
   (a) is proportional to the cross-sectional area of the plates
   (b) is proportional to the distance between the plates
   (c) depends on the number of plates
   (d) is proportional to the relative permittivity of the dielectric

7. Which of the following statement is false?
   (a) An air capacitor is normally a variable type
   (b) A paper capacitor generally has a shorter service life than most other types of capacitor
   (c) An electrolytic capacitor must be used only on a.c. supplies
   (d) Plastic capacitors generally operate satisfactorily under conditions of high temperature

8. The energy stored in a $10\,\mu F$ capacitor when charged to $500\,V$ is
   (a) $1.25\,mJ$ (b) $0.025\,\mu J$ (c) $1.25\,J$ (d) $1.25\,C$

9. The capacitance of a variable air capacitor is at maximum when
   (a) the movable plates half overlap the fixed plates
   (b) the movable plates are most widely separated from the fixed plates
   (c) both sets of plates are exactly meshed
   (d) the movable plates are closer to one side of the fixed plate than to the other

10. When a voltage of $1\,kV$ is applied to a capacitor, the charge on the capacitor is $500\,nC$. The capacitance of the capacitor is:
    (a) $2 \times 10^9\,F$ (b) $0.5\,pF$ (c) $0.5\,mF$ (d) $0.5\,nF$

11. When a magnetic flux of $10\,Wb$ links with a circuit of 20 turns in $2\,s$, the induced e.m.f. is:
    (a) 1 V   (b) 4 V   (c) 100 V   (d) 400 V

12. A current of $10\,A$ in a coil of 1000 turns produces a flux of $10\,mWb$ linking with the coil. The coil inductance is:
    (a) $10^6\,H$ (b) $1\,H$ (c) $1\,\mu H$ (d) $1\,mH$

13. Which of the following statements is false?
    The inductance of an inductor increases:
    (a) with a short, thick, coil
    (b) when wound on an iron core
    (c) as the number of turns increases
    (d) as the cross-sectional area of the coil decreases

14. Self-inductance occurs when:
    (a) the current is changing
    (b) the circuit is changing
    (c) the flux is changing
    (d) the resistance is changing

16

Heat energy

At the end of this chapter you should be able to:

- distinguish between heat and temperature
- appreciate that temperature is measured on the Celsius or the thermodynamic scale
- convert temperatures from Celsius into kelvin and vice versa
- recognise several temperature measuring devices
- define specific heat capacity, c and recognize typical values
- calculate the quantity of heat energy Q using $Q = mc(t_2 - t_1)$
- understand change of state from solid to liquid to gas, and vice versa
- distinguish between sensible and latent heat
- define specific latent heat of fusion
- define specific latent heat of vaporisation
- recognise typical values of latent heats of fusion and vaporisation
- calculate quantity of heat Q using $Q = mL$
- describe the principle of operation of a simple refrigerator

16.1 Introduction

Heat is a form of energy and is measured in joules.

Temperature is the degree of hotness or coldness of a substance. Heat and temperature are thus not the same thing. For example, twice the heat energy is needed to boil a full container of water than half a container – that is, different amounts of heat energy are needed to cause an equal rise in the temperature of different amounts of the same substance.

Temperature is measured either (i) on the **Celsius** (°C) scale (formerly Centigrade), where the temperature at which ice melts, i.e. the freezing point of water, is taken as 0°C and the point at which water boils under normal atmospheric pressure is taken as 100°C, or (ii) on the **thermodynamic scale**, in which the unit of temperature is the **kelvin (K)**. The kelvin scale uses the same temperature interval as the Celsius scale but as its zero takes the 'absolute zero of temperature' which is at about −273°C. Hence,

kelvin temperature = degree Celsius + 273

i.e.

$$\boxed{\mathbf{K = (°C) + 273}}$$

Thus, for example, $0°C = 273\,K$, $25°C = 298\,K$ and $100°C = 373\,K$.

Problem 1. Convert the following temperatures into the kelvin scale:

(a) 37°C (b) −28°C

From above, kelvin temperature = degree Celsius + 273

(a) 37°C corresponds to a kelvin temperature of $37 + 273$, i.e. **310 K**.

(b) −28°C corresponds to a kelvin temperature of $-28 + 273$, i.e. **245 K**.

Problem 2. Convert the following temperatures into the Celsius scale:

(a) 365 K (b) 213 K

From above, $K = (°C) + 273$

Hence, degree Celsius = kelvin temperature − 273

(a) 365 K corresponds to 365 − 273, i.e. **92°C**.

(b) 213 K corresponds to 213 − 273, i.e. **−60°C**.

Now try the following exercise

**Exercise 68   Further problems on temperature scales
(Answers on page 304)**

1. Convert the following temperatures into the kelvin
   scale:
   (a) 51°C   (b) −78°C   (c) 183°C

2. Convert the following temperatures into the Celsius
   scale:
   (a) 307 K   (b) 237 K   (c) 415 K

16.2   The measurement of temperature

A **thermometer** is an instrument which measures temperature. Any substance which possesses one or more properties which vary with temperature can be used to measure temperature. These properties include changes in length, area or volume, electrical resistance or in colour. Examples of temperature measuring devices include:

(i) **liquid-in-glass thermometer**, which uses the expansion of a liquid with increase in temperature as its principle of operation

(ii) **thermocouples**, which use the e.m.f. set up when the junction of two dissimilar metals is heated

(iii) **resistance thermometer**, which uses the change in electrical resistance caused by temperature change, and

(iv) **pyrometers**, which are devices for measuring very high temperatures, using the principle that all substances emit radiant energy when hot, the rate of emission depending on their temperature.

Each of these temperature measuring devices, together with others, are described in Chapter 36.

16.3   Specific heat capacity

The **specific heat capacity** of a substance is the quantity of heat energy required to raise the temperature of 1 kg of the substance by 1°C. The symbol used for specific heat capacity is c and the units are J/(kg°C) or J/(kg K). (Note that these units may also be written as $J\,kg^{-1}\,°C^{-1}$ or $J\,kg^{-1}\,K^{-1}$) Some typical values of specific heat capacity for the range of temperature 0°C to 100°C include:

Water	4190 J/(kg°C)
Aluminium	950 J/(kg°C)
Iron	500 J/(kg°C)
Ice	2100 J/(kg°C)
Copper	390 J/(kg°C)
Lead	130 J/(kg°C)

Hence to raise the temperature of 1 kg of iron by 1°C requires 500 J of energy, to raise the temperature of 5 kg of iron by 1°C requires (500×5) J of energy, and to raise the temperature of 5 kg of iron by 40°C requires $(500 \times 5 \times 40)$ J of energy, i.e. 100 kJ.

In general, the quantity of heat energy, Q, required to raise a mass m kg of a substance with a specific heat capacity c J/(kg°C) from temperature t_1°C to t_2°C is given by:

$$Q = mc\,(t_2 - t_1)\ \text{joules}$$

Problem 3.   Calculate the quantity of heat required to raise the temperature of 5 kg of water from 0°C to 100°C. Assume the specific heat capacity of water is 4200 J/(kg°C).

Quantity of heat energy,

$Q = mc(t_2 - t_1)$

$= 5\,\text{kg} \times 4200\,\text{J/(kg°C)} \times (100 - 0)°\text{C}$

$= 5 \times 4200 \times 100$

$= \textbf{2\,100\,000\,J}$ or   **2100 kJ**   or   **2.1 MJ**

Problem 4.   A block of cast iron having a mass of 10 kg cools from a temperature of 150°C to 50°C. How much energy is lost by the cast iron? Assume the specific heat capacity of iron is 500 J/(kg°C).

Quantity of heat energy,

$Q = mc(t_2 - t_1)$

$= 10\,\text{kg} \times 500\,\text{J/(kg°C)} \times (50 - 150)°\text{C}$

$= 10 \times 500 \times (-100)$

$= \textbf{−500\,000\,J}$ or   **−500 kJ**   or   **−0.5 MJ**

(Note that the minus sign indicates that heat is given out or lost.)

Problem 5.   Some lead having a specific heat capacity of 130 J/(kg°C) is heated from 27°C to its melting point at 327°C. If the quantity of heat required is 780 kJ, determine the mass of the lead.

Quantity of heat, $Q = mc(t_2 - t_1)$, hence,

$$780 \times 10^3 \text{ J} = m \times 130 \text{ J/(kg°C)} \times (327 - 27)\text{°C}$$

i.e. $780\,000 = m \times 130 \times 300$

from which, **mass, m** $= \dfrac{780\,000}{130 \times 300} \text{ kg} = \textbf{20 kg}$.

Problem 6.   5.7 MJ of heat energy are supplied to 30 kg of aluminium which is initially at a temperature of 20°C. If the specific heat capacity of aluminium is 950 J/(kg°C), determine its final temperature.

Quantity of heat, $Q = mc(t_2 - t_1)$, hence,

$$5.7 \times 10^6 \text{ J} = 30 \text{ kg} \times 950 \text{ J/(kg°C)}$$
$$\times (t_2 - 20)\text{°C}$$

from which, $(t_2 - 20) = \dfrac{5.7 \times 10^6}{30 \times 950} = 200$

Hence the **final temperature**, $t_2 = 200 + 20 = \textbf{220°C}$.

Problem 7.   A copper container of mass 500 g contains 1 litre of water at 293 K. Calculate the quantity of heat required to raise the temperature of the water and container to boiling point assuming there are no heat losses. Assume that the specific heat capacity of copper is 390 J/(kg K), the specific heat capacity of water is 4.2 kJ/(kg K) and 1 litre of water has a mass of 1 kg.

Heat is required to raise the temperature of the water, and also to raise the temperature of the copper container.

For the water: $m = 1$ kg, $t_1 = 293$ K,

$$t_2 = 373 \text{ K (i.e. boiling point) and}$$

$$c = 4.2 \text{ kJ/(kg K)}.$$

Quantity of heat required for the water is given by:

$$Q_W = mc(t_2 - t_1) = (1 \text{ kg}) \left(4.2 \frac{\text{kg}}{\text{kg K}}\right)(373 - 293)\text{ K}$$

$$= 4.2 \times 80 \text{ kJ}$$

i.e. $Q_W = \textbf{336 kJ}$

For the copper container: $m = 500$ g $= 0.5$ kg, $t_1 = 293$ K, $t_2 = 373$ K and $c = 390$ J/(kg K) $= 0.39$ kJ/(kg K).

Quantity of heat required for the copper container is given by:

$$Q_C = mc(t_2 - t_1) = (0.5 \text{ kg})(0.39 \text{ kJ/(kg K)})(80 \text{ K})$$

i.e. $Q_C = \textbf{15.6 kJ}$

Total quantity of heat required,

$$Q = Q_W + Q_C = 336 + 15.6 = \textbf{351.6 kJ}$$

Now try the following exercise

Exercise 69   Further problems on specific heat capacity (Answers on page 304)

1. Determine the quantity of heat energy (in megajoules) required to raise the temperature of 10 kg of water from 0°C to 50°C. Assume the specific heat capacity of water is 4200 J/(kg°C).

2. Some copper, having a mass of 20 kg, cools from a temperature of 120°C to 70°C. If the specific heat capacity of copper is 390 J/(kg°C), how much heat energy is lost by the copper?

3. A block of aluminium having a specific heat capacity of 950 J/(kg°C) is heated from 60°C to its melting point at 660°C. If the quantity of heat required is 2.85 MJ, determine the mass of the aluminium block.

4. 20.8 kJ of heat energy is required to raise the temperature of 2 kg of lead from 16°C to 96°C. Determine the specific heat capacity of lead.

5. 250 kJ of heat energy is supplied to 10 kg of iron which is initially at a temperature of 15°C. If the specific heat capacity of iron is 500 J/(kg°C) determine its final temperature.

16.4   Change of state

A material may exist in any one of three states – solid, liquid or gas. If heat is supplied at a constant rate to some ice initially at, say, −30°C, its temperature rises as shown in Fig. 16.1. Initially the temperature increases from −30°C to 0°C as shown by the line *AB*. It then remains constant at 0°C for the time *BC* required for the ice to melt into water.

When melting commences the energy gained by continual heating is offset by the energy required for the change of state and the temperature remains constant even though heating is continued. When the ice is completely melted to water, continual heating raises the temperature to 100°C, as shown by *CD* in Fig. 16.1. The water then begins to boil and the temperature again remains constant at 100°C, shown as *DE*, until all the water has vaporised.

Continual heating raises the temperature of the steam as shown by *EF* in the region where the steam is termed superheated.

Changes of state from solid to liquid or liquid to gas occur without change of temperature and such changes are reversible processes. When heat energy flows to or from a substance and causes a change of temperature, such as between *A* and *B*, between *C* and *D* and between *E* and *F* in Fig. 16.1, it is called **sensible heat** (since it can be 'sensed' by a thermometer).

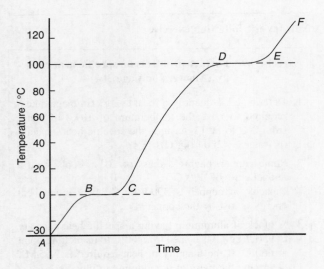

Fig. 16.1

Heat energy which flows to or from a substance while the temperature remains constant, such as between *B* and *C* and between *D* and *E* in Fig. 16.1, is called **latent heat** (latent means concealed or hidden).

Problem 8. Steam initially at a temperature of 130°C is cooled to a temperature of 20°C below the freezing point of water, the loss of heat energy being at a constant rate. Make a sketch, and briefly explain, the expected temperature/time graph representing this change.

A temperature/time graph representing the change is shown in Fig. 16.2. Initially steam cools until it reaches the boiling point of water at 100°C. Temperature then remains constant, i.e. between *A* and *B*, even though it is still giving off heat

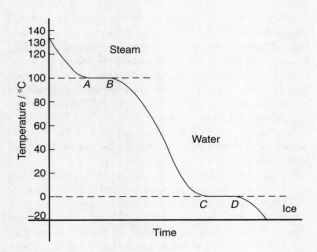

Fig. 16.2

(i.e. latent heat). When all the steam at 100°C has changed to water at 100°C it starts to cool again until it reaches the freezing point of water at 0°C. From *C* to *D* the temperature again remains constant until all the water is converted to ice. The temperature of the ice then decreases as shown.

Now try the following exercise

Exercise 70   A further problem on change of state (Answer on page 304)

1. Some ice, initially at −40°C, has heat supplied to it at a constant rate until it becomes superheated steam at 150°C. Sketch a typical temperature/time graph expected and use it to explain the difference between sensible and latent heat.

16.5 Latent heats of fusion and vaporisation

The **specific latent heat of fusion** is the heat required to change 1 kg of a substance from the solid state to the liquid state (or vice versa) at constant temperature.

The **specific latent heat of vaporisation** is the heat required to change 1 kg of a substance from a liquid to a gaseous state (or vice versa) at constant temperature.

The units of the specific latent heats of fusion and vaporisation are J/kg, or more often kJ/kg, and some typical values are shown in Table 16.1.

Table 16.1

	Latent heat of fusion (kJ/kg)	Melting point (°C)
Mercury	11.8	−39
Lead	22	327
Silver	100	957
Ice	335	0
Aluminium	387	660

	Latent heat of vaporisation (kJ/kg)	Boiling point (°C)
Oxygen	214	−183
Mercury	286	357
Ethyl alcohol	857	79
Water	2257	100

The quantity of heat Q supplied or given out during a change of state is given by:

$$Q = mL$$

where m is the mass in kilograms and L is the specific latent heat.

Thus, for example, the heat required to convert 10 kg of ice at 0°C to water at 0°C is given by 10 kg × 335 kJ/kg = 3350 kJ or 3.35 MJ.

Besides changing temperature, the effects of supplying heat to a material can involve changes in dimensions, as well as in colour, state and electrical resistance. Most substances expand when heated and contract when cooled, and there are many practical applications and design implications of thermal movement (see Chapter 35).

Problem 9. How much heat is needed to melt completely 12 kg of ice at 0°C? Assume the latent heat of fusion of ice is 335 kJ/kg.

Quantity of heat required,

$$Q = mL = 12 \text{ kg} \times 335 \text{ kJ/kg}$$

$$= \mathbf{4020\,kJ} \quad \text{or} \quad \mathbf{4.02\,MJ}$$

Problem 10. Calculate the heat required to convert 5 kg of water at 100°C to superheated steam at 100°C. Assume the latent heat of vaporisation of water is 2260 kJ/kg.

Quantity of heat required,

$$Q = mL = 5 \text{ kg} \times 2260 \text{ kJ/kg}$$

$$= \mathbf{11\,300\,kJ} \quad \text{or} \quad \mathbf{11.3\,MJ}$$

Problem 11. Determine the heat energy needed to convert 5 kg of ice initially at −20°C completely to water at 0°C. Assume the specific heat capacity of ice is 2100 J/(kg°C) and the specific latent heat of fusion of ice is 335 kJ/kg.

Quantity of heat energy needed,

$$Q = \text{sensible heat} + \text{latent heat}.$$

The quantity of heat needed to raise the temperature of ice from −20°C to 0°C, i.e. sensible heat,

$$Q_1 = mc(t_2 - t_1) = 5 \text{ kg} \times 2100 \text{ J/(kg°C)}$$

$$\times (0 - -20)°\text{C}$$

$$= (5 \times 2100 \times 20) \text{ J} = \mathbf{210\,kJ}$$

The quantity of heat needed to melt 5 kg of ice at 0°C, i.e. the latent heat,

$$Q_2 = mL = 5 \text{ kg} \times 335 \text{ kJ/kg} = \mathbf{1675\,kJ}$$

Total heat energy needed,

$$Q = Q_1 + Q_2 = 210 + 1675 = \mathbf{1885\,kJ}$$

Problem 12. Calculate the heat energy required to convert completely 10 kg of water at 50°C into steam at 100°C, given that the specific heat capacity of water is 4200 J/(kg°C) and the specific latent heat of vaporisation of water is 2260 kJ/kg

---

Quantity of heat required = sensible heat + latent heat.

Sensible heat, $Q_1 = mc(t_2 - t_1)$

$$= 10 \text{ kg} \times 4200 \text{ J/(kg°C)} \times (100 - 50)°\text{C}$$

$$= \mathbf{2100\,kJ}$$

Latent heat, $Q_2 = mL = 10 \text{ kg} \times 2260 \text{ kJ/kg}$

$$= \mathbf{22\,600\,kJ}$$

Total heat energy required,

$$Q = Q_1 + Q_2 = (2100 + 22\,600) \text{ kJ}$$

$$= \mathbf{24\,700\,kJ} \quad \text{or} \quad \mathbf{24.70\,MJ}$$

Problem 13. Determine the amount of heat energy needed to change 400 g of ice, initially at −20°C, into steam at 120°C. Assume the following: latent heat of fusion of ice = 335 kJ/kg, latent heat of vaporisation of water = 2260 kJ/kg, specific heat capacity of ice = 2.14 kJ/(kg°C), specific heat capacity of water = 4.2 kJ/(kg°C) and specific heat capacity of steam = 2.01 kJ/(kg°C).

The energy needed is determined in five stages:

(i) Heat energy needed to change the temperature of ice from −20°C to 0°C is given by:

$$Q_1 = mc(t_2 - t_1)$$

$$= 0.4 \text{ kg} \times 2.14 \text{ kJ/(kg°C)} \times (0 - -20)°\text{C}$$

$$= \mathbf{17.12\,kJ}$$

(ii) Latent heat needed to change ice at 0°C into water at 0°C is given by:

$$Q_2 = mL_f = 0.4 \text{ kg} \times 335 \text{ kJ/kg} = \mathbf{134\,kJ}$$

(iii) Heat energy needed to change the temperature of water from 0°C (i.e. melting point) to 100°C (i.e. boiling point) is given by:

$$Q_3 = mc(t_2 - t_1)$$

$$= 0.4 \text{ kg} \times 4.2 \text{ kJ/(kg°C)} \times 100°\text{C}$$

$$= \mathbf{168\,kJ}$$

(iv) Latent heat needed to change water at 100°C into steam at 100°C is given by:

$$Q_4 = mL_v = 0.4 \text{ kg} \times 2260 \text{ kJ/kg} = \mathbf{904\,kJ}$$

(v) Heat energy needed to change steam at 100°C into steam at 120°C is given by:

$$Q_5 = mc(t_1 - t_2)$$

$$= 0.4\,\text{kg} \times 2.01\,\text{kJ/(kg°C)} \times 20°\text{C}$$

$$= \mathbf{16.08\,kJ}$$

Total heat energy needed,

$$Q = Q_1 + Q_2 + Q_3 + Q_4 + Q_5$$

$$= 17.12 + 134 + 168 + 904 + 16.08$$

$$= \mathbf{1239.2\,kJ}$$

Now try the following exercise

> **Exercise 71   Further problems on the latent heats of fusion and vaporisation (Answers on page 304)**
>
> 1. How much heat is needed to melt completely 25 kg of ice at 0°C. Assume the specific latent heat of fusion of ice is 335 kJ/kg.
>
> 2. Determine the heat energy required to change 8 kg of water at 100°C to superheated steam at 100°C. Assume the specific latent heat of vaporisation of water is 2260 kJ/kg.
>
> 3. Calculate the heat energy required to convert 10 kg of ice initially at −30°C completely into water at 0°C. Assume the specific heat capacity of ice is 2.1 kJ/(kg°C) and the specific latent heat of fusion of ice is 335 kJ/kg.
>
> 4. Determine the heat energy needed to convert completely 5 kg of water at 60°C to steam at 100°C, given that the specific heat capacity of water is 4.2 kJ/(kg°C) and the specific latent heat of vaporisation of water is 2260 kJ/kg.

16.6   A simple refrigerator

The boiling point of most liquids may be lowered if the pressure is lowered. In a simple refrigerator a working fluid, such as ammonia or freon, has the pressure acting on it reduced. The resulting lowering of the boiling point causes the liquid to vaporise. In vaporising, the liquid takes in the necessary latent heat from its surroundings, i.e. the freezer, which thus becomes cooled. The vapour is immediately removed by a pump to a condenser which is outside of the cabinet, where it is compressed and changed back into a liquid, giving out latent heat. The cycle is repeated when the liquid is pumped back to the freezer to be vaporised.

Now try the following exercises

> **Exercise 72   Short answer questions on heat energy**
>
> 1. Differentiate between temperature and heat.
>
> 2. Name two scales on which temperature is measured.
>
> 3. Name any four temperature measuring devices.
>
> 4. Define specific heat capacity and name its unit.
>
> 5. Differentiate between sensible and latent heat.
>
> 6. The quantity of heat, Q, required to raise a mass m kg from temperature $t_1°$C to $t_2°$C, the specific heat capacity being c, is given by $Q = \ldots$
>
> 7. What is meant by the specific latent heat of fusion?
>
> 8. Define the specific latent heat of vaporisation.
>
> 9. Explain briefly the principle of operation of a simple refrigerator.

> **Exercise 73   Multi-choice questions on heat energy (Answers on page 305)**
>
> 1. Heat energy is measured in:
>   (a) kelvin   (b) watts   (c) kilograms   (d) joules
>
> 2. A change of temperature of 20°C is equivalent to a change in thermodynamic temperature of:
>   (a) 293 K   (b) 20 K   (c) 80 K   (d) 120 K
>
> 3. A temperature of 20°C is equivalent to:
>   (a) 293 K   (b) 20 K   (c) 80 K   (d) 120 K
>
> 4. The unit of specific heat capacity is:
>   (a) joules per kilogram   (b) joules
>   (c) joules per kilogram kelvin   (d) cubic metres
>
> 5. The quantity of heat required to raise the temperature of 500 g of iron by 2°C, given that the specific heat capacity is 500 J/(kg°C), is:
>   (a) 500 kJ   (b) 0.5 kJ   (c) 2 J   (d) 250 kJ
>
> 6. The heat energy required to change 1 kg of a substance from a liquid to a gaseous state at the same temperature is called:
>   (a) specific heat capacity
>   (b) specific latent heat of vaporisation
>   (c) sensible heat
>   (d) specific latent heat of fusion
>
> 7. The temperature of pure melting ice is:
>   (a) 373 K   (b) 273 K   (c) 100 K   (d) 0 K

8. 1.95 kJ of heat is required to raise the temperature of 500 g of lead from 15°C to its final temperature. Taking the specific heat capacity of lead to be 130 J/(kg°C), the final temperature is:

    (a) 45°C   (b) 37.5°C   (c) 30°C   (d) 22.5°C

9. Which of the following temperatures is absolute zero?

    (a) 0°C   (b) −173°C   (c) −273°C   (d) −373°C

10. When two wire of different metals are twisted together and heat applied to the junction, an e.m.f. is produced. This effect is used in a thermocouple to measure:

    (a) e.m.f.   (b) temperature
    (c) expansion   (d) heat

Assignment 5

This assignment covers the material contained in Chapters 13 to 16. The marks for each question are shown in brackets at the end of each question.

1. A 100 W electric light bulb is connected to a 200 V supply. Calculate (a) the current flowing in the bulb, and (b) the resistance of the bulb.   (4)

2. Determine the charge transferred when a current of 5 mA flows for 10 minutes.   (4)

3. A current of 12 A flows in the element of an electric fire of resistance 10 Ω. Determine the power dissipated by the element. If the fire is on for 5 hours every day, calculate for a one week period (a) the energy used, and (b) cost of using the fire if electricity cost 7p per unit.   (6)

4. Four cells, each with an internal resistance of 0.40 Ω and an e.m.f. of 2.5 V are connected in series to a load of 38.4 Ω. Determine the current flowing in the circuit and the p.d. at the battery terminals.   (5)

5. The charge on the plates of a capacitor is 8 mC when the potential between them is 4 kV. Determine the capacitance of the capacitor.   (3)

6. Two parallel rectangular plates measuring 80 mm by 120 mm are separated by 4 mm of mica and carry an electric charge of 0.48 µC. The voltage between the plates is 500 V. Calculate (a) the electric flux density (b) the electric field strength, and (c) the capacitance of the capacitor, in picofarads, if the relative permittivity of mica is 5.   (7)

7. An e.m.f. of 2.5 kV is induced in a coil when a current of 2 A collapses to zero in 5 ms. Calculate the inductance of the coil.   (4)

8. A flux of 15 mWb links with a 2000 turn coil when a current of 2 A passes through the coil. Calculate (a) the inductance of the coil, (b) the energy stored in the magnetic field, and (c) the average e.m.f. induced if the current falls to zero in 300 ms.   (6)

9. A block of aluminium having a mass of 20 kg cools from a temperature of 250°C to 80°C. How much energy is lost by the aluminium? Assume the specific heat capacity of aluminium is 950 J/(kg°C).   (4)

10. Calculate the heat energy required to convert completely 12 kg of water at 30°C to superheated steam at 100°C. Assume that the specific heat capacity of water is 4200 J/(kg°C) and the specific latent heat of vaporisation of water is 2260 kJ/(kg°C).   (7)

Formulae for energy applications

Formula	Formula Symbols	Units
Density = $\dfrac{\text{mass}}{\text{volume}}$	$\rho = \dfrac{m}{V}$	kg/m^3
Work done = force × distance moved	$W = Fs$	J
Efficiency = $\dfrac{\text{useful output energy}}{\text{input energy}}$		
Power = $\dfrac{\text{energy used (or work done)}}{\text{time taken}}$ = force × velocity	$P = \dfrac{E}{t} = Fv$	W
Potential energy = weight × change in height	$E_p = mgh$	J
kinetic energy = $\dfrac{1}{2}$ × mass × (speed)^2	$E_k = \dfrac{1}{2}mv^2$	J
Charge = current × time	$Q = It$	C
Resistance = $\dfrac{\text{potential difference}}{\text{current}}$	$R = \dfrac{V}{I}$	Ω
Electrical power = potential difference × current	$P = VI = I^2R = \dfrac{V^2}{R}$	W
Terminal p.d. = source e.m.f. − (current) (resistance)	$V = E - Ir$	V
Electric field strength = $\dfrac{\text{p.d. across dielectric}}{\text{thickness of dielectric}}$	$E = \dfrac{V}{d}$	V/m
Electric flux density = $\dfrac{\text{charge}}{\text{area}}$	$D = \dfrac{Q}{A}$	C/m^2
Charge = capacitance × potential difference	$Q = C \times V$	C
Parallel plate capacitor	$C = \dfrac{\varepsilon_0 \varepsilon_r A(n-1)}{d}$	F
Total capacitance of capacitors in series:	$\dfrac{1}{C} = \dfrac{1}{C_1} + \dfrac{1}{C_2} + \cdots$	

Total capacitance of capacitors in parallel: $\qquad C = C_1 + C_2 + \cdots \qquad$ F

Energy stored in capacitor $\qquad W = \dfrac{1}{2}CV^2 \qquad$ J

Induced e.m.f. = number of coil turns × rate of change of flux $\qquad E = -N\dfrac{d\Phi}{dt} \qquad$ V

Induced e.m.f. = inductance × rate of change of current $\qquad E = -L\dfrac{dI}{dt} \qquad$ V

Inductance = $\dfrac{\text{number of coil turns} \times \text{flux}}{\text{current}} \qquad L = \dfrac{N\Phi}{I} \qquad$ H

Energy stored in magnetic field of inductor $\qquad W = \dfrac{1}{2}LI^2 \qquad$ J

Kelvin temperature = degrees Celsius + 273

Quantity of heat energy = mass × specific heat capacity
× change in temperature $\qquad Q = mc(t_2 - t_1) \qquad$ J

Section 3

Electrical Applications

17

Resistance variation

At the end of this chapter you should be able to:

- appreciate that electrical resistance depends on four factors

- appreciate that resistance $R = \dfrac{\rho l}{a}$, where ρ is the resistivity

- recognise typical values of resistivity and its unit

- perform calculations using $R = \dfrac{\rho l}{a}$

- define the temperature coefficient of resistance, α

- recognise typical values for α

- perform calculations using $R_\theta = R_0(1 + \alpha\theta)$

17.1 Resistance and resistivity

The resistance of an electrical conductor depends on four factors, these being: (a) the length of the conductor, (b) the cross-sectional area of the conductor, (c) the type of material and (d) the temperature of the material. Resistance, R, is directly proportional to length, l, of a conductor, i.e. $R \propto l$. Thus, for example, if the length of a piece of wire is doubled, then the resistance is doubled.

Resistance, R, is inversely proportional to cross-sectional area, a, of a conductor, i.e. $R \propto \dfrac{1}{a}$. Thus, for example, if the cross-sectional area of a piece of wire is doubled then the resistance is halved.

Since $R \propto l$ and $R \propto \dfrac{1}{a}$ then $R \propto \dfrac{l}{a}$. By inserting a constant of proportionality into this relationship the type of material used may be taken into account. The constant of proportionality is known as the **resistivity** of the material and is given the symbol ρ (Greek rho).

Thus, resistance $\boxed{R = \dfrac{\rho l}{a} \text{ ohms}}$

ρ is measured in ohm metres (Ωm).

The value of the resistivity is that resistance of a unit cube of the material measured between opposite faces of the cube.

Resistivity varies with temperature and some typical values of resistivities measured at about room temperature are given below:

Copper $1.7 \times 10^{-8} \Omega$m (or $0.017 \mu\Omega$m)
Aluminium $2.6 \times 10^{-8} \Omega$m (or $0.026 \mu\Omega$m)
Carbon (graphite) $10 \times 10^{-8} \Omega$m ($0.10 \mu\Omega$m)
Glass $1 \times 10^{10} \Omega$m (or $10^4 \mu\Omega$m)
Mica $1 \times 10^{13} \Omega$m (or $10^7 \mu\Omega$m)

Note that good conductors of electricity have a low value of resistivity and good insulators have a high value of resistivity.

Problem 1. The resistance of a 5 m length of wire is 600Ω. Determine (a) the resistance of an 8 m length of the same wire, and (b) the length of the same wire when the resistance is 420Ω.

(a) Resistance, R, is directly proportional to length, l, i.e. $R \propto 1$. Hence, $600 \Omega \propto 5$ m or $600 = (k)$ (5), where k is the coefficient of proportionality. Hence, $k = \dfrac{600}{5} = 120$.

When the length l is 8 m, then resistance

$$R = kl = (120)(8) = \textbf{960}\,\boldsymbol{\Omega}$$

(b) When the resistance is 420Ω, $420 = kl$, from which,

$$\text{length } l = \dfrac{420}{k} = \dfrac{420}{120} = \textbf{3.5 m}$$

Problem 2. A piece of wire of cross-sectional area $2\,mm^2$ has a resistance of $300\,\Omega$. Find (a) the resistance of a wire of the same length and material if the cross-sectional area is $5\,mm^2$, (b) the cross-sectional area of a wire of the same length and material of resistance $750\,\Omega$.

Resistance R is inversely proportional to cross-sectional area, a, i.e. $R \propto \dfrac{1}{a}$

Hence $300\,\Omega \propto \dfrac{1}{2\,mm^2}$ or $300 = (k)\left(\dfrac{1}{2}\right)$

from which, the coefficient of proportionality,

$$k = 300 \times 2 = 600$$

(a) When the cross-sectional area $a = 5\,mm^2$

then $R = (k)\left(\dfrac{1}{5}\right) = (600)\left(\dfrac{1}{5}\right) = \mathbf{120\,\Omega}$

(Note that resistance has decreased as the cross-sectional is increased)

(b) When the resistance is $750\,\Omega$,

then $750 = (k)\left(\dfrac{1}{a}\right)$

from which cross-sectional area,

$$a = \dfrac{k}{750} = \dfrac{600}{750} = \mathbf{0.8\,mm^2}$$

Problem 3. A wire of length $8\,m$ and cross-sectional area $3\,mm^2$ has a resistance of $0.16\,\Omega$. If the wire is drawn out until its cross-sectional area is $1\,mm^2$, determine the resistance of the wire.

Resistance R is directly proportional to length l, and inversely proportional to the cross-sectional area, a,

i.e. $R \propto \dfrac{l}{a}$ or $R = k\left(\dfrac{l}{a}\right)$

where k is the coefficient of proportionality.

Since $R = 0.16$, $l = 8$ and $a = 3$, then $0.16 = (k)\left(\dfrac{8}{3}\right)$

from which, $k = 0.16 \times \dfrac{3}{8} = 0.06$

If the cross-sectional area is reduced to $\frac{1}{3}$ of its original area then the length must be tripled to 3×8, i.e. $24\,m$.

New resistance $R = k\left(\dfrac{l}{a}\right) = 0.06\left(\dfrac{24}{1}\right) = \mathbf{1.44\,\Omega}$

Problem 4. Calculate the resistance of a $2\,km$ length of aluminium overhead power cable if the cross-sectional area of the cable is $100\,mm^2$. Take the resistivity of aluminium to be $0.03 \times 10^{-6}\,\Omega m$.

Length, $l = 2\,km = 2000\,m$, area, $a = 100\,mm^2 = 100 \times 10^{-6}\,m^2$ and resistivity, $\rho = 0.03 \times 10^{-6}\,\Omega m$.

Resistance $R = \dfrac{\rho l}{a} = \dfrac{(0.03 \times 10^{-6}\,\Omega m)(2000\,m)}{(100 \times 10^{-6}\,m^2)}$

$$= \dfrac{0.03 \times 2000}{100}\,\Omega = \mathbf{0.6\,\Omega}$$

Problem 5. Calculate the cross-sectional area, in mm^2, of a piece of copper wire, $40\,m$ in length and having a resistance of $0.25\,\Omega$. Take the resistivity of copper as $0.02 \times 10^{-6}\,\Omega m$.

Resistance $R = \dfrac{\rho l}{a}$ hence

cross-sectional area, $a = \dfrac{\rho l}{R} = \dfrac{(0.02 \times 10^{-6}\,\Omega m)(40\,m)}{0.25\,\Omega}$

$$= 3.2 \times 10^{-6}\,m^2$$

$$= (3.2 \times 10^{-6}) \times 10^6\,mm^2$$

$$= \mathbf{3.2\,mm^2}$$

Problem 6. The resistance of $1.5\,km$ of wire of cross-sectional area $0.17\,mm^2$ is $150\,\Omega$. Determine the resistivity of the wire.

Resistance, $R = \dfrac{\rho l}{a}$ hence

resistivity, $\rho = \dfrac{Ra}{l} = \dfrac{(150\,\Omega)(0.17 \times 10^{-6}\,m^2)}{(1500\,m)}$

$$= \mathbf{0.017 \times 10^{-6}\,\Omega m} \quad \text{or} \quad \mathbf{0.017\,\mu\Omega m}$$

Problem 7. Determine the resistance of $1200\,m$ of copper cable having a diameter of $12\,mm$ if the resistivity of copper is $1.7 \times 10^{-8}\,\Omega m$.

Cross-sectional area of cable,

$$a = \pi r^2 = \pi\left(\dfrac{12}{2}\right)^2 = 36\pi\,mm^2 = 36\pi \times 10^{-6}\,m^2$$

Resistance, $R = \dfrac{\rho l}{a} = \dfrac{(1.7 \times 10^{-8}\,\Omega m)(1200\,m)}{(36\pi \times 10^{-6}\,m^2)}$

$$= \dfrac{1.7 \times 1200 \times 10^6}{10^8 \times 36\pi}\,\Omega$$

$$= \dfrac{1.7 \times 12}{36\pi}\,\Omega = \mathbf{0.180\,\Omega}$$

Now try the following exercise

Exercise 74   Further problems on resistance and resistivity (Answers on page 305)

1. The resistance of a 2 m length of cable is 2.5 Ω. Determine (a) the resistance of a 7 m length of the same cable, and (b) the length of the same wire when the resistance is 6.25 Ω.

2. Some wire of cross-sectional area 1 mm² has a resistance of 20 Ω. Determine (a) the resistance of a wire of the same length and material if the cross-sectional area is 4 mm², and (b) the cross-sectional area of a wire of the same length and material if the resistance is 32 Ω.

3. Some wire of length 5 m and cross-sectional area 2 mm² has a resistance of 0.08 Ω. If the wire is drawn out until its cross-sectional area is 1 mm², determine the resistance of the wire.

4. Find the resistance of 800 m of copper cable of cross-sectional area 20 mm². Take the resistivity of copper as 0.02 μΩm.

5. Calculate the cross-sectional area, in mm², of a piece of aluminium wire 100 m long and having a resistance of 2 Ω. Take the resistivity of aluminium as 0.03×10^{-6} Ωm.

6. The resistance of 500 m of wire of cross-sectional area 2.6 mm² is 5 Ω. Determine the resistivity of the wire in μΩm.

7. Find the resistance of 1 km of copper cable having a diameter of 10 mm if the resistivity of copper is 0.017×10^{-6} Ωm.

17.2   Temperature coefficient of resistance

In general, as the temperature of a material increases, most conductors increase in resistance, insulators decrease in resistance, whilst the resistance of some special alloys remain almost constant.

The **temperature coefficient of resistance** of a material is the increase in the resistance of a 1 Ω resistor of that material when it is subjected to a rise of temperature of 1°C. The symbol used for the temperature coefficient of resistance is α (Greek alpha). Thus, if some copper wire of resistance 1 Ω is heated through 1°C and its resistance is then measured as 1.0043 Ω then $\alpha = 0.0043$ Ω/Ω°C for copper. The units are usually expressed only as 'per °C', i.e. $\alpha = 0.0043$/°C for copper. If the 1 Ω resistor of copper is heated through 100°C then the resistance at 100°C would be $1 + 100 \times 0.0043 = 1.43$ Ω.

Some typical values of temperature coefficient of resistance measured at 0°C are given below:

Copper	0.0043/°C
Nickel	0.0062/°C
Constantan	0
Aluminium	0.0038/°C
Carbon	−0.00048/°C
Eureka	0.00001/°C

(Note that the negative sign for carbon indicates that its resistance falls with increase of temperature.)

If the resistance of a material at 0°C is known the resistance at any other temperature can be determined from:

$$R_\theta = R_0(1 + \alpha_0\theta)$$

where R_0 = resistance at 0°C,
        R_θ = resistance at temperature θ°C
and α_0 = temperature coefficient of resistance at 0°C.

Problem 8.   A coil of copper wire has a resistance of 100 Ω when its temperature is 0°C. Determine its resistance at 70°C if the temperature coefficient of resistance of copper at 0°C is 0.0043/°C.

Resistance $R_\theta = R_0(1 + \alpha_0\theta)$

Hence resistance at 100°C,

$$R_{100} = 100[1 + (0.0043)(70)]$$
$$= 100[1 + 0.301] = 100(1.301) = \mathbf{130.1\,\Omega}$$

Problem 9.   An aluminium cable has a resistance of 27 Ω at a temperature of 35°C. Determine its resistance at 0°C. Take the temperature coefficient of resistance at 0°C to be 0.0038/°C.

Resistance at θ°C, $R_\theta = R_0(1 + \alpha_0\theta)$

Hence resistance at 0°C,

$$R_0 = \frac{R_\theta}{(1 + \alpha_0\theta)} = \frac{27}{[1 + (0.0038)(35)]}$$
$$= \frac{27}{1 + 0.133} = \frac{27}{1.133} = \mathbf{23.83\,\Omega}$$

Problem 10.   A carbon resistor has a resistance of 1 kΩ at 0°C. Determine its resistance at 80°C. Assume that the temperature coefficient of resistance for carbon at 0°C is −0.0005/°C.

Resistance at temperature θ°C,

$$R_\theta = R_0(1 + \alpha_0\theta)$$

i.e. $R_\theta = 1000[1 + (-0.0005)(80)]$

$$= 1000[1 - 0.040] = 1000(0.96) = \mathbf{960\,\Omega}$$

If the resistance of a material at room temperature (approximately 20°C), R_{20}, and the temperature coefficient of resistance at 20°C, α_{20}, are known then the resistance R_θ at temperature θ°C is given by:

$$R_\theta = R_{20}[1 + \alpha_{20}(\theta - 20)]$$

Problem 11. A coil of copper wire has a resistance of 10 Ω at 20°C. If the temperature coefficient of resistance of copper at 20°C is 0.004/°C determine the resistance of the coil when the temperature rises to 100°C.

Resistance at θ°C, $R_\theta = R_{20}[1 + \alpha_{20}(\theta - 20)]$

Hence resistance at 100°C,

$$R_{100} = 10[1 + (0.004)(100 - 20)]$$

$$= 10[1 + (0.004)(80)]$$

$$= 10[1 + 0.32]$$

$$= 10(1.32) = \textbf{13.2 Ω}$$

Problem 12. The resistance of a coil of aluminium wire at 18°C is 200 Ω. The temperature of the wire is increased and the resistance rises to 240 Ω. If the temperature coefficient of resistance of aluminium is 0.0039/°C at 18°C determine the temperature to which the coil has risen.

Let the temperature rise to θ°C

Resistance at θ°C, $R_\theta = R_{18}[1 + \alpha_{18}(\theta - 18)]$

i.e. $\qquad 240 = 200[1 + (0.0039)(\theta - 18)]$

$$240 = 200 + (200)(0.0039)(\theta - 18)$$

$$240 - 200 = 0.78(\theta - 18)$$

$$40 = 0.78(\theta - 18)$$

$$\frac{40}{0.78} = \theta - 18$$

$$51.28 = \theta - 18$$

from which, $\qquad \theta = 51.28 + 18 = 69.28$°C

Hence the temperature of the coil increases to 69.28°C

If the resistance at 0°C is not known, but is known at some other temperature θ_1, then the resistance at any temperature can be found as follows:

$$R_1 = R_0(1 + \alpha_0\theta_1) \quad \text{and} \quad R_2 = R_0(1 + \alpha_0\theta_2)$$

Dividing one equation by the other gives:

$$\frac{R_1}{R_2} = \frac{1 + \alpha_0\theta_1}{1 + \alpha_0\theta_2}$$

where R_2 = resistance at temperature θ_2

Problem 13. Some copper wire has a resistance of 200 Ω at 20°C. A current is passed through the wire and the temperature rises to 90°C. Determine the resistance of the wire at 90°C, correct to the nearest ohm, assuming that the temperature coefficient of resistance is 0.004/°C at 0°C.

$R_{20} = 200$ Ω, $\alpha_0 = 0.004$/°C $\quad$ and $\quad \dfrac{R_{20}}{R_{90}} = \dfrac{[1 + \alpha_0(20)]}{[1 + \alpha_0(90)]}$

Hence $\quad R_{90} = \dfrac{R_{20}[1 + 90\alpha_0]}{[1 + 20\alpha_0]} = \dfrac{200[1 + 90(0.004)]}{[1 + 20(0.004)]}$

$$= \dfrac{200[1 + 0.36]}{[1 + 0.08]}$$

$$= \dfrac{200(1.36)}{(1.08)} = \textbf{251.85 Ω}$$

i.e. the resistance of the wire at 90°C is 252 Ω, correct to the nearest ohm.

Now try the following exercises

Exercise 75 Further problems on the temperature coefficient of resistance (Answers on page 305)

1. A coil of aluminium wire has a resistance of 50 Ω when its temperature is 0°C. Determine its resistance at 100°C if the temperature coefficient of resistance of aluminium at 0°C is 0.0038/°C.

2. A copper cable has a resistance of 30 Ω at a temperature of 50°C. Determine its resistance at 0°C. Take the temperature coefficient of resistance of copper at 0°C as 0.0043/°C.

3. The temperature coefficient of resistance for carbon at 0°C is −0.00048/°C. What is the significance of the minus sign? A carbon resistor has a resistance of 500 Ω at 0°C. Determine its resistance at 50°C.

4. A coil of copper wire has a resistance of 20 Ω at 18°C. If the temperature coefficient of resistance of copper at 18°C is 0.004/°C, determine the resistance of the coil when the temperature rises to 98°C.

5. The resistance of a coil of nickel wire at 20°C is 100 Ω. The temperature of the wire is increased and the resistance rises to 130 Ω. If the temperature coefficient

of resistance of nickel is 0.006/°C at 20°C, determine the temperature to which the coil has risen.

6. Some aluminium wire has a resistance of $50\,\Omega$ at 20°C. The wire is heated to a temperature of 100°C. Determine the resistance of the wire at 100°C, assuming that the temperature coefficient of resistance at 0°C is 0.004/°C.

7. A copper cable is 1.2 km long and has a cross-sectional area of $5\,\text{mm}^2$. Find its resistance at 80°C if at 20°C the resistivity of copper is $0.02 \times 10^{-6}\,\Omega\text{m}$ and its temperature coefficient of resistance is 0.004/°C.

Exercise 76 Short answer questions on resistance variation

1. Name four factors which can effect the resistance of a conductor.

2. If the length of a piece of wire of constant cross-sectional area is halved, the resistance of the wire is......

3. If the cross-sectional area of a certain length of cable is trebled, the resistance of the cable is......

4. What is resistivity? State its unit and the symbol used.

5. Complete the following: Good conductors of electricity have a......value of resistivity and good insulators have a......value of resistivity.

6. What is meant by the 'temperature coefficient of resistance?' State its units and the symbols used.

7. If the resistance of a metal at 0°C is R_0, R_θ is the resistance at θ°C and α_0 is the temperature coefficient of resistance at 0°C then: $R_\theta = \ldots\ldots$

Exercise 77 Multi-choice questions on resistance variation (Answers on page 305)

1. The unit of resistivity is:

    (a) ohms   (b) ohm millimetre
    (c) ohm metre   (d) ohm/metre

2. The length of a certain conductor of resistance $100\,\Omega$ is doubled and its cross-sectional area is halved. Its new resistance is:

    (a) $100\,\Omega$ (b) $200\,\Omega$ (c) $50\,\Omega$ (d) $400\,\Omega$

3. The resistance of a 2 km length of cable of cross-sectional area $2\,\text{mm}^2$ and resistivity of $2 \times 10^{-8}\,\Omega\text{m}$ is:

    (a) $0.02\,\Omega$ (b) $20\,\Omega$ (c) $0.02\,\text{m}\Omega$ (d) $200\,\Omega$

4. A piece of graphite has a cross-sectional area of $10\,\text{mm}^2$. If its resistance is $0.1\,\Omega$ and its resistivity $10 \times 10^{-8}\,\Omega\text{m}$, its length is:

    (a) 10 km (b) 10 cm (c) 10 mm (d) 10 m

5. The symbol for the unit of temperature coefficient of resistance is:

    (a) $\Omega/°C$ (b) Ω (c) $°C$ (d) $\Omega/\Omega°C$

6. A coil of wire has a resistance of $10\,\Omega$ at 0°C. If the temperature coefficient of resistance for the wire is 0.004/°C, its resistance at 100°C is:

    (a) $0.4\,\Omega$ (b) $1.4\,\Omega$ (c) $14\,\Omega$ (d) $10\,\Omega$

7. A nickel coil has a resistance of $13\,\Omega$ at 50°C. If the temperature coefficient of resistance at 0°C is 0.006/°C, the resistance at 0°C is:

    (a) $16.9\,\Omega$ (b) $10\,\Omega$ (c) $43.3\,\Omega$ (d) $0.1\,\Omega$

18

Series and parallel networks

At the end of this chapter you should be able to:

- calculate unknown voltages, current and resistance's in a series circuit
- understand voltage division in a series circuit
- calculate unknown voltages, currents and resistance's in a parallel network
- calculate unknown voltages, currents and resistance's in series-parallel networks
- understand current division in a two-branch parallel network
- describe the advantages and disadvantages of series and parallel connection of lamps

18.1 Series circuits

Figure 18.1 shows three resistors R_1, R_2 and R_3 connected end to end, i.e. in series, with a battery source of V volts. Since the circuit is closed a current I will flow and the p.d. across each resistor may be determined from the voltmeter readings V_1, V_2 and V_3

In a series circuit

(a) the current I is the same in all parts of the circuit and hence the same reading is found on each of the ammeters shown, and

(b) the sum of the voltages V_1, V_2 and V_3 is equal to the total applied voltage, V,

i.e. $$V = V_1 + V_2 + V_3$$

From Ohm's law: $V_1 = IR_1$, $V_2 = IR_2$, $V_3 = IR_3$ and $V = IR$ where R is the total circuit resistance. Since $V = V_1 + V_2 + V_3$ then $IR = IR_1 + IR_2 + IR_3$
Dividing throughout by I gives

$$R = R_1 + R_2 + R_3$$

Thus for a series circuit, the total resistance is obtained by adding together the values of the separate resistances.

Problem 1. For the circuit shown in Fig. 18.2, determine (a) the battery voltage V, (b) the total resistance of the circuit, and (c) the values of resistors R_1, R_2 and R_3, given that the p.d.'s across R_1, R_2 and R_3 are 5 V, 2 V and 6 V respectively.

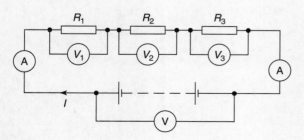

Fig. 18.1

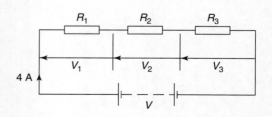

Fig. 18.2

(a) Battery voltage

$$V = V_1 + V_2 + V_3 = 5 + 2 + 6 = \mathbf{13\,V}$$

(b) Total circuit resistance

$$R = \frac{V}{I} = \frac{13}{4} = \mathbf{3.25\,\Omega}$$

(c) Resistance $R_1 = \dfrac{V_1}{I} = \dfrac{5}{4} = \mathbf{1.25\,\Omega}$,

resistance $R_2 = \dfrac{V_2}{I} = \dfrac{2}{4} = \mathbf{0.5\,\Omega}$ and

resistance $R_3 = \dfrac{V_3}{I} = \dfrac{6}{4} = \mathbf{1.5\,\Omega}$.

(Check: $R_1 + R_2 + R_3 = 1.25 + 0.5 + 1.5 = 3.25\,\Omega = R$)

Problem 2. For the circuit shown in Fig. 18.3, determine the p.d. across resistor R_3. If the total resistance of the circuit is $100\,\Omega$, determine the current flowing through resistor R_1. Find also the value of resistor R_2

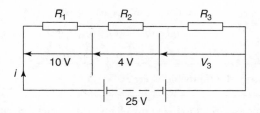

Fig. 18.3

P.d. across R_3, $V_3 = 25 - 10 - 4 = \mathbf{11\,V}$

$$\text{Current } I = \frac{V}{R} = \frac{25}{100} = \mathbf{0.25\,A},$$

which is the current flowing in each resistor.

$$\text{Resistance } R_2 = \frac{V_2}{I} = \frac{4}{0.25} = \mathbf{16\,\Omega}$$

Problem 3. A 12 V battery is connected to a circuit having three series-connected resistors having resistance's of $4\,\Omega$, $9\,\Omega$ and $11\,\Omega$. Determine the current flowing through, and the p.d. across the $9\,\Omega$ resistor. Find also the power dissipated in the $11\,\Omega$ resistor.

The circuit diagram is shown in Fig. 18.4.

Total resistance $R = 4 + 9 + 11 = 24\,\Omega$

$$\text{Current } I = \frac{V}{R} = \frac{12}{24} = \mathbf{0.5\,A},$$

which is the current in the $9\,\Omega$ resistor.
P.d. across the $9\,\Omega$ resistor,

$$V_1 = I \times 9 = 0.5 \times 9 = \mathbf{4.5\,V}$$

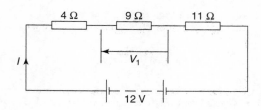

Fig. 18.4

Power dissipated in the $11\,\Omega$ resistor,

$$P = I^2 R = (0.5)^2 (11) = (0.25)(11) = \mathbf{2.75\,W}$$

18.2 Potential divider

The voltage distribution for the circuit shown in Fig. 18.5(a) is given by:

$$V_1 = \left(\frac{R_1}{R_1 + R_2}\right) V \text{ and } V_2 = \left(\frac{R_2}{R_1 + R_2}\right) V$$

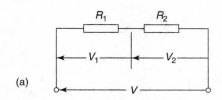

(a)

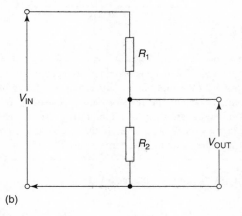

(b)

Fig. 18.5

The circuit shown in Fig. 18.5(b) is often referred to as a **potential divider** circuit. Such a circuit can consist of a number of similar elements in series connected across a voltage source, voltages being taken from connections between the elements. Frequently the divider consists of two

138   *Science for Engineering*

resistors as shown in Fig. 18.5(b), where

$$V_{OUT} = \left(\frac{R_2}{R_1 + R_2}\right) V_{IN}$$

Problem 4.   Determine the value of voltage V shown in Fig. 18.6

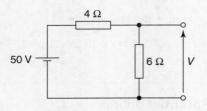

Fig. 18.6

Figure 18.6 may be redrawn as shown in Fig. 18.7, and voltage

$$V = \left(\frac{6}{6+4}\right)(50) = 30\,\text{V}$$

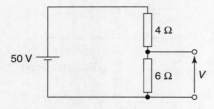

Fig. 18.7

Problem 5.   Two resistors are connected in series across a 24 V supply and a current of 3 A flows in the circuit. If one of the resistors has a resistance of $2\,\Omega$ determine (a) the value of the other resistor, and (b) the p.d. across the $2\,\Omega$ resistor. If the circuit is connected for 50 hours, how much energy is used?

The circuit diagram is shown in Fig. 18.8

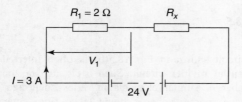

Fig. 18.8

(a)   Total circuit resistance

$$R = \frac{V}{I} = \frac{24}{3} = 8\,\Omega$$

Value of unknown resistance,

$$R_x = 8 - 2 = \mathbf{6\,\Omega}$$

(b)   P.d. across $2\,\Omega$ resistor,

$$V_1 = IR_1 = 3 \times 2 = \mathbf{6\,V}$$

Alternatively, from above,

$$V_1 = \left(\frac{R_1}{R_1 + R_x}\right) V$$

$$= \left(\frac{2}{2+6}\right)(24) = 6\,\text{V}$$

Energy used = power × time

$$= (V \times I) \times t$$

$$= (24 \times 3\,\text{W})(50\,\text{h})$$

$$= 3600\,\text{Wh} = \mathbf{3.6\,kWh}$$

Now try the following exercise

Exercise 78   Further problems on series circuits (Answers on page 305)

1.   The p.d's measured across three resistors connected in series are 5 V, 7 V and 10 V, and the supply current is 2 A. Determine (a) the supply voltage, (b) the total circuit resistance and (c) the values of the three resistors.

2.   For the circuit shown in Fig. 18.9, determine the value of V_1. If the total circuit resistance is $36\,\Omega$ determine the supply current and the value of resistors R_1, R_2 and R_3

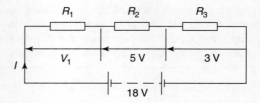

Fig. 18.9

3.   When the switch in the circuit in Fig. 18.10 is closed the reading on voltmeter 1 is 30 V and that on voltmeter 2 is 10 V. Determine the reading on the ammeter and the value of resistor R_x

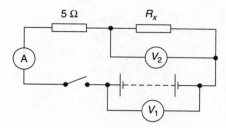

Fig. 18.10

4. Calculate the value of voltage V in Fig. 18.11

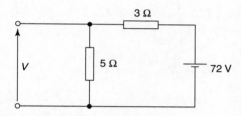

Fig. 18.11

5. Two resistors are connected in series across an 18 V supply and a current of 5 A flows. If one of the resistors has a value of 2.4 Ω determine (a) the value of the other resistor and (b) the p.d. across the 2.4 Ω resistor.

18.3 Parallel networks

Figure 18.12 shows three resistors, R_1, R_2 and R_3 connected across each other, i.e. in parallel, across a battery source of V volts.

In a parallel circuit:

(a) the sum of the currents I_1, I_2 and I_3 is equal to the total circuit current, I,

i.e. $\boxed{I = I_1 + I_2 + I_3}$ and

(b) the source p.d., V volts, is the same across each of the resistors.

From Ohm's law: $I_1 = \dfrac{V}{R_1}$, $I_2 = \dfrac{V}{R_2}$, $I_3 = \dfrac{V}{R_3}$ and $I = \dfrac{V}{R}$
where R is the total circuit resistance. Since $I = I_1 + I_2 + I_3$
then, $\dfrac{V}{R} = \dfrac{V}{R_1} + \dfrac{V}{R_2} + \dfrac{V}{R_3}$

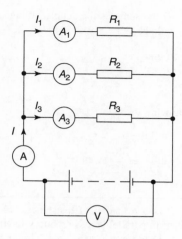

Fig. 18.12

Dividing throughout by V gives:

$$\frac{1}{R} = \frac{1}{R_1} + \frac{1}{R_2} + \frac{1}{R_3}$$

This equation must be used when finding the total resistance R of a parallel circuit.

For the special case of **two resistors in parallel**

$$\frac{1}{R} = \frac{1}{R_1} + \frac{1}{R_2} = \frac{R_2 + R_1}{R_1 R_2}$$

Hence $\boxed{R = \dfrac{R_1 R_2}{R_1 + R_2}}$ $\left(\text{i.e.} \ \dfrac{\text{product}}{\text{sum}} \right)$

Problem 6. For the circuit shown in Fig. 18.13, determine (a) the reading on the ammeter, and (b) the value of resistor R_2

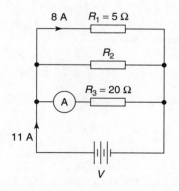

Fig. 18.13

P.d. across R_1 is the same as the supply voltage V.
Hence supply voltage, $V = 8 \times 5 = 40$ V

(a) Reading on ammeter,

$$I = \frac{V}{R_3} = \frac{40}{20} = 2\,\mathbf{A}$$

(b) Current flowing through R_2

$$= 11 - 8 - 2 = 1\,\text{A}$$

Hence $R_2 = \dfrac{V}{I_2} = \dfrac{40}{1} = \mathbf{40\,\Omega}$

Problem 7. Two resistors, of resistance $3\,\Omega$ and $6\,\Omega$, are connected in parallel across a battery having a voltage of 12 V. Determine (a) the total circuit resistance and (b) the current flowing in the $3\,\Omega$ resistor.

The circuit diagram is shown in Fig. 18.14

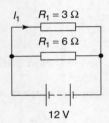

Fig. 18.14

(a) The total circuit resistance R is given by

$$\frac{1}{R} = \frac{1}{R_1} + \frac{1}{R_2}$$

$$= \frac{1}{3} + \frac{1}{6} = \frac{2+1}{6} = \frac{3}{6}$$

Since $\dfrac{1}{R} = \dfrac{3}{6}$ then $R = 2\,\Omega$

(Alternatively, $R = \dfrac{R_1 R_2}{R_1 + R_2}$

$$= \frac{3 \times 6}{3 + 6} = \frac{18}{9} = \mathbf{2\,\Omega})$$

(b) Current in the $3\,\Omega$ resistance,

$$I_1 = \frac{V}{R_1} = \frac{12}{3} = 4\,\mathbf{A}$$

Problem 8. For the circuit shown in Fig. 18.15, find (a) the value of the supply voltage V and (b) the value of current I

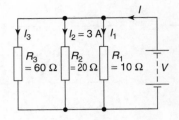

Fig. 18.15

(a) p.d. across $20\,\Omega$ resistor $= I_2 R_2 = 3 \times 20 = 60$ V, hence supply voltage $V = \mathbf{60\,V}$ since the circuit is connected in parallel.

(b) Current $I_1 = \dfrac{V}{R_1} = \dfrac{60}{10} = 6$ A,

$$I_2 = 3\,\text{A and}$$

$$I_3 = \frac{V}{R_3} = \frac{60}{60} = 1\,\text{A}$$

Current $I = I_1 + I_2 + I_3$

hence, $I = 6 + 3 + 1 = \mathbf{10\,A}$

Alternatively,

$$\frac{1}{R} = \frac{1}{60} + \frac{1}{20} + \frac{1}{10} = \frac{1 + 3 + 6}{60} = \frac{10}{60}$$

Hence total resistance

$$R = \frac{60}{10} = 6\,\Omega$$

and current

$$I = \frac{V}{R} = \frac{60}{6} = \mathbf{10\,A}$$

Problem 9. Given four $1\,\Omega$ resistors, state how they must be connected to give an overall resistance of (a) $\frac{1}{4}\,\Omega$ (b) $1\,\Omega$ (c) $1\frac{1}{3}\,\Omega$ (d) $2\frac{1}{2}\,\Omega$, all four resistors being connected in each case.

(a) **All four in parallel** (see Fig. 18.16),

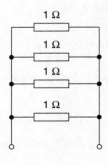

Fig. 18.16

since $\dfrac{1}{R} = \dfrac{1}{1} + \dfrac{1}{1} + \dfrac{1}{1} + \dfrac{1}{1} = \dfrac{4}{1}$

i.e. $R = \dfrac{1}{4}\,\Omega$

(b) **Two in series, in parallel with another two in series** (see Fig. 18.17), since $1\,\Omega$ and $1\,\Omega$ in series gives $2\,\Omega$, and $2\,\Omega$ in parallel with $2\,\Omega$ gives

$$\dfrac{2 \times 2}{2 + 2} = \dfrac{4}{4} = 1\,\Omega$$

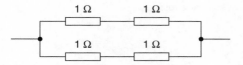

Fig. 18.17

(c) **Three in parallel, in series with one** (see Fig. 18.18), since for the three in parallel,

$$\dfrac{1}{R} = \dfrac{1}{1} + \dfrac{1}{1} + \dfrac{1}{1} = \dfrac{3}{1}$$

i.e. $R = \frac{1}{3}\,\Omega$ and $\frac{1}{3}\,\Omega$ in series with $1\,\Omega$ gives $1\frac{1}{3}\,\Omega$.

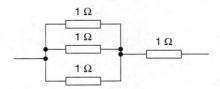

Fig. 18.18

(d) **Two in parallel, in series with two in series** (see Fig. 18.19), since for the two in parallel

$$R = \dfrac{1 \times 1}{1 + 1} = \dfrac{1}{2}\,\Omega,$$

and $\frac{1}{2}\,\Omega$, $1\,\Omega$ and $1\,\Omega$ in series gives $2\frac{1}{2}\,\Omega$.

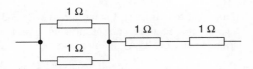

Fig. 18.19

Problem 10. Find the equivalent resistance for the circuit shown in Fig. 18.20.

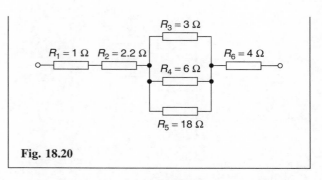

Fig. 18.20

R_3, R_4 and R_5 are connected in parallel and their equivalent resistance R is given by:

$$\dfrac{1}{R} = \dfrac{1}{3} + \dfrac{1}{6} + \dfrac{1}{18} = \dfrac{6 + 3 + 1}{18} = \dfrac{10}{18}$$

hence $R = \dfrac{18}{10} = 1.8\,\Omega$

The circuit is now equivalent to four resistors in series and the **equivalent circuit resistance** $= 1 + 2.2 + 1.8 + 4 = \mathbf{9\,\Omega}$

Problem 11. Resistance's of $10\,\Omega$, $20\,\Omega$ and $30\,\Omega$ are connected (a) in series and (b) in parallel to a $240\,\text{V}$ supply. Calculate the supply current in each case.

(a) The series circuit is shown in Fig. 18.21.

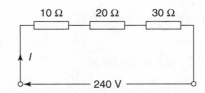

Fig. 18.21

The equivalent resistance

$$R_T = 10\,\Omega + 20\,\Omega + 30\,\Omega = 60\,\Omega$$

Supply current

$$I = \dfrac{V}{R_T} = \dfrac{240}{60} = \mathbf{4\,A}$$

(b) The parallel circuit is shown in Fig. 18.22.
The equivalent resistance R_T of $10\,\Omega$, $20\,\Omega$ and $30\,\Omega$ resistance's connected in parallel is given by:

$$\dfrac{1}{R_T} = \dfrac{1}{10} + \dfrac{1}{20} + \dfrac{1}{30} = \dfrac{6 + 3 + 2}{60} = \dfrac{11}{60}$$

hence $R_T = \dfrac{60}{11}\,\Omega.$

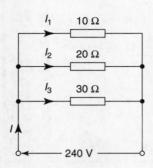

Fig. 18.22

Supply current

$$I = \frac{V}{R_T} = \frac{240}{\dfrac{60}{11}} = \frac{240 \times 11}{60} = \textbf{44 A}$$

(Check:

$$I_1 = \frac{V}{R_1} = \frac{240}{10} = 24\,\text{A},$$

$$I_2 = \frac{V}{R_2} = \frac{240}{20} = 12\,\text{A and}$$

$$I_3 = \frac{V}{R_3} = \frac{240}{30} = 8\,\text{A}$$

For a parallel circuit $I = I_1 + I_2 + I_3 = 24 + 12 + 8 = \textbf{44 A}$, as above)

18.4 Current division

For the circuit shown in Fig. 18.23, the total circuit resistance, R_T is given by

$$R_T = \frac{R_1 R_2}{R_1 + R_2}$$

and

$$V = IR_T = I\left(\frac{R_1 R_2}{R_1 + R_2}\right)$$

Current $I_1 = \dfrac{V}{R_1} = \dfrac{I}{R_1}\left(\dfrac{R_1 R_2}{R_1 + R_2}\right) = \left(\dfrac{R_2}{R_1 + R_2}\right)(I)$

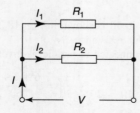

Fig. 18.23

Similarly,

current $I_2 = \dfrac{V}{R_2} = \dfrac{I}{R_2}\left(\dfrac{R_1 R_2}{R_1 + R_2}\right) = \left(\dfrac{R_1}{R_1 + R_2}\right)(I)$

Summarising, with reference to Fig. 18.23:

$$I_1 = \left(\frac{R_2}{R_1 + R_2}\right)(I) \quad \text{and} \quad I_2 = \left(\frac{R_1}{R_1 + R_2}\right)(I)$$

Problem 12. For the series-parallel arrangement shown in Fig. 18.24, find (a) the supply current, (b) the current flowing through each resistor and (c) the p.d. across each resistor.

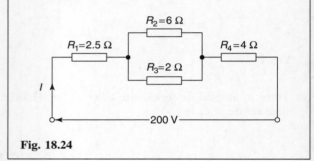

Fig. 18.24

(a) The equivalent resistance R_x of R_2 and R_3 in parallel is:

$$R_x = \frac{6 \times 2}{6 + 2} == 1.5\,\Omega$$

The equivalent resistance R_T of R_1, R_x and R_4 in series is:

$$R_T = 2.5 + 1.5 + 4 = 8\,\Omega$$

Supply current $I = \dfrac{V}{R_T} = \dfrac{200}{8} = \textbf{25 A}$

(b) The current flowing through R_1 and R_4 is 25 A
The current flowing through R_2

$$= \left(\frac{R_3}{R_2 + R_3}\right)I = \left(\frac{2}{6 + 2}\right)25 = \textbf{6.25 A}$$

The current flowing through R_3

$$= \left(\frac{R_2}{R_2 + R_3}\right)I = \left(\frac{6}{6 + 2}\right)25 = \textbf{18.75 A}$$

(Note that the currents flowing through R_2 and R_3 must add up to the total current flowing into the parallel arrangement, i.e. 25 A)

(c) The equivalent circuit of Fig. 18.24 is shown in Fig. 18.25

p.d. across R_1, i.e. $V_1 = IR_1 = (25)(2.5) = \textbf{62.5 V}$

p.d. across R_x, i.e. $V_x = IR_x = (25)(1.5) = \textbf{37.5 V}$

p.d. across R_4, i.e. $V_4 = IR_4 = (25)(4) = \textbf{100 V}$

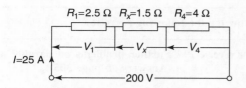

Fig. 18.25

Hence the p.d. across R_2 = p.d. across R_3 = **37.5 V**

Problem 13. For the circuit shown in Fig. 18.26 calculate (a) the value of resistor R_x such that the total power dissipated in the circuit is 2.5 kW, (b) the current flowing in each of the four resistors.

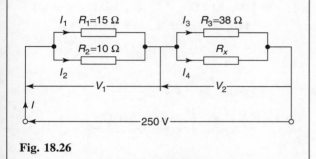

Fig. 18.26

(a) Power dissipated $P = VI$ watts, hence $2500 = (250)(I)$

i.e. $I = \dfrac{2500}{250} = 10\,\text{A}$

From Ohm's law,

$$R_T = \frac{V}{I} = \frac{250}{10} = 25\,\Omega,$$

where R_T is the equivalent circuit resistance.
The equivalent resistance of R_1 and R_2 in parallel is

$$\frac{15 \times 10}{15 + 10} = \frac{150}{25} = 6\,\Omega$$

The equivalent resistance of resistors R_3 and R_x in parallel is equal to $25\,\Omega - 6\,\Omega$, i.e. $19\,\Omega$.

There are three methods whereby R_x can be determined.

Method 1

The voltage $V_1 = IR$, where R is $6\,\Omega$, from above,

i.e. $V_1 = (10)(6) = 60\,\text{V}$

Hence $V_2 = 250\,\text{V} - 60\,\text{V} = 190\,\text{V} = $ p.d. across R_3

$\qquad\qquad\qquad\qquad\qquad\quad = $ p.d. across R_x

$$I_3 = \frac{V_2}{R_3} = \frac{190}{38} = 5\,\text{A}$$

Thus $I_4 = 5\,\text{A}$ also, since $I = 10\,\text{A}$

Thus $\boldsymbol{R_x} = \dfrac{V_2}{I_4} = \dfrac{190}{5} = 38\,\Omega$

Method 2

Since the equivalent resistance of R_3 and R_x in parallel is $19\,\Omega$,

$$19 = \frac{38R_x}{38 + R_x} \left(\text{i.e. } \frac{\text{product}}{\text{sum}}\right)$$

Hence $19(38 + R_x) = 38R_x$

$$722 + 19R_x = 38R_x$$

$$722 = 38R_x - 19R_x = 19R_x$$

Thus $\boldsymbol{R_x} = \dfrac{722}{19} = 38\,\Omega$

Method 3

When two resistors having the same value are connected in parallel the equivalent resistance is always half the value of one of the resistors. Thus, in this case, since $R_T = 19\,\Omega$ and $R_3 = 38\,\Omega$, then $R_x = 38\,\Omega$ could have been deduced on sight.

(b) Current $I_1 = \left(\dfrac{R_2}{R_1 + R_2}\right) I = \left(\dfrac{10}{15 + 10}\right)(10)$

$$= \left(\frac{2}{5}\right)(10) = \mathbf{4\,A}$$

Current $I_2 = \left(\dfrac{R_1}{R_1 + R_2}\right) I = \left(\dfrac{15}{15 + 10}\right)(10)$

$$= \left(\frac{3}{5}\right)(10) = \mathbf{6\,A}$$

From part (a), method 1, $\boldsymbol{I_3 = I_4 = 5\,A}$

Problem 14. For the arrangement shown in Fig. 18.27, find the current I_x

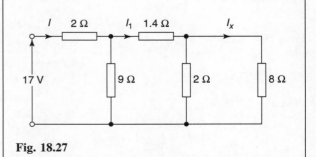

Fig. 18.27

Commencing at the right-hand side of the arrangement shown in Fig. 18.27, the circuit is gradually reduced in stages as shown in Fig. 18.28(a)–(d).

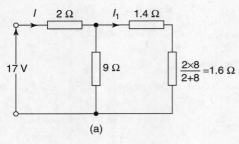

(a)

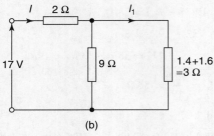

(b)

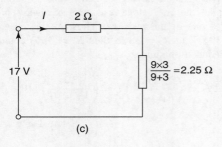

(c)

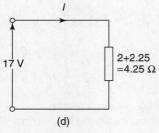

(d)

Fig. 18.28

From Fig. 18.28(d),

$$I = \frac{17}{4.25} = 4\,\text{A}$$

From Fig. 18.28(b),

$$I_1 = \left(\frac{9}{9+3}\right)(I) = \left(\frac{9}{12}\right)(4) = 3\,\text{A}$$

From Fig. 18.27,

$$I_x = \left(\frac{2}{2+8}\right)(I_1) = \left(\frac{2}{10}\right)(3) = \mathbf{0.6\,A}$$

Now try the following exercise

Exercise 79   Further problems on parallel networks (Answers on page 305)

1. Resistance's of $4\,\Omega$ and $12\,\Omega$ are connected in parallel across a 9 V battery. Determine (a) the equivalent circuit resistance, (b) the supply current, and (c) the current in each resistor.

2. For the circuit shown in Fig. 18.29 determine (a) the reading on the ammeter, and (b) the value of resistor R.

3. Find the equivalent resistance when the following resistance's are connected (a) in series (b) in parallel

   (i) $3\,\Omega$ and $2\,\Omega$ (ii) $20\,\text{k}\Omega$ and $40\,\text{k}\Omega$

   (iii) $4\,\Omega$, $8\,\Omega$ and $16\,\Omega$ (iv) $800\,\Omega$, $4\,\text{k}\Omega$ and 1500Ω

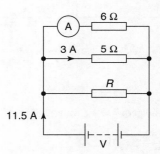

Fig. 18.29

4. Find the total resistance between terminals A and B of the circuit shown in Fig. 18.30(a).

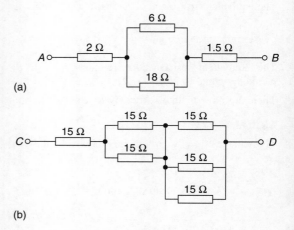

(a)

(b)

Fig. 18.30

5. Find the equivalent resistance between terminals C and D of the circuit shown in Fig. 18.30(b).

6. Resistors of $20\,\Omega$, $20\,\Omega$ and $30\,\Omega$ are connected in parallel. What resistance must be added in series with the combination to obtain a total resistance of $10\,\Omega$. If the complete circuit expends a power of $0.36\,\text{kW}$, find the total current flowing.

7. (a) Calculate the current flowing in the $30\,\Omega$ resistor shown in Fig. 18.31. (b) What additional value of resistance would have to be placed in parallel with the $20\,\Omega$ and $30\,\Omega$ resistors to change the supply current to $8\,\text{A}$, the supply voltage remaining constant.

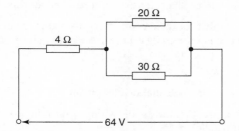

Fig. 18.31

8. For the circuit shown in Fig. 18.32, find (a) V_1 (b) V_2, without calculating the current flowing.

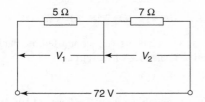

Fig. 18.32

9. Determine the currents and voltages indicated in the circuit shown in Fig. 18.33.

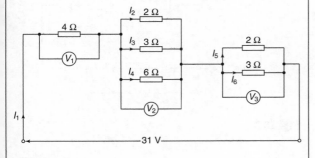

Fig. 18.33

10. Find the current I in Fig. 18.34.

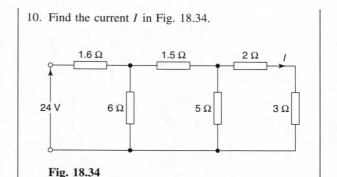

Fig. 18.34

18.5 Wiring lamps in series and in parallel

Series connection

Figure 18.35 shows three lamps, each rated at 240 V, connected in series across a 240 V supply.

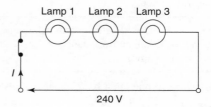

Fig. 18.35

(i) Each lamp has only $\dfrac{240}{3}$ V, i.e., 80 V across it and thus each lamp glows dimly.

(ii) If another lamp of similar rating is added in series with the other three lamps then each lamp now has $\dfrac{240}{4}$ V, i.e., 60 V across it and each now glows even more dimly.

(iii) If a lamp is removed from the circuit or if a lamp develops a fault (i.e. an open circuit) or if the switch is opened, then the circuit is broken, no current flows, and the remaining lamps will not light up.

(iv) Less cable is required for a series connection than for a parallel one.

The series connection of lamps is usually limited to decorative lighting such as for Christmas tree lights.

Parallel connection

Figure 18.36 shows three similar lamps, each rated at 240 V, connected in parallel across a 240 V supply.

(i) Each lamp has 240 V across it and thus each will glow brilliantly at their rated voltage.

(ii) If any lamp is removed from the circuit or develops a fault (open circuit) or a switch is opened, the remaining lamps are unaffected.

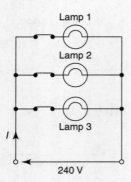

Fig. 18.36

(iii) The addition of further similar lamps in parallel does not affect the brightness of the other lamps.
(iv) More cable is required for parallel connection than for a series one.

The parallel connection of lamps is the most widely used in electrical installations.

Problem 15. If three identical lamps are connected in parallel and the combined resistance is 150 Ω, find the resistance of one lamp.

Let the resistance of one lamp be R, then,

$$\frac{1}{150} = \frac{1}{R} + \frac{1}{R} + \frac{1}{R} = \frac{3}{R}$$

from which,

$$R = 3 \times 150 = \mathbf{450\ \Omega}$$

Problem 16. Three identical lamps A, B and C are connected in series across a 150 V supply. State (a) the voltage across each lamp, and (b) the effect of lamp C failing.

(a) Since each lamp is identical and they are connected in series there is $\dfrac{150}{3}$ V, i.e. 50 V across each.

(b) If lamp C fails, i.e., open circuits, no current will flow and **lamps A and B will not operate**.

Now try the following exercises

Exercise 80   Further problems on wiring lamps in series and in parallel (Answers on page 305)

1. If four identical lamps are connected in parallel and the combined resistance is 100 Ω, find the resistance of one lamp.

---

2. Three identical filament lamps are connected (a) in series, (b) in parallel across a 210 V supply. State for each connection the p.d. across each lamp.

Exercise 81   Short answer questions on series and parallel networks

1. Name three characteristics of a series circuit.
2. Show that for three resistors R_1, R_2 and R_3 connected in series the equivalent resistance R is given by:
   $R = R_1 + R_2 + R_3$
3. Name three characteristics of a parallel network.
4. Show that for three resistors R_1, R_2 and R_3 connected in parallel the equivalent resistance R is given by:
   $$\frac{1}{R} = \frac{1}{R_1} + \frac{1}{R_2} + \frac{1}{R_3}$$
5. Explain the potential divider circuit.
6. Compare the merits of wiring lamps in (a) series (b) parallel.

Exercise 82   Multi-choice questions on series and parallel networks (Answers on page 305)

1. If two 4 Ω resistors are connected in series the effective resistance of the circuit is:

   (a) 8 Ω   (b) 4 Ω   (c) 2 Ω   (d) 1 Ω

2. If two 4 Ω resistors are connected in parallel the effective resistance of the circuit is:

   (a) 8 Ω   (b) 4 Ω   (c) 2 Ω   (d) 1 Ω

3. With the switch in Fig. 18.37 closed, the ammeter reading will indicate:

   (a) $1\frac{2}{3}$ A   (b) 75 A   (c) $\frac{1}{3}$ A   (d) 3 A

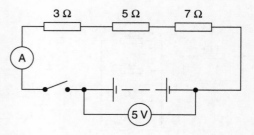

Fig. 18.37

4. The effect of connecting an additional parallel load to an electrical supply source is to increase the

   (a) resistance of the load

   (b) voltage of the source

   (c) current taken from the source

   (d) p.d. across the load

5. The equivalent resistance when a resistor of $\frac{1}{3}\,\Omega$ is connected in parallel with a $\frac{1}{4}\,\Omega$ resistance is:

   (a) $\frac{1}{7}\,\Omega$ (b) $7\,\Omega$ (c) $\frac{1}{12}\,\Omega$ (d) $\frac{3}{4}\,\Omega$

6. A $6\,\Omega$ resistor is connected in parallel with the three resistors of Fig. 18.38. With the switch closed the ammeter reading will indicate:

   (a) $\frac{3}{4}\,A$ (b) $4\,A$ (c) $\frac{1}{4}\,A$ (d) $1\frac{1}{3}\,A$

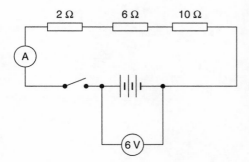

Fig. 18.38

7. A $10\,\Omega$ resistor is connected in parallel with a $15\,\Omega$ resistor and the combination is connected in series with a $12\,\Omega$ resistor. The equivalent resistance of the circuit is:

   (a) $37\,\Omega$ (b) $18\,\Omega$ (c) $27\,\Omega$ (d) $4\,\Omega$

8. When three $3\,\Omega$ resistors are connected in parallel, the total resistance is:

   (a) $3\,\Omega$ (b) $9\,\Omega$ (c) $1\,\Omega$ (d) $0.333\,\Omega$

9. The total resistance of two resistors R_1 and R_2 when connected in parallel is given by:

   (a) $R_1 + R_2$

   (b) $\dfrac{1}{R_1} + \dfrac{1}{R_2}$

   (c) $\dfrac{R_1 + R_2}{R_1 R_2}$

   (d) $\dfrac{R_1 R_2}{R_1 + R_2}$

Assignment 6

This assignment covers the material contained in chapters 17 and 18. The marks for each question are shown in brackets at the end of each question.

1. Calculate the resistance of 1200 m of copper cable of cross-sectional area $15\,\text{mm}^2$. Take the resistivity of copper as $0.02\,\mu\Omega\text{m}$. (5)

2. At a temperature of $40\,°C$, an aluminium cable has a resistance of $25\,\Omega$. If the temperature coefficient of resistance at $0\,°C$ is $0.0038/°C$, calculate it's resistance at $0\,°C$. (6)

3. The resistance of a coil of copper wire at $20\,°C$ is $150\,\Omega$. The temperature of the wire is increased and the resistance rises to $200\,\Omega$. If the temperature coefficient of resistance of copper is $0.004/°C$ at $20\,°C$ determine the temperature to which the coil has risen, correct to the nearest degree. (8)

4. Resistance's of $5\,\Omega$, $7\,\Omega$, and $8\,\Omega$ are connected in series. If a $10\,V$ supply voltage is connected across the arrangement determine the current flowing through and the p.d. across the $7\,\Omega$ resistor. Calculate also the power dissipated in the $8\,\Omega$ resistor. (6)

5. For the series-parallel network shown in Fig. A6.1, find (a) the supply current, (b) the current flowing through each resistor, (c) the p.d. across each resistor, (d) the total power dissipated in the circuit, (e) the

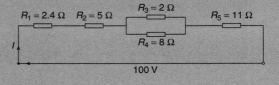

Fig. A6.1

cost of energy if the circuit is connected for 80 hours. Assume electrical energy costs 7.2 p per unit.   (15)

6. For the arrangement shown in Fig. A6.2, determine the current I_x (8)

7. Four identical filament lamps are connected (a) in series, (b) in parallel across a 240 V supply. State for each connection the p.d. across each lamp.   (2)

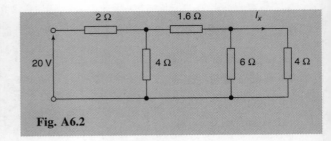

Fig. A6.2

19

Kirchhoff's laws

At the end of this chapter you should be able to:

- state Kirchhoff's current and voltage laws
- evaluate unknown currents and e.m.f.'s in series/parallel circuits, using Kirchhoff's laws

19.1 Introduction

Complex d.c. circuits cannot always be solved by Ohm's law and the formulae for series and parallel resistors alone. Kirchhoff (a German physicist) developed two laws which further help the determination of unknown currents and voltages in d.c. series/parallel networks.

19.2 Kirchhoff's current and voltage laws

Current law

At any junction in an electric circuit the total current flowing towards that junction is equal to the total current flowing away from the junction, i.e. $\Sigma I = 0$

Thus referring to Fig. 19.1:

$$I_1 + I_2 + I_3 = I_4 + I_5$$

or $\boxed{I_1 + I_2 + I_3 - I_4 - I_5 = 0}$

Voltage law

In any closed loop in a network, the algebraic sum of the voltage drops (i.e. products of current and resistance) taken around the loop is equal to the resultant e.m.f. acting in that loop.

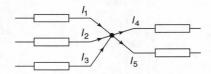

Fig. 19.1

Thus referring to Fig. 19.2:

$$\boxed{E_1 - E_2 = IR_1 + IR_2 + IR_3}$$

(Note that if current flows away from the positive terminal of a source, that source is considered by convention to be positive. Thus moving anticlockwise around the loop of Fig. 19.2, E_1 is positive and E_2 is negative)

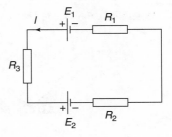

Fig. 19.2

19.3 Worked problems on Kirchhoff's laws

Problem 1. Determine the value of the unknown currents marked in Fig. 19.3.

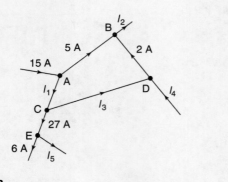

Fig. 19.3

Applying Kirchhoff's current law to each junction in turn gives:

For junction A: $15 = 5 + I_1$

Hence $I_1 = 10\,\text{A}$

For junction B: $5 + 2 = I_2$

Hence $I_2 = 7\,\text{A}$

For junction C: $I_1 = 27 + I_3$

i.e. $10 = 27 + I_3$

Hence $I_3 = 10 - 27 = -17\,\text{A}$

(i.e. in the opposite direction to that shown in Fig. 19.3)

For junction D: $I_3 + I_4 = 2$

i.e. $-17 + I_4 = 2$

Hence $I_4 = 17 + 2 = 19\,\text{A}$

For junction E: $27 = 6 + I_5$

Hence $I_5 = 27 - 6 = 21\,\text{A}$

Problem 2. Determine the value of e.m.f. E in Fig. 19.4.

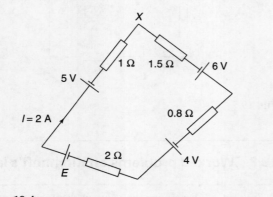

Fig. 19.4

Applying Kirchhoff's voltage law and moving clockwise around the loop of Fig. 19.4 starting at point X gives:

$$6 + 4 + E - 5 = I(1.5) + I(0.8) + I(2) + I(1)$$

$$5 + E = I(5.3) = 2(5.3) \text{ since current } I \text{ is } 2\,\text{A}$$

Hence $5 + E = 10.6$

and   **e.m.f. $E = 10.6 - 5 = 5.6\,\text{V}$**

Problem 3. Use Kirchhoff's laws to determine the current flowing in the $4\,\Omega$ resistance of the network shown in Fig. 19.5.

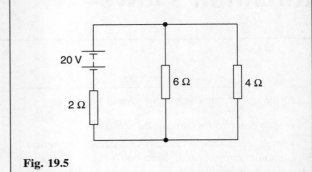

Fig. 19.5

Step 1

Label current I_1 flowing from the positive terminal of the 20 V source and current I_2 flowing through the 6 V resistance as shown in Fig. 19.6. By **Kirchhoff's current law**, the current flowing in the 4 V resistance must be $(I_1 - I_2)$

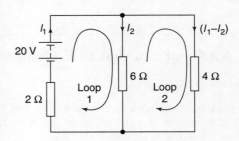

Fig. 19.6

Step 2

Label loops 1 and 2 as shown in Fig. 19.6 (both loops have been shown clockwise, although they do not need to be in the same direction). **Kirchhoff's voltage law** is now applied to each loop in turn:

For loop 1: $20 = 2I_1 + 6I_2$ $\qquad\qquad$ (1)

For loop 2: $0 = 4(I_1 - I_2) - 6I_2$ $\qquad\qquad$ (2)

Note the zero on the left-hand side of equation (2) since there is no voltage source in loop 2. Note also the minus sign in front of $6I_2$. This is because loop 2 is moving through the 6 V resistance in the opposite direction to current I_2

Equation (2) simplifies to:

$$0 = 4I_1 - 10I_2 \qquad (3)$$

Step 3

Solve the simultaneous equations (1) and (3) for currents I_1 and I_2 (see Chapter 8):

$$20 = 2I_1 + 6I_2 \qquad (1)$$

$$0 = 4I_1 - 10I_2 \qquad (3)$$

2 × equation (1) gives:

$$40 = 4I_1 + 12I_2 \qquad (4)$$

Equation (4) – equation (3) gives:

$$40 = 0 + (12I_2 - -10I_2)$$

i.e. $\qquad\qquad 40 = 22I_2$

Hence, **current $I_2 = \dfrac{40}{22} = \mathbf{1.818\,A}$**

Substituting $I_2 = 1.818$ into equation (1) gives:

$$20 = 2I_1 + 6(1.818)$$

$$20 = 2I_1 + 10.908$$

and $\qquad\qquad 20 - 10.908 = 2I_1$

from which, $\qquad I_1 = \dfrac{20 - 10.908}{2}$

$$= \dfrac{9.092}{2} = \mathbf{4.546\,A}$$

Hence the current flowing in the 4 V resistance is $I_1 - I_2$, i.e. $(4.546 - 1.818) = \mathbf{2.728\,A}$.

The currents and their directions are as shown in Fig. 19.7.

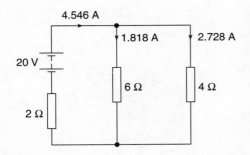

Fig. 19.7

Problem 4. Use Kirchhoff's laws to determine the current flowing in each branch of the network shown in Fig. 19.8.

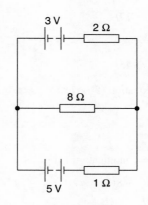

Fig. 19.8

Step 1

The currents I_1 and I_2 are labelled as shown in Fig. 19.9 and by Kirchhoff's current law the current in the 8 Ω resistance is $(I_1 + I_2)$

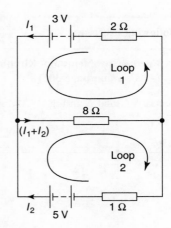

Fig. 19.9

Step 2

Loops 1 and 2 are labelled as shown in Fig. 19.9. Kirchhoff's voltage law is now applied to each loop in turn.

For loop 1: $\quad 3 = 2I_1 + 8(I_1 + I_2) \qquad (1)$

For loop 2: $\quad 5 = 8(I_1 + I_2) + (1)(I_2) \qquad (2)$

Equation (1) simplifies to:

$$3 = 10I_1 + 8I_2 \qquad (3)$$

Equation (2) simplifies to:

$$5 = 8I_1 + 9I_2 \quad\quad (4)$$

$4 \times$ equation (3) gives:

$$12 = 40I_1 + 32I_2 \quad\quad (5)$$

$5 \times$ equation (4) gives:

$$25 = 40I_1 + 45I_2 \quad\quad (6)$$

Equation (6) − equation (5) gives:

$$13 = 13I_2$$

and $I_2 = 1\,\mathrm{A}$

Substituting $I_2 = 1$ in equation (3) gives:

$$3 = 10I_1 + 8(1)$$

$$3 - 8 = 10I_1$$

and $I_1 = \dfrac{-5}{10} = -0.5\,\mathrm{A}$

(i.e. I_1 is flowing in the opposite direction to that shown in Fig. 19.9).

The **current in the 8 Ω resistance** is $(I_1 + I_2) = (-0.5 + 1)$

$$= 0.5\,\mathrm{A}$$

Now try the following exercises

Exercise 83   Further problems on Kirchhoff's laws (Answers on page 305)

1. Find currents I_3, I_4 and I_6 in Fig. 19.10.

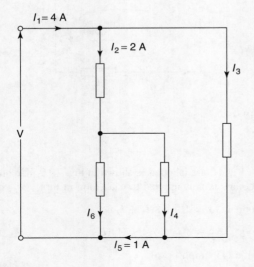

$I_1 = 4\,\mathrm{A}$

$I_2 = 2\,\mathrm{A}$

I_3

V

I_6 I_4

$I_5 = 1\,\mathrm{A}$

Fig. 19.10

2. For the networks shown in Fig. 19.11, find the values of the currents marked.

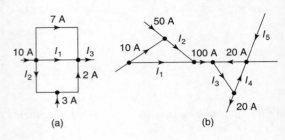

(a)   (b)

Fig. 19.11

3. Use Kirchhoff's laws to find the current flowing in the 6 Ω resistor of Fig. 19.12 and the power dissipated in the 4 Ω resistor.

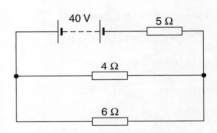

40 V   5 Ω

4 Ω

6 Ω

Fig. 19.12

4. Find the current flowing in the 3 Ω resistor for the network shown in Fig. 19.13(a). Find also the p.d. across the 10 Ω and 2 Ω resistors.

5. For the network shown in Fig. 19.13(b), find: (a) the current in the battery, (b) the current in the 300 Ω resistor, (c) the current in the 90 Ω resistor, and (d) the power dissipated in the 150 Ω resistor.

6. For the bridge network shown in Fig. 19.13(c), find the currents I_1 to I_5

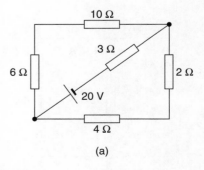

10 Ω

3 Ω

6 Ω   2 Ω

20 V

4 Ω

(a)

Fig. 19.13

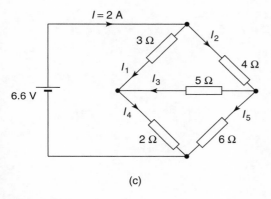

(b)

(c)

Fig. 19.13 (*Continued*)

Exercise 84   Short answer questions on Kirchhoff's laws

1. State Kirchhoff's current law.
2. State Kirchhoff's voltage law.

Exercise 85   Multi-choice questions on Kirchhoff's laws (Answers on page 306)

1. The current flowing in the branches of a d.c. circuit may be determined using:
   (a) Kirchhoff's laws   (b) Lenz's law
   (c) Faraday's laws   (d) Fleming's left-hand rule

2. Which of the following statements is true?
   For the junction in the network shown in Fig. 19.14:
   (a) $I_5 - I_4 = I_3 - I_2 + I_1$
   (b) $I_1 + I_2 + I_3 = I_4 + I_5$
   (c) $I_2 + I_3 + I_5 = I_1 + I_4$
   (d) $I_1 - I_2 - I_3 - I_4 + I_5 = 0$

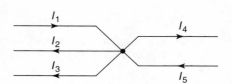

Fig. 19.14

3. Which of the following statements is true?
   For the circuit shown in Fig. 19.15:
   (a) $E_1 + E_2 + E_3 = Ir_1 + Ir_2 + Ir_3$
   (b) $E_2 + E_3 - E_1 - I(r_1 + r_2 + r_3) = 0$
   (c) $I(r_1 + r_2 + r_3) = E_1 - E_2 - E_3$
   (d) $E_2 + E_3 - E_1 = Ir_1 + Ir_2 + Ir_3$

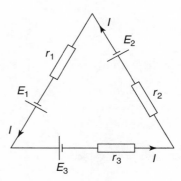

Fig. 19.15

4. The current I flowing in resistor R in the circuit shown in Fig. 19.16 is:

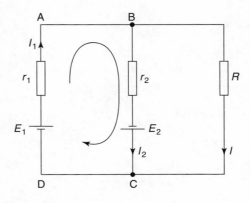

Fig. 19.16

(a) $I_2 - I_1$ (b) $I_1 - I_2$

(c) $I_1 + I_2$ (d) I_1

5. Applying Kirchhoff's voltage law clockwise around loop *ABCD* in the circuit shown in Fig. 19.16 gives:

(a) $E_1 - E_2 = I_1 r_1 + I_2 r_2$

(b) $E_2 = E_1 + I_1 r_1 + I_2 r_2$

(c) $E_1 + E_2 = I_1 r_1 + I_2 r_2$

(d) $E_1 + I_1 r_1 = E_2 + I_2 r_2$

20

Electromagnetism

20.1 Magnetic field due to an electric current

A permanent magnet is a piece of ferromagnetic material (such as iron, nickel or cobalt) which has properties of attracting other pieces of these materials.

The area around a magnet is called the **magnetic field** and it is in this area that the effects of the **magnetic force** produced by the magnet can be detected. The magnetic field of a bar magnet can be represented pictorially by the 'lines of force' (or lines of 'magnetic flux' as they are called) as shown in Fig. 20.1. Such a field pattern can be produced by placing iron filings in the vicinity of the magnet.

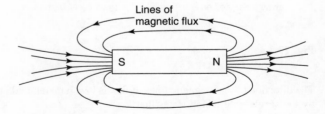

Fig. 20.1

The field direction at any point is taken as that in which the north-seeking pole of a compass needle points when suspended in the field. External to the magnet the direction of the field is north to south.

The laws of magnetic attraction and repulsion can be demonstrated by using two bar magnets. In Fig. 20.2(a), with **unlike pole** adjacent, **attraction** occurs. In Fig. 20.2(b), with **like poles** adjacent, **repulsion** occurs.

Magnetic fields are produced by electric currents as well as by permanent magnets. The field forms a circular pattern

(a)

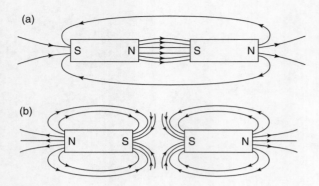

(b)

Fig. 20.2

with the current-carrying conductor at the centre. The effect is portrayed in Fig. 20.3 where the convention adopted is:

(i) Current flowing **away** from the viewer, i.e. into the paper, is indicated by $\oplus$. This may be thought of as the feathered end of the shaft of an arrow. See Fig. 20.3(a).

(ii) Current flowing **towards** the viewer, i.e. out of the paper, is indicated by $\bullet$. This may be thought of as the point of an arrow. See Fig. 20.3(b).

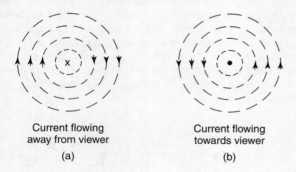

Current flowing
away from viewer

(a)

Current flowing
towards viewer

(b)

Fig. 20.3

The direction of the magnetic lines of flux is best remembered by the **screw rule** which states that:

If a normal right-hand thread screw is screwed along the conductor in the direction of the current, the direction of rotation of the screw is in the direction of the magnetic field.

For example, with current flowing away from the viewer (Fig. 20.3(a)) a right-hand thread screw driven into the paper has to be rotated clockwise. Hence the direction of the magnetic field is clockwise.

A magnetic field set up by a long coil, or **solenoid**, is shown in Fig. 20.4 and is seen to be similar to that of a bar magnet. If the solenoid is wound on an iron bar an even stronger magnetic field is produced, the iron becoming

magnetised and behaving like a permanent magnet. The **direction** of the magnetic field produced by the current I in the solenoid may be found by either of two methods, i.e. the screw rule or the grip rule.

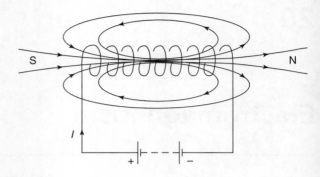

Fig. 20.4

(a) **The screw rule** states that if a normal right-hand thread screw is placed along the axis of the solenoid and is screwed in the direction of the current it moves in the direction of the magnetic field **inside** the solenoid (i.e. points in the direction of the north pole).

(b) **The grip rule** states that if the coil is gripped with the **right** hand, with the fingers pointing in the direction of the current, then the thumb, outstretched parallel to the axis of the solenoid, points in the direction of the magnetic field **inside** the solenoid (i.e. points in the direction of the north pole).

Problem 1. Figure 20.5 shows a coil of wire wound on an iron core connected to a battery. Sketch the magnetic field pattern associated with the current carrying coil and determine the polarity of the field.

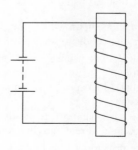

Fig. 20.5

The magnetic field associated with the solenoid in Fig. 20.5 is similar to the field associated with a bar magnet and is as shown in Fig. 20.6. The polarity of the field is determined either by the screw rule or by the grip rule. Thus the north pole is at the bottom and the south pole at the top.

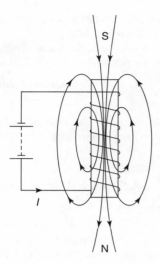

Fig. 20.6

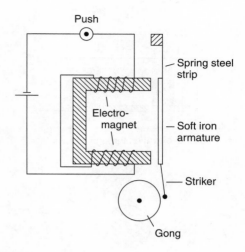

Fig. 20.7

20.2 Electromagnets

The solenoid is very important in electromagnetic theory since the magnetic field inside the solenoid is practically uniform for a particular current, and is also versatile, inasmuch that a variation of the current can alter the strength of the magnetic field. An electromagnet, based on the solenoid, provides the basis of many items of electrical equipment, examples of which include electric bells, relays, lifting magnets and telephone receivers.

(i) Electric bell

There are various types of electric bell, including the single-stroke bell, the trembler bell, the buzzer and a continuously ringing bell, but all depend on the attraction exerted by an electromagnet on a soft iron armature. A typical single stroke bell circuit is shown in Fig. 20.7. When the push button is operated a current passes through the coil. Since the iron-cored coil is energized the soft iron armature is attracted to the electromagnet. The armature also carries a striker which hits the gong. When the circuit is broken the coil becomes demagnetised and the spring steel strip pulls the armature back to its original position. The striker will only operate when the push is operated.

(ii) Relay

A relay is similar to an electric bell except that contacts are opened or closed by operation instead of a gong being struck. A typical simple relay is shown in Fig. 20.8, which consists of a coil wound on a soft iron core. When the coil is energised the hinged soft iron armature is attracted to the electromagnet and pushes against two fixed contacts

so that they are connected together, thus closing some other electrical circuit.

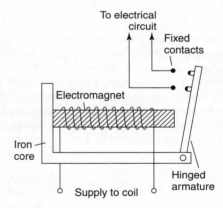

Fig. 20.8

(iii) Lifting magnet

Lifting magnets, incorporating large electromagnets, are used in iron and steel works for lifting scrap metal. A typical robust lifting magnet, capable of exerting large attractive forces, is shown in the elevation and plan view of Fig. 20.9 where a coil, *C*, is wound round a central core, *P*, of the iron casting. Over the face of the electromagnet is placed a protective non-magnetic sheet of material, *R*. The load, *Q*, which must be of magnetic material is lifted when the coils are energised, the magnetic flux paths, *M*, being shown by the broken lines.

(iv) Telephone receiver

Whereas a transmitter or microphone changes sound waves into corresponding electrical signals, a telephone receiver

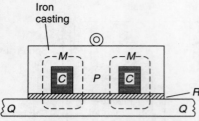

Iron casting

Sectional elevation through a diameter

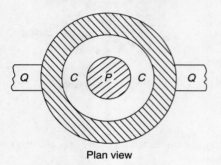

Plan view

Fig. 20.9

converts the electrical waves back into sound waves. A typical telephone receiver is shown in Fig. 20.10 and consists of a permanent magnet with coils wound on its poles. A thin, flexible diaphragm of magnetic material is held in position near to the magnetic poles but not touching them. Variation in current from the transmitter varies the magnetic field and the diaphragm consequently vibrates. The vibration produces sound variations corresponding to those transmitted.

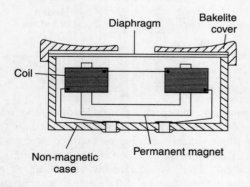

Fig. 20.10

20.3   Magnetic flux and flux density

Magnetic flux is the amount of magnetic field (or the number of lines of force) produced by a magnetic source. The symbol

for magnetic flux is Φ (Greek letter 'phi'). The unit of magnetic flux is the **weber, Wb**.

Magnetic flux density is the amount of flux passing through a defined area that is perpendicular to the direction of the flux:

$$\textbf{Magnetic flux density} = \frac{\textbf{magnetic flux}}{\textbf{area}}$$

The symbol for magnetic flux density is B. The unit of magnetic flux density is the **tesla, T**, where $1\,T = 1\,Wb/m^2$. Hence

$$\boxed{B = \frac{\Phi}{A}\ \textbf{tesla}} \quad \text{where } A(m^2) \text{ is the area}$$

Problem 2.   A magnetic pole face has a rectangular section having dimensions 200 mm by 100 mm. If the total flux emerging from the pole is $150\,\mu Wb$, calculate the flux density.

Flux, $\Phi = 150\,\mu Wb = 150 \times 10^{-6}\,Wb$, and cross sectional area, $A = 200 \times 100 = 20\,000\,mm^2 = 20\,000 \times 10^{-6}\,m^2$

$$\text{Flux density, } \textbf{\textit{B}} = \frac{\Phi}{A} = \frac{150 \times 10^{-6}}{20\,000 \times 10^{-6}}$$

$$= \textbf{0.0075 T} \quad \text{or} \quad \textbf{7.5 mT}$$

Problem 3.   The maximum working flux density of a lifting electromagnet is 1.8 T and the effective area of a pole face is circular in cross-section. If the total magnetic flux produced is 353 mWb, determine the radius of the pole face.

Flux density $B = 1.8\,T$ and

flux $\Phi 353\,mWb = 353 \times 10^{-3}\,Wb$.

Since $B = \dfrac{\Phi}{A}$, cross-sectional area

$$A = \frac{\Phi}{B} = \frac{353 \times 10^{-3}}{1.8}\,m^2 = 0.1961\,m^2$$

The pole face is circular, hence area $= \pi r^2$, where r is the radius.

Hence $\pi r^2 = 0.1961$, from which,

$$r^2 = \frac{0.1961}{\pi} \quad \text{and radius}$$

$$r = \sqrt{\frac{0.1961}{\pi}} = 0.250\,m$$

i.e. **the radius of the pole face is 250 mm**.

Now try the following exercise

Exercise 86 Further problems on magnetic circuits
(Answers on page 306)

1. What is the flux density in a magnetic field of cross-sectional area 20 cm² having a flux of 3 mWb?

2. Determine the total flux emerging from a magnetic pole face having dimensions 5 cm by 6 cm, if the flux density is 0.9 T

3. The maximum working flux density of a lifting electromagnet is 1.9 T and the effective area of a pole face is circular in cross-section. If the total magnetic flux produced is 611 mWb determine the radius of the pole face.

4. An electromagnet of square cross-section produces a flux density of 0.45 T. If the magnetic flux is 720 μWb find the dimensions of the electromagnet cross-section.

20.4 Force on a current-carrying conductor

If a current-carrying conductor is placed in a magnetic field produced by permanent magnets, then the fields due to the current-carrying conductor and the permanent magnets interact and cause a force to be exerted on the conductor. The force on the current-carrying conductor in a magnetic field depends upon:

(a) the flux density of the field, B teslas

(b) the strength of the current, I amperes,

(c) the length of the conductor perpendicular to the magnetic field, l metres, and

(d) the directions of the field and the current.

When the magnetic field, the current and the conductor are mutually at right angles then:

$$\boxed{\text{Force } F = BIl \text{ newtons}}$$

When the conductor and the field are at an angle $\theta°$ to each other then:

$$\boxed{\text{Force } F = BIl \sin \theta \text{ newtons}}$$

Since when the magnetic field, current and conductor are mutually at right angles, $F = BIl$, the magnetic flux density B may be defined by $B = \dfrac{F}{Il}$, i.e. the flux density is 1 T if the force exerted on 1 m of a conductor when the conductor carries a current of 1 A is 1 N.

Loudspeaker

A simple application of the above force is the moving-coil loudspeaker. The loudspeaker is used to convert electrical signals into sound waves.

Figure 20.11 shows a typical loudspeaker having a magnetic circuit comprising a permanent magnet and soft iron pole pieces so that a strong magnetic field is available in the short cylindrical air-gap. A moving coil, called the voice or speech coil, is suspended from the end of a paper or plastic cone so that it lies in the gap. When an electric current flows through the coil it produces a force which tends to move the cone backwards and forwards according to the direction of the current. The cone acts as a piston, transferring this force to the air, and producing the required sound waves.

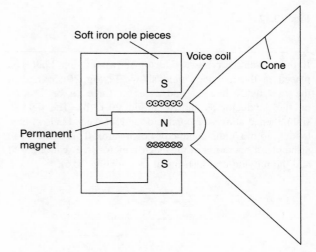

Fig. 20.11

Problem 4. A conductor carries a current of 20 A and is at right-angles to a magnetic field having a flux density of 0.9 T. If the length of the conductor in the field is 30 cm, calculate the force acting on the conductor. Determine also the value of the force if the conductor is inclined at an angle of 30° to the direction of the field.

$B = 0.9$ T, $I = 20$ A and $l = 30$ cm $= 0.30$ m.

Force $F = BIl = (0.9)(20)(0.30)$ newtons when the conductor is at right-angles to the field, as shown in Fig. 20.12(a), i.e. **$F = 5.4$ N**

When the conductor is inclined at 30° to the field, as shown in Fig. 20.12(b),

then force, $F = BIl \sin \theta$

$$= (0.9)(20)(0.30) \sin 30° = \textbf{2.7 N}$$

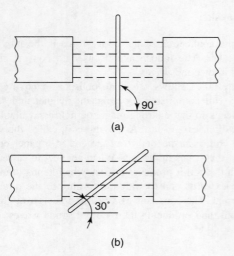

Fig. 20.12

If the current-carrying conductor shown in Fig. 20.3(a) is placed in the magnetic field shown in Fig. 20.13(a), then the two fields interact and cause a force to be exerted on the conductor as shown in Fig. 20.13(b). The field is strengthened above the conductor and weakened below, thus tending to move the conductor downwards. This is the basic principle of operation of the electric motor (see section 20.5) and the moving-coil instrument (see section 20.6).

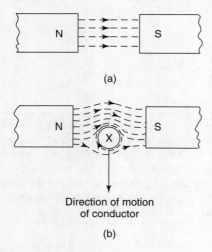

Fig. 20.13

The direction of the force exerted on a conductor can be pre-determined by using **Fleming's left-hand rule** (often called the motor rule) which states:

Let the thumb, first finger and second finger of the left hand be extended such that they are all at right-angles to each other, (as shown in Fig. 20.14). If the first finger points in the direction of the magnetic field, the second finger points in the direction of the current, then the thumb will point in the direction of the motion of the conductor.

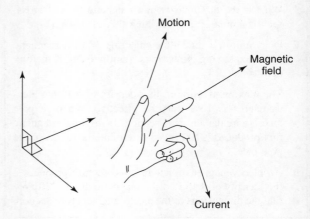

Fig. 20.14

Summarising:

First finger − Field

SeCond finger − Current

ThuMb − Motion

> *Problem 5*. Determine the current required in a 400 mm length of conductor of an electric motor, when the conductor is situated at right-angles to a magnetic field of flux density 1.2 T, if a force of 1.92 N is to be exerted on the conductor. If the conductor is vertical, the current flowing downwards and the direction of the magnetic field is from left to right, what is the direction of the force?

Force = 1.92 N, $l = 400$ mm $= 0.40$ m and $B = 1.2$ T.

Since $F = BIl$, then $I = \dfrac{F}{Bl}$ hence

current $I = \dfrac{1.92}{(1.2)(0.4)} = \mathbf{4\,A}$

If the current flows downwards, the direction of its magnetic field due to the current alone will be clockwise when viewed from above. The lines of flux will reinforce (i.e. strengthen) the main magnetic field at the back of the conductor and will be in opposition in the front (i.e. weaken the field). **Hence the force on the conductor will be from back to front (i.e. toward the viewer).** This direction may also have been deduced using Fleming's left-hand rule.

Problem 6. A conductor 350 mm long carries a current of 10 A and is at right-angles to a magnetic field lying between two circular pole faces each of radius 60 mm. If the total flux between the pole faces is 0.5 mWb, calculate the magnitude of the force exerted on the conductor.

$l = 350\,\text{mm} = 0.35\,\text{m}$, $I = 10\,\text{A}$, area of pole face $A = \pi r^2 = \pi(0.06)^2\,\text{m}^2$ and $\Phi = 0.5\,\text{mWb} = 0.5 \times 10^{-3}\,\text{Wb}$.

Force $F = BIl$, and $B = \dfrac{\Phi}{A}$ hence force,

$$F = \frac{\Phi}{A}Il = \frac{(0.5 \times 10^{-3})}{\pi(0.06)^2}(10)(0.35)\ \text{newtons}$$

i.e. **force = 0.155 N**

Problem 7. With reference to Fig. 20.15 determine (a) the direction of the force on the conductor in Fig. 20.15(a), (b) the direction of the force on the conductor in Fig. 20.15(b), (c) the direction of the current in Fig. 20.15(c), (d) the polarity of the magnetic system in Fig. 20.15(d).

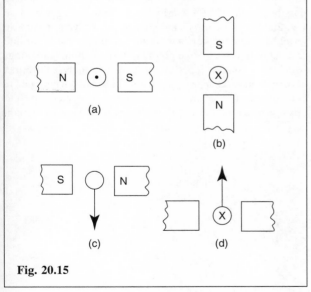

Fig. 20.15

(a) The direction of the main magnetic field is from north to south, i.e. left to right. The current is flowing towards the viewer, and using the screw rule, the direction of the field is anticlockwise. Hence either by Fleming's left-hand rule, or by sketching the interacting magnetic field as shown in Fig. 20.16(a), the direction of the force on the conductor is seen to be upward.

(b) Using a similar method to part (a) it is seen that the force on the conductor is to the right – see Fig. 20.16(b).

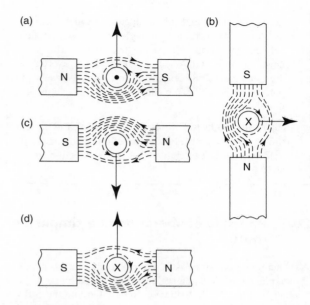

Fig. 20.16

(c) Using Fleming's left-hand rule, or by sketching as in Fig. 20.16(c), it is seen that the current is toward the viewer, i.e. out of the paper.

(d) Similar to part (c), the polarity of the magnetic system is as shown in Fig. 20.16(d).

Now try the following exercise

Exercise 87 Further problems on the force on a current-carrying conductor (Answers on page 306)

1. A conductor carries a current of 70 A at right-angles to a magnetic field having a flux density of 1.5 T. If the length of the conductor in the field is 200 mm calculate the force acting on the conductor. What is the force when the conductor and field are at an angle of 45°?

2. Calculate the current required in a 240 mm length of conductor of a d.c. motor when the conductor is situated at right-angles to the magnetic field of flux density 1.25 T, if a force of 1.20 N is to be exerted on the conductor.

3. A conductor 30 cm long is situated at right-angles to a magnetic field. Calculate the strength of the magnetic field if a current of 15 A in the conductor produces a force on it of 3.6 N.

4. A conductor 300 mm long carries a current of 13 A and is at right-angles to a magnetic field between two circular pole faces, each of diameter 80 mm. If the total flux between the pole faces is 0.75 mWb calculate the force exerted on the conductor.

5. (a) A 400 mm length of conductor carrying a current of 25 A is situated at right-angles to a magnetic field between two poles of an electric motor. The poles have a circular cross-section. If the force exerted on the conductor is 80 N and the total flux between the pole faces is 1.27 mWb, determine the diameter of a pole face.

(b) If the conductor in part (a) is vertical, the current flowing downwards and the direction of the magnetic field is from left to right, what is the direction of the 80 N force?

20.5 Principle of operation of a simple d.c. motor

A rectangular coil which is free to rotate about a fixed axis is shown placed inside a magnetic field produced by permanent magnets in Fig. 20.17. A direct current is fed into the coil via carbon brushes bearing on a commutator, which consists of a metal ring split into two halves separated by insulation. When current flows in the coil a magnetic field is set up around the coil which interacts with the magnetic field produced by the magnets. This causes a force F to be exerted on the current-carrying conductor which, by Fleming's left-hand rule, is downwards between points A and B and upward between C and D for the current direction shown. This causes a torque and the coil rotates anticlockwise. When the coil has turned through 90° from the position shown in Fig. 20.17 the brushes connected to the positive and negative terminals of the supply make contact with different halves of the commutator ring, thus reversing the direction of the current

flow in the conductor. If the current is not reversed and the coil rotates past this position the forces acting on it change direction and it rotates in the opposite direction thus never making more than half a revolution. The current direction is reversed every time the coil swings through the vertical position and thus the coil rotates anti-clockwise for as long as the current flows. This is the principle of operation of a d.c. motor which is thus a device that takes in electrical energy and converts it into mechanical energy.

20.6 Principle of operation of a moving-coil instrument

A moving-coil instrument operates on the motor principle. When a conductor carrying current is placed in a magnetic field, a force F is exerted on the conductor, given by $F = BIl$. If the flux density B is made constant (by using permanent magnets) and the conductor is a fixed length (say, a coil) then the force will depend only on the current flowing in the conductor.

In a moving-coil instrument a coil is placed centrally in the gap between shaped pole pieces as shown by the front elevation in Fig. 20.18(a). (The airgap is kept as small as possible, although for clarity it is shown exaggerated in Fig. 20.18). The coil is supported by steel pivots, resting in jewel bearings, on a cylindrical iron core. Current is led into and out of the coil by two phosphor bronze spiral hairsprings which are wound in opposite directions to minimise the effect of temperature change and to limit the coil swing (i.e. to **control** the movement) and return the movement to zero position when no current flows. Current flowing in the coil

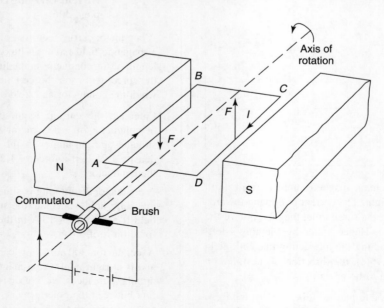

Fig. 20.17

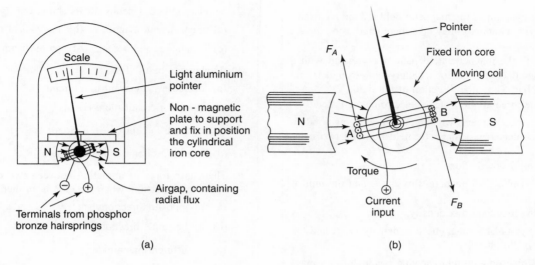

Fig. 20.18

produces forces as shown in Fig. 20.18(b), the directions being obtained by Fleming's left-hand rule. The two forces, F_A and F_B, produce a torque which will move the coil in a clockwise direction, i.e. move the pointer from left to right. Since force is proportional to current the scale is linear.

When the aluminium frame, on which the coil is wound, is rotated between the poles of the magnet, small currents (called eddy currents) are induced into the frame, and this provides automatically the necessary **damping** of the system due to the reluctance of the former to move within the magnetic field. The moving-coil instrument will measure only direct current or voltage and the terminals are marked positive and negative to ensure that the current passes through the coil in the correct direction to deflect the pointer 'up the scale'.

The range of this sensitive instrument is extended by using shunts and multipliers (see Chapter 23).

20.7 Force on a charge

When a charge of Q coulombs is moving at a velocity of v m/s in a magnetic field of flux density B teslas, the charge moving perpendicular to the field, then the magnitude of the force F exerted on the charge is given by:

$$F = QvB \text{ newtons}$$

Problem 8. An electron in a television tube has a charge of 1.6×10^{-19} coulombs and travels at 3×10^7 m/s perpendicular to a field of flux density $18.5\,\mu\text{T}$. Determine the force exerted on the electron in the field.

From above, force $F = QvB$ newtons, where

$$Q = \text{charge in coulombs} = 1.6 \times 10^{-19}\,\text{C},$$

$$v = \text{velocity of charge} = 3 \times 10^7\,\text{m/s},$$

and $B = \text{flux density} = 18.5 \times 10^{-6}\,\text{T}$

Hence force on electron,

$$F = 1.6 \times 10^{-19} \times 3 \times 10^7 \times 18.5 \times 10^{-6}$$
$$= 1.6 \times 3 \times 18.5 \times 10^{-18}$$
$$= 88.8 \times 10^{-18} = \mathbf{8.88 \times 10^{-17}\,N}$$

Now try the following exercises

Exercise 88 Further problems on the force on a charge (Answers on page 306)

1. Calculate the force exerted on a charge of 2×10^{-18} C travelling at 2×10^6 m/s perpendicular to a field of density 2×10^{-7} T

2. Determine the speed of a 10^{-19} C charge travelling perpendicular to a field of flux density 10^{-7} T, if the force on the charge is 10^{-20} N.

Exercise 89 Short answer questions on electromagnetism

1. What is a permanent magnet?

2. Sketch the pattern of the magnetic field associated with a bar magnet. Mark the direction of the field.

3. The direction of the magnetic field around a current-carrying conductor may be remembered using the ... rule.

4. Sketch the magnetic field pattern associated with a solenoid connected to a battery and wound on an iron bar. Show the direction of the field.

5. Name three applications of electromagnetism.

6. State what happens when a current-carrying conductor is placed in a magnetic field between two magnets.

7. Define magnetic flux.

8. The symbol for magnetic flux is ... and the unit of flux is the ....

9. Define magnetic flux density.

10. The symbol for magnetic flux density is ... and the unit of flux density is ....

11. The force on a current-carrying conductor in a magnetic field depends on four factors. Name them.

12. The direction of the force on a conductor in a magnetic field may be predetermined using Fleming's ... rule.

13. State three applications of the force on a current-carrying conductor.

14. Figure 20.19 shows a simplified diagram of a section through the coil of a moving-coil instrument. For the direction of current flow shown in the coil determine the direction that the pointer will move.

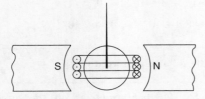

Fig. 20.19

15. Explain, with the aid of a sketch, the action of a simplified d.c. motor.

16. Sketch and label the movement of a moving-coil instrument. Briefly explain the principle of operation of such an instrument.

Exercise 90   Multi-choice questions on electro-magetism (Answers on page 306)

1. The unit of magnetic flux density is the:
   (a) weber   (b) weber per metre
   (c) ampere per metre   (d) tesla

2. An electric bell depends for its action on:
   (a) a permanent magnet   (b) reversal of current
   (c) a hammer and a gong   (d) an electromagnet

3. A relay can be used to:
   (a) decrease the current in a circuit
   (b) control a circuit more readily
   (c) increase the current in a circuit
   (d) control a circuit from a distance

4. There is a force of attraction between two current-carrying conductors when the current in them is:
   (a) in opposite directions
   (b) in the same direction
   (c) of different magnitude
   (d) of the same magnitude

5. The magnetic field due to a current-carrying conductor takes the form of:
   (a) rectangles
   (b) concentric circles
   (c) wavy lines
   (d) straight lines radiating outwards

6. The total flux in the core of an electrical machine is 20 mWb and its flux density is 1 T. The cross-sectional area of the core is:
   (a) $0.05 \, m^2$ (b) $0.02 \, m^2$
   (c) $20 \, m^2$ (d) $50 \, m^2$

7. A conductor carries a current of 10 A at right-angles to a magnetic field having a flux density of 500 mT. If the length of the conductor in the field is 20 cm, the force on the conductor is:
   (a) 100 kN   (b) 1 kN   (c) 100 N   (d) 1 N

8. If a conductor is horizontal, the current flowing from left to right and the direction of the surrounding magnetic field is from above to below, the force exerted on the conductor is:
   (a) from left to right
   (b) from below to above
   (c) away from the viewer
   (d) towards the viewer

9. For the current-carrying conductor lying in the magnetic field shown in Fig. 20.20(a), the direction of the force on the conductor is:
   (a) to the left   (b) upwards
   (c) to the right   (d) downwards

10. For the current-carrying conductor lying in the magnetic field shown in Fig. 20.20(b), the direction of the current in the conductor is:
    (a) towards the viewer
    (b) away from the viewer

    (c) remain in the vertical position
    (d) experience a force towards the north pole

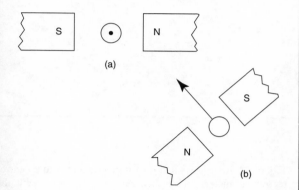

(a)

(b)

Fig. 20.20

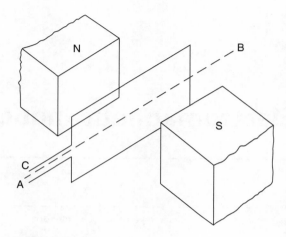

Fig. 20.21

11. Figure 20.21 shows a rectangular coil of wire placed in a magnetic field and free to rotate about axis *AB*. If the current flows into the coil at *C*, the coil will:
    (a) commence to rotate anti-clockwise
    (b) commence to rotate clockwise

12. The force on an electron travelling at 10^7 m/s in a magnetic field of density $10\,\mu$T is 1.6×10^{-17} N. The electron has a charge of:
    (a) 1.6×10^{-28} C (b) 1.6×10^{-15} C
    (c) 1.6×10^{-19} C (d) 1.6×10^{-25} C

21

Electromagnetic induction

At the end of this chapter you should be able to:

- understand how an e.m.f. may be induced in a conductor
- state Faraday's laws of electromagnetic induction
- state Lenz's law
- use Fleming's right-hand rule for relative directions
- appreciate that the induced e.m.f., $E = Blv$ or $E = Blv\sin\theta$
- calculate induced e.m.f. given B, l, v and θ and determine relative directions
- define inductance L and state its unit
- define mutual inductance
- calculate mutual inductance using $E_2 = -M\dfrac{\mathrm{d}I_1}{\mathrm{d}t}$
- describe the principle of operation of a transformer
- perform calculations using $\dfrac{V_1}{V_2} = \dfrac{N_1}{N_2} = \dfrac{I_2}{I_1}$ for an ideal transformer
- define the 'rating' of a transformer

21.1 Introduction to electromagnetic induction

When a conductor is moved across a magnetic field so as to cut through the lines of force (or flux), an electromotive force (e.m.f.) is produced in the conductor. If the conductor forms part of a closed circuit then the e.m.f. produced causes an electric current to flow round the circuit. Hence an e.m.f. (and thus current) is 'induced' in the conductor as a result of its movement across the magnetic field. This effect is known as **'electromagnetic induction'**.

Figure 21.1(a) shows a coil of wire connected to a centre-zero galvanometer, which is a sensitive ammeter with the zero-current position in the centre of the scale.

(a) When the magnet is moved at constant speed towards the coil (Fig. 21.1(a)), a deflection is noted on the galvanometer showing that a current has been produced in the coil.

(b) When the magnet is moved at the same speed as in (a) but away from the coil the same deflection is noted but is in the opposite direction (see Fig. 21.1(b)).

(c) When the magnet is held stationary, even within the coil, no deflection is recorded.

(d) When the coil is moved at the same speed as in (a) and the magnet held stationary the same galvanometer deflection is noted.

(e) When the relative speed is, say, doubled, the galvanometer deflection is doubled.

(f) When a stronger magnet is used, a greater galvanometer deflection is noted.

(g) When the number of turns of wire of the coil is increased, a greater galvanometer deflection is noted.

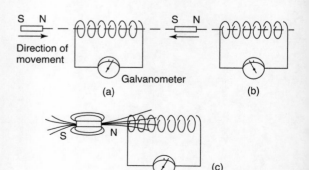

Fig. 21.1

Figure 21.1(c) shows the magnetic field associated with the magnet. As the magnet is moved towards the coil, the magnetic flux of the magnet moves across, or cuts, the coil. **It is the relative movement of the magnetic flux and the coil that causes an e.m.f. and thus current, to be induced in the coil**. This effect is known as **electromagnetic induction**. The laws of electromagnetic induction stated in Section 21.2 evolved from experiments such as those described above.

21.2 Laws of electromagnetic induction

Faraday's laws of electromagnetic induction state:

(i) *An induced e.m.f. is set up whenever the magnetic field linking that circuit changes.*

(ii) *The magnitude of the induced e.m.f. in any circuit is proportional to the rate of change of the magnetic flux linking the circuit.*

Lenz's law states:

The direction of an induced e.m.f. is always such that it tends to set up a current opposing the motion or the change of flux responsible for inducing that e.m.f.

An alternative method to Lenz's law of determining relative directions is given by **Fleming's right-hand rule** (often called the generator rule) which states:

Let the thumb, first finger and second finger of the right hand be extended such that they are all at right angles to each other (as shown in Fig. 21.2). If the first finger points in the direction of the magnetic field and the thumb points in the direction of motion of the conductor relative to the magnetic field, then the second finger will point in the direction of the induced e.m.f.

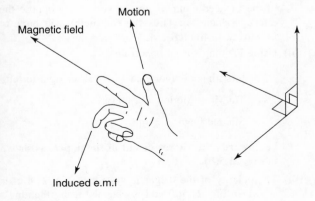

Fig. 21.2

Summarising:

First finger – Field

Thumb – Motion

SEcond finger – E.m.f.

In a generator, conductors forming an electric circuit are made to move through a magnetic field. By Faraday's law an e.m.f. is induced in the conductors and thus a source of e.m.f. is created. A generator converts mechanical energy into electrical energy. (The action of a simple a.c. generator is described in Chapter 22)

The induced e.m.f. E set up between the ends of the conductor shown in Fig. 21.3 is given by:

$$E = Blv \text{ volts}$$

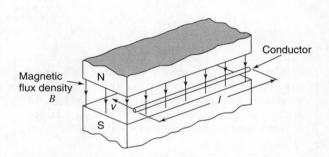

Fig. 21.3

where B, the flux density, is measured in teslas, l, the length of conductor in the magnetic field, is measured in metres, and v, the conductor velocity, is measured in metres per second.

If the conductor moves at an angle $\theta°$ to the magnetic field (instead of at 90° as assumed above) then

$$E = Blv \sin \theta \text{ volts}$$

Problem 1. A conductor 300 mm long moves at a uniform speed of 4 m/s at right-angles to a uniform magnetic field of flux density 1.25 T. Determine the current flowing in the conductor when (a) its ends are open-circuited, (b) its ends are connected to a load of 20 Ω resistance.

When a conductor moves in a magnetic field it will have an e.m.f. induced in it but this e.m.f. can only produce a current if there is a closed circuit.

Induced e.m.f. $E = Blv = (1.25) \left(\dfrac{300}{1000} \right) (4) = 1.5 \text{ V}$

(a) If the ends of the conductor are open circuited **no current will flow** even though 1.5 V has been induced.

(b) From Ohm's law, current,

$$I = \frac{E}{R} = \frac{1.5}{20} = 0.075 \text{ A or } 75 \text{ mA}$$

Problem 2. At what velocity must a conductor 75 mm long cut a magnetic field of flux density 0.6 T if an e.m.f. of 9 V is to be induced in it? Assume the conductor, the field and the direction of motion are mutually perpendicular.

Induced e.m.f. $E = Blv$, hence velocity $v = \dfrac{E}{Bl}$

Hence velocity,

$$v = \frac{9}{(0.6)(75 \times 10^{-3})}$$

$$= \frac{9 \times 10^3}{0.6 \times 75} = 200 \text{ m/s}$$

Problem 3. A conductor moves with a velocity of 15 m/s at an angle of (a) 90° (b) 60° and (c) 30° to a magnetic field produced between two square-faced poles of side length 2 cm. If the flux leaving a pole face is 5 μWb, find the magnitude of the induced e.m.f. in each case.

$v = 15$ m/s, length of conductor in magnetic field, $l = 2 \text{ cm} = 0.02 \text{ m}$, $A = (2 \times 2) \text{ cm}^2 = 4 \times 10^{-4} \text{ m}^2$ and $\Phi = 5 \times 10^{-6}$ Wb

(a) $E_{90} = Blv \sin 90° = \left(\dfrac{\Phi}{A}\right) lv \sin 90°$

$$= \left(\frac{5 \times 10^{-6}}{4 \times 10^{-4}}\right)(0.02)(15)(1) = 3.75 \text{ mV}$$

(b) $E_{60} = Blv \sin 60° = E_{90} \sin 60°$

$$= 3.75 \sin 60° = 3.25 \text{ mV}$$

(c) $E_{30} = Blv \sin 30° = E_{90} \sin 30°$

$$= 3.75 \sin 30° = 1.875 \text{ mV}$$

Problem 4. The wing span of a metal aeroplane is 36 m. If the aeroplane is flying at 400 km/h, determine the e.m.f. induced between its wing tips. Assume the vertical component of the earth's magnetic field is 40 μT.

Induced e.m.f. across wing tips, $E = Blv$

$$B = 40 \text{ μT} = 40 \times 10^{-6} \text{ T}, \quad l = 36 \text{ m}$$

and $v = 400 \dfrac{\text{km}}{\text{h}} \times 1000 \dfrac{\text{m}}{\text{km}} \times \dfrac{1 \text{ h}}{60 \times 60 \text{ s}}$

$$= \frac{(400)(1000)}{3600} = \frac{4000}{36} \text{ m/s}$$

Hence

$$E = Blv = (40 \times 10^{-6})(36)\left(\frac{4000}{36}\right) = 0.16 \text{ V}$$

Problem 5. The diagram shown in Fig. 21.4 represents the generation of e.m.f's. Determine (i) the direction in which the conductor has to be moved in Fig. 21.4(a), (ii) the direction of the induced e.m.f. in Fig. 21.4(b), (iii) the polarity of the magnetic system in Fig. 21.4(c).

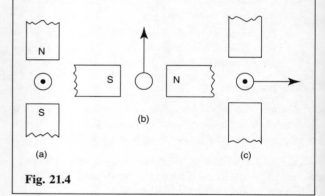

Fig. 21.4

The direction of the e.m.f., and thus the current due to the e.m.f. may be obtained by either Lenz's law or Fleming's Right-hand rule (i.e. GeneRator rule).

(i) Using Lenz's law: The field due to the magnet and the field due to the current-carrying conductor are shown in Fig. 21.5(a) and are seen to reinforce to the left of the conductor. Hence the force on the conductor is to the right. However Lenz's law states that the direction of the induced e.m.f. is always such as to oppose the effect producing it. **Thus the conductor will have to be moved to the left**.

(ii) Using Fleming's right-hand rule:

First finger − Field, i.e. $N \to S$, or right to left;

ThuMb − Motion, i.e. upwards;

SEcond finger − E.m.f.

i.e. **towards the viewer or out of the paper**, as shown in Fig. 21.5(b).

(iii) The polarity of the magnetic system of Fig. 21.4(c) is shown in Fig. 21.5(c) and is obtained using Fleming's right-hand rule.

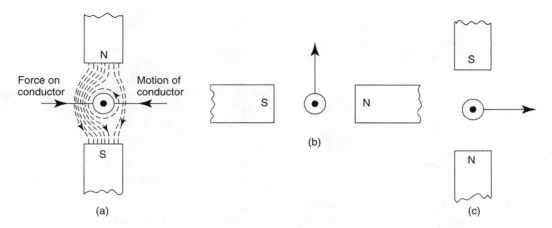

Fig. 21.5

Now try the following exercise

**Exercise 91   Further problems on induced e.m.f.
(Answers on page 306)**

1. A conductor of length 15 cm is moved at 750 mm/s at right-angles to a uniform flux density of 1.2 T. Determine the e.m.f. induced in the conductor.

2. Find the speed that a conductor of length 120 mm must be moved at right angles to a magnetic field of flux density 0.6 T to induce in it an e.m.f. of 1.8 V

3. A 25 cm long conductor moves at a uniform speed of 8 m/s through a uniform magnetic field of flux density 1.2 T. Determine the current flowing in the conductor when (a) its ends are open-circuited, (b) its ends are connected to a load of 15 ohms resistance.

4. A straight conductor 500 mm long is moved with constant velocity at right angles both to its length and to a uniform magnetic field. Given that the e.m.f. induced in the conductor is 2.5 V and the velocity is 5 m/s, calculate the flux density of the magnetic field. If the conductor forms part of a closed circuit of total resistance 5 ohms, calculate the force on the conductor.

5. A car is travelling at 80 km/h. Assuming the back axle of the car is 1.76 m in length and the vertical component of the earth's magnetic field is $40\,\mu T$, find the e.m.f. generated in the axle due to motion.

6. A conductor moves with a velocity of 20 m/s at an angle of (a) 90° (b) 45° (c) 30°, to a magnetic field produced between two square-faced poles of side length 2.5 cm. If the flux on the pole face is 60 mWb, find the magnitude of the induced e.m.f. in each case.

21.3   Self inductance

Inductance is the name given to the property of a circuit whereby there is an e.m.f. induced into the circuit by the change of flux linkages produced by a current change, as stated in Chapter 15, page 114.

When the e.m.f. is induced in the same circuit as that in which the current is changing, the property is called **self inductance, L**.

The unit of inductance is the **henry, H**.

21.4   Mutual inductance

When an e.m.f. is induced in a circuit by a change of flux due to current changing in an adjacent circuit, the property is called **mutual inductance, M**.

Mutually induced e.m.f. in the second coil,

$$E_2 = -M\frac{dI_1}{dt}\ \text{volts}$$

where M is the **mutual inductance** between two coils, in henrys, and $\dfrac{dI_1}{dt}$ is the rate of change of current in the first coil.

Problem 6. Calculate the mutual inductance between two coils when a current changing at 200 A/s in one coil induces an e.m.f. of 1.5 V in the other.

Induced e.m.f. $|E_2| = M\dfrac{dI_1}{dt}$, i.e. $1.5 = M(200)$.

Thus **mutual inductance**,

$$M = \frac{1.5}{200} = 0.0075\,\text{H or } 7.5\,\text{mH}$$

Problem 7. The mutual inductance between two coils is 18 mH. Calculate the steady rate of change of current in one coil to induce an e.m.f. of 0.72 V in the other.

Induced e.m.f. $|E_2| = M \dfrac{dI_1}{dt}$

Hence rate of change of current,

$$\frac{dI_1}{dt} = \frac{|E_2|}{M} = \frac{0.72}{0.018} = \textbf{40 A/s}$$

Problem 8. Two coils have a mutual inductance of 0.2 H. If the current in one coil is changed from 10 A to 4 A in 10 ms, calculate (a) the average induced e.m.f. in the second coil, (b) the change of flux linked with the second coil if it is wound with 500 turns.

(a) Induced e.m.f. $E_2 = -M \dfrac{dI_1}{dt} = -(0.2) \left(\dfrac{10 - 4}{10 \times 10^{-3}} \right)$

$$= -\textbf{120 V}$$

(b) From Section 15.8, induced e.m.f. $|E_2| = N \dfrac{d\Phi}{dt}$, hence

$$d\Phi = \frac{|E_2| \, dt}{N}$$

Thus the change of flux,

$$d\Phi = \frac{(120)(10 \times 10^{-3})}{500} = \textbf{2.4 mWb}$$

Now try the following exercise

Exercise 92   Further problems on mutual inductance
(Answers on page 306)

1. The mutual inductance between two coils is 150 mH. Find the magnitude of the e.m.f. induced in one coil when the current in the other is increasing at a rate of 30 A/s

2. Determine the mutual inductance between two coils when a current changing at 50 A/s in one coil induces an e.m.f. of 80 mV in the other.

3. Two coils have a mutual inductance of 0.75 H. Calculate the magnitude of the e.m.f. induced in one coil when a current of 2.5 A in the other coil is reversed in 15 ms.

4. The mutual inductance between two coils is 240 mH. If the current in one coil changes from 15 A to 6 A in 12 ms, calculate (a) the average e.m.f. induced in the other coil, (b) the change of flux linked with the other coil if it is wound with 400 turns.

21.5   The transformer

A transformer is a device which uses the phenomenon of mutual induction to change the values of alternating voltages and currents. In fact, one of the main advantages of a.c. transmission and distribution is the ease with which an alternating voltage can be increased or decreased by transformers.

Losses in transformers are generally low and thus efficiency is high. Being static they have a long life and are very stable.

Transformers range in size from the miniature units used in electronic applications to the large power transformers used in power stations. The principle of operation is the same for each.

A transformer is represented in Fig. 21.6(a) as consisting of two electrical circuits linked by a common ferromagnetic core. One coil is termed the **primary winding** which is connected to the supply of electricity, and the other the **secondary winding**, which may be connected to a load. A circuit diagram symbol for a transformer is shown in Fig. 21.6(b).

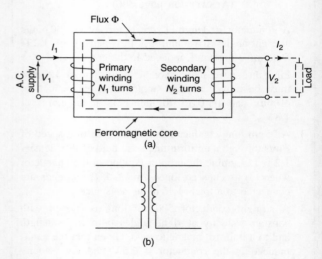

Fig. 21.6

Transformer principle of operation

When the secondary is an open-circuit and an alternating voltage V_1 is applied to the primary winding, a small current - called the no-load current I_0 - flows, which sets up a magnetic flux in the core. This alternating flux links with both primary and secondary coils and induces in them e.m.f.'s of E_1 and E_2 respectively by mutual induction.

The induced e.m.f. E in a coil of N turns is given by $E = -N \dfrac{d\Phi}{dt}$ volts, where $\dfrac{d\Phi}{dt}$ is the rate of change of flux (see Chapter 15, page 115). In an ideal transformer, the rate of change of flux is the same for both primary and secondary and thus $\dfrac{E_1}{N_1} = \dfrac{E_2}{N_2}$ i.e. **the induced e.m.f. per turn is constant**.

Assuming no losses, $E_1 = V_1$ and $E_2 = V_2$

Hence $\qquad \dfrac{V_1}{N_1} = \dfrac{V_2}{N_2}$ or $\dfrac{V_1}{V_2} = \dfrac{N_1}{N_2}$ $\qquad$ (1)

$\dfrac{V_1}{V_2}$ is called the voltage ratio and $\dfrac{N_1}{N_2}$ the turns ratio, or the **'transformation ratio'** of the transformer. If N_2 is less than N_1 then V_2 is less than V_1 and the device is termed a **step-down transformer**. If N_2 is greater then N_1 then V_2 is greater than V_1 and the device is termed a **step-up transformer**.

When a load is connected across the secondary winding, a current I_2 flows. In an ideal transformer losses are neglected and a transformer is considered to be 100% efficient.

Hence input power = output power, or $V_1 I_1 = V_2 I_2$ i.e. in an ideal transformer, the **primary and secondary ampere-turns are equal**

Thus $\qquad \dfrac{V_1}{V_2} = \dfrac{I_2}{I_1}$ $\qquad$ (2)

Combining equations (1) and (2) gives:

$$\boxed{\dfrac{V_1}{V_2} = \dfrac{N_1}{N_2} = \dfrac{I_2}{I_1}}$$ $\qquad$ (3)

The **rating** of a transformer is stated in terms of the volt-amperes that it can transform without overheating. With reference to Fig. 21.6(a), the transformer rating is either $V_1 I_1$ or $V_2 I_2$, where I_2 is the full-load secondary current.

Problem 9. A transformer has 500 primary turns and 3000 secondary turns. If the primary voltage is 240 V, determine the secondary voltage, assuming an ideal transformer.

For an ideal transformer, voltage ratio = turns ratio

i.e. $\dfrac{V_1}{V_2} = \dfrac{N_1}{N_2}$ hence $\dfrac{240}{V_2} = \dfrac{500}{3000}$

Thus secondary voltage,

$$V_2 = \dfrac{(240)(3000)}{500} = \mathbf{1440\,V} \quad \text{or} \quad \mathbf{1.44\,kV}$$

Problem 10. An ideal transformer with a turns ratio of 2:7 is fed from a 240 V supply. Determine its output voltage.

A turns ratio of 2:7 means that the transformer has 2 turns on the primary for every 7 turns on the secondary (i.e. a step-up transformer); thus $\dfrac{N_1}{N_2} = \dfrac{2}{7}$

For an ideal transformer,

$$\dfrac{N_1}{N_2} = \dfrac{V_1}{V_2} \quad \text{hence} \quad \dfrac{2}{7} = \dfrac{240}{V_2}$$

Thus the secondary voltage,

$$V_2 = \dfrac{(240)(7)}{2} = \mathbf{840\,V}$$

Problem 11. An ideal transformer has a turns ratio of 8:1 and the primary current is 3 A when it is supplied at 240 V. Calculate the secondary voltage and current.

A turns ratio of 8:1 means $\dfrac{N_1}{N_2} = \dfrac{8}{1}$ i.e. a step-down transformer.

$$\dfrac{N_1}{N_2} = \dfrac{V_1}{V_2}$$

hence, secondary voltage

$$V_2 = V_1 \left(\dfrac{N_2}{N_1} \right) = 240 \left(\dfrac{1}{8} \right) = \mathbf{30\ volts}$$

Also, $\dfrac{N_1}{N_2} = \dfrac{I_2}{I_1}$ hence, secondary current

$$I_2 = I_1 \left(\dfrac{N_1}{N_2} \right) = 3 \left(\dfrac{8}{1} \right) = \mathbf{24\,A}$$

Problem 12. An ideal transformer, connected to a 240 V mains, supplies a 12 V, 150 W lamp. Calculate the transformer turns ratio and the current taken from the supply.

$V_1 = 240\,V$, $V_2 = 12\,V$, and since power $P = VI$, then

$$I_2 = \dfrac{P}{V_2} = \dfrac{150}{12} = 12.5\,A$$

Turns ratio $= \dfrac{N_1}{N_2} = \dfrac{V_1}{V_2} = \dfrac{240}{12} = \mathbf{20}$

$$\dfrac{V_1}{V_2} = \dfrac{I_2}{I_1} \text{ from which, } I_1 = I_2 \left(\dfrac{V_2}{V_1} \right)$$

$$= 12.5 \left(\dfrac{12}{240} \right)$$

Hence the current taken from the supply,

$$I_1 = \dfrac{12.5}{20} = \mathbf{0.625\,A}$$

Problem 13.   A $12\,\Omega$ resistor is connected across the secondary winding of an ideal transformer whose secondary voltage is 120 V. Determine the primary voltage if the supply current is 4 A.

Secondary current

$$I_2 = \frac{V_2}{R_2} = \frac{120}{12} = 10\,\text{A}$$

$$\frac{V_1}{V_2} = \frac{I_2}{I_1}$$

from which the primary voltage

$$V_1 = V_2\left(\frac{I_2}{I_1}\right) = 120\left(\frac{10}{4}\right) = \textbf{300 volts}$$

Now try the following exercises

Exercise 93   Further problems on the transformer principle of operation (Answers on page 306)

1. A transformer has 600 primary turns connected to a 1.5 kV supply. Determine the number of secondary turns for a 240 V output voltage, assuming no losses.

2. An ideal transformer with a turns ratio of 2:9 is fed from a 220 V supply. Determine its output voltage.

3. An ideal transformer has a turns ratio of 12:1 and is supplied at 192 V. Calculate the secondary voltage.

4. A transformer primary winding connected across a 415 V supply has 750 turns. Determine how many turns must be wound on the secondary side if an output of 1.66 kV is required.

5. An ideal transformer has a turns ratio of 12:1 and is supplied at 180 V when the primary current is 4 A. Calculate the secondary voltage and current.

6. A step-down transformer having a turns ratio of 20:1 has a primary voltage of 4 kV and a load of 10 kW. Neglecting losses, calculate the value of the secondary current.

7. A transformer has a primary to secondary turns ratio of 1:15. Calculate the primary voltage necessary to supply a 240 V load. If the load current is 3 A determine the primary current. Neglect any losses.

8. A $20\,\Omega$ resistance is connected across the secondary winding of a single-phase power transformer whose secondary voltage is 150 V. Calculate the primary voltage and the turns ratio if the supply current is 5 A, neglecting losses.

Exercise 94   Short answer questions on electromagnetic induction

1. What is electromagnetic induction?
2. State Faraday's laws of electromagnetic induction
3. State Lenz's law
4. Explain briefly the principle of the generator
5. The direction of an induced e.m.f. in a generator may be determined using Fleming's .... rule
6. The e.m.f. E induced in a moving conductor may be calculated using the formula $E = Blv$. Name the quantities represented and their units
7. What is self-inductance? State its symbol and unit
8. What is mutual inductance? State its symbol and unit
9. The mutual inductance between two coils is M. The e.m.f. E_2 induced in one coil by the current changing at $\frac{dI_1}{dt}$ in the other is given by: $E_2 = \ldots\ldots$ volts
10. What is a transformer?
11. Explain briefly how a voltage is induced in the secondary winding of a transformer
12. Draw the circuit diagram symbol for a transformer
13. State the relationship between turns and voltage ratios for a transformer
14. How is a transformer rated?
15. Briefly describe the principle of operation of a transformer

Exercise 95   Multi-choice questions on electromagnetic induction (Answers on page 306)

1. A current changing at a rate of 5 A/s in a coil of inductance 5 H induces an e.m.f. of:
    (a) 25 V in the same direction as the applied voltage
    (b) 1 V in the same direction as the applied voltage
    (c) 25 V in the opposite direction to the applied voltage
    (d) 1 V in the opposite direction to the applied voltage

2. A bar magnet is moved at a steady speed of 1.0 m/s towards a coil of wire which is connected to a centre-zero galvanometer. The magnet is now withdrawn along the same path at 0.5 m/s. The deflection of the galvanometer is in the:
    (a) same direction as previously, with the magnitude of the deflection doubled
    (b) opposite direction as previously, with the magnitude of the deflection halved

(c) same direction as previously, with the magnitude of the deflection halved

(d) opposite direction as previously, with the magnitude of the deflection doubled

3. An e.m.f. of 1 V is induced in a conductor moving at 10 cm/s in a magnetic field of 0.5 T. The effective length of the conductor in the magnetic field is:

(a) 20 cm   (b) 5 m   (c) 20 m   (d) 50 m

4. Which of the following is false?

(a) Fleming's left-hand rule or Lenz's law may be used to determine the direction of an induced e.m.f.

(b) An induced e.m.f. is set up whenever the magnetic field linking that circuit changes

(c) The direction of an induced e.m.f. is always such as to oppose the effect producing it

(d) The induced e.m.f. in any circuit is proportional to the rate of change of the magnetic flux linking the circuit

5. A strong permanent magnet is plunged into a coil and left in the coil. What is the effect produced on the coil after a short time?

(a) The coil winding becomes hot

(b) The insulation of the coil burns out

(c) A high voltage is induced

(d) There is no effect

6. Self-inductance occurs when:

(a) the current is changing

(b) the circuit is changing

(c) the flux is changing

(d) the resistance is changing

7. Faraday's laws of electromagnetic induction are related to:

(a) the e.m.f. of a chemical cell

(b) the e.m.f. of a generator

(c) the current flowing in a conductor

(d) the strength of a magnetic field

8. The mutual inductance between two coils, when a current changing at 20 A/s in one coil induces an e.m.f. of 10 mV in the other, is:

(a) 0.5 H   (b) 200 mH   (c) 0.5 mH   (d) 2 H

9. A transformer has 800 primary turns and 100 secondary turns. To obtain 40 V from the secondary winding the voltage applied to the primary winding must be:

(a) 5 V   (b) 20 V   (c) 2.5 V   (d) 320 V

10. A step-up transformer has a turns ratio of 10. If the output current is 5 A, the input current is:

(a) 50 A   (b) 5 A   (c) 2.5 A   (d) 0.5 A

11. A 440 V/110 V transformer has 1000 turns on the primary winding. The number of turns on the secondary is:

(a) 550   (b) 250   (c) 4000   (d) 25

12. A 1 kV/250 V transformer has 500 turns on the secondary winding. The number of turns on the primary is:

(a) 2000   (b) 125   (c) 1000   (d) 250

13. The power input to a mains transformer is 200 W. If the primary current is 2.5 A, the secondary voltage is 2 V and assuming no losses in the transformer, the turns ratio is:

(a) 80:1 step up   (b) 40:1 step up

(c) 80:1 step down   (d) 40:1 step down

14. An ideal transformer has a turns ratio of 1:5 and is supplied at 200 V when the primary current is 3 A. Which of the following statements is false?

(a) The turns ratio indicates a step-up transformer

(b) The secondary voltage is 40 V

(c) The secondary current is 15 A

(d) The transformer rating is 0.6 kVA

(e) The secondary voltage is 1 kV

(f) The secondary current is 0.6 A

Assignment 7

This assignment covers the material contained Chapter 19 to 21. The marks for each question are shown in brackets at the end of each question.

1. Determine the value of currents I_1 to I_5 shown in Fig. A7.1.   (5)

2. Find the current flowing in the 5 Ω resistor of the circuit shown in Fig. A7.2 using Kirchhoff's laws. Find also the current flowing in each of the other two branches of the circuit.   (11)

3. The maximum working flux density of a lifting electromagnet is 1.7 T and the effective area of a pole face is circular in cross-section. If the total magnetic flux produced is 214 mWb, determine the radius of the pole face.   (6)

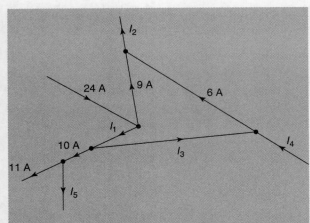

Fig. A7.1

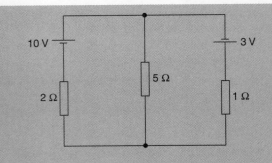

Fig. A7.2

4. A conductor, 25 cm long is situtated at right angles to a magnetic field. Determine the strength of the magnetic field if a current of 12 A in the conductor produces a force on it of 4.5 N.   (5)

5. An electron in a television tube has a charge of 1.5×10^{-19} C and travels at 3×10^7 m/s perpendicular to a field of flux density 20 μT. Calculate the force exerted on the electron in the field.   (5)

6. A lorry is travelling at 100 km/h. Assuming the vertical component of the earth's magnetic field is 40 μT and the back axle of the lorry is 1.98 m, find the e.m.f. generated in the axle due to motion.   (6)

7. Two coils, *P* and *Q*, have a mutual inductance of 100 mH. If a current 3 A in coil *P* is reversed in 20 ms, determine (a) the average e.m.f. induced in coil *Q*, and (b) the flux change linked with coil *Q* if it wound with 200 turns.   (6)

8. An ideal transformer connected to a 250 V mains, supplies a 25 V, 200 W lamp. Calculate the transformer turns ratio and the current taken from the supply. (6)

22

Alternating voltages and currents

At the end of this chapter you should be able to:

- appreciate why a.c. is used in preference to d.c.
- describe the principle of operation of an a.c. generator
- distinguish between unidirectional and alternating waveforms
- define cycle, period or periodic time T and frequency f of a waveform
- perform calculations involving $T = \dfrac{1}{f}$
- define instantaneous, peak, mean and r.m.s. values, and form and peak factors for a sine wave
- calculate mean and r.m.s. values and form and peak factors for given waveforms

22.1 Introduction

Electricity is produced by generators at power stations and then distributed by a vast network of transmission lines (called the National Grid system) to industry and for domestic use. It is easier and cheaper to generate **alternating current (a.c.)** than direct current (d.c.) and a.c. is more conveniently distributed than d.c. since its voltage can be readily altered using transformers. Whenever d.c. is needed in preference to a.c., devices called **rectifiers** are used for conversion.

22.2 The a.c. generator

Let a single turn coil be free to rotate at constant angular velocity symmetrically between the poles of a magnet system as shown in Fig. 22.1.

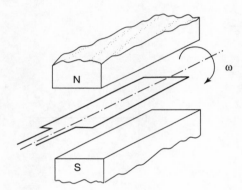

Fig. 22.1

An e.m.f. is generated in the coil (from Faraday's Laws) which varies in magnitude and reverses its direction at regular intervals. The reason for this is shown in Fig. 22.2. In positions (a), (e) and (i) the conductors of the loop are effectively moving along the magnetic field, no flux is cut and hence no e.m.f. is induced. In position (c) maximum flux is cut and hence maximum e.m.f. is induced. In position (g), maximum flux is cut and hence maximum e.m.f. is again induced. However, using Fleming's right-hand rule, the induced e.m.f. is in the opposite direction to that in position (c) and is thus shown as $-E$. In positions (b), (d), (f) and (h) some flux is cut and hence some e.m.f. is induced. If all such positions of the coil are considered, in one revolution of the coil, one cycle of alternating e.m.f. is produced as shown. This is the principle of operation of the **a.c. generator** (i.e. the **alternator**).

22.3 Waveforms

If values of quantities which vary with time t are plotted to a base of time, the resulting graph is called a **waveform**. Some typical waveforms are shown in Fig. 22.3. Waveforms (a)

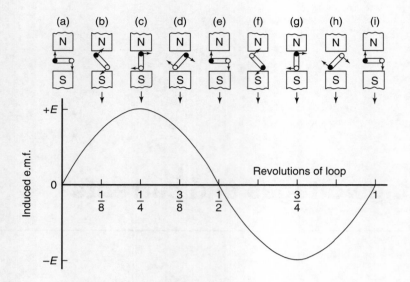

Fig. 22.2

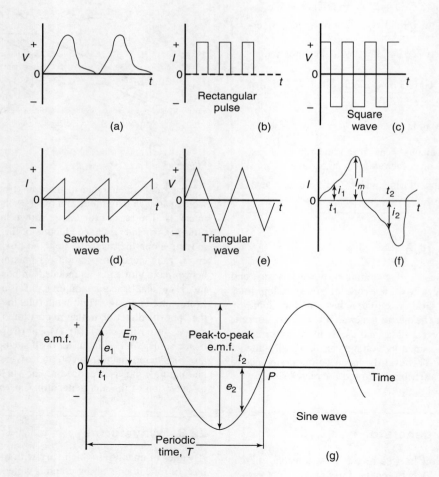

Fig. 22.3

and (b) are **unidirectional waveforms**, for, although they vary considerably with time, they flow in one direction only (i.e. they do not cross the time axis and become negative). Waveforms (c) to (g) are called **alternating waveforms** since their quantities are continually changing in direction (i.e. alternately positive and negative).

A waveform of the type shown in Fig. 22.3(g) is called a **sine wave**. It is the shape of the waveform of e.m.f. produced by an alternator and thus the mains electricity supply is of 'sinusoidal' form.

One complete series of values is called a **cycle** (i.e. from O to P in Fig. 22.3(g)).

The time taken for an alternating quantity to complete one cycle is called the **period** or the **periodic time**, T, of the waveform.

The number of cycles completed in one second is called the **frequency**, f, of the supply and is measured in **hertz, Hz**. The standard frequency of the electricity supply in Great Britain is 50 Hz.

$$T = \frac{1}{f} \quad \text{or} \quad f = \frac{1}{T}$$

Problem 1. Determine the periodic time for frequencies of (a) 50 Hz and (b) 20 kHz.

(a) Periodic time $T = \dfrac{1}{f} = \dfrac{1}{50} = 0.02\,\text{s}$ or **20 ms**

(b) Periodic time $T = \dfrac{1}{f} = \dfrac{1}{20\,000} = 0.00005\,\text{s}$ or **50 µs**

Problem 2. Determine the frequencies for periodic times of (a) 4 ms (b) 8 µs.

(a) Frequency $f = \dfrac{1}{T} = \dfrac{1}{4 \times 10^{-3}} = \dfrac{1000}{4} = \textbf{250 Hz}$

(b) Frequency $f = \dfrac{1}{T} = \dfrac{1}{8 \times 10^{-6}} = \dfrac{1\,000\,000}{8}$

$$= \textbf{125\,000 Hz} \quad \text{or} \quad \textbf{125 kHz} \quad \text{or}$$
$$\textbf{0.125 MHz}$$

Problem 3. An alternating current completes 5 cycles in 8 ms. What is its frequency?

Time for one cycle $= \frac{8}{5}$ ms $= 1.6$ ms $=$ periodic time T.

Frequency $f = \dfrac{1}{T} = \dfrac{1}{1.6 \times 10^{-3}} = \dfrac{1000}{1.6} = \dfrac{10\,000}{16}$

$$= \textbf{625 Hz}$$

Now try the following exercise

Exercise 96 Further problems on frequency and periodic time (Answers on page 306)

1. Determine the periodic time for the following frequencies:

   (a) 2.5 Hz   (b) 100 Hz   (c) 40 kHz

2. Calculate the frequency for the following periodic times:

   (a) 5 ms   (b) 50 µs   (c) 0.2 s

3. An alternating current completes 4 cycles in 5 ms. What is its frequency?

22.4 A.C. values

Instantaneous values are the values of the alternating quantities at any instant of time. They are represented by small letters, i, v, e, etc., (see Figs. 22.3(f) and (g)).

The largest value reached in a half cycle is called the **peak value** or the **maximum value** or the **crest value** or the **amplitude** of the waveform. Such values are represented by V_m, I_m, E_m, etc. (see Figs. 22.3(f) and (g)). A **peak-to-peak** value of e.m.f. is shown in Fig. 22.3(g) and is the difference between the maximum and minimum values in a cycle.

The **average** or **mean value** of a symmetrical alternating quantity, (such as a sine wave), is the average value measured over a half cycle, (since over a complete cycle the average value is zero).

$$\text{Average or mean value} = \frac{\text{area under the curve}}{\text{length of base}}$$

The area under the curve is found by approximate methods such as the trapezoidal rule, the mid-ordinate rule or Simpson's rule. Average values are represented by V_{AV}, I_{AV}, E_{AV}, etc.

For a sine wave:

average value $= 0.637 \times$ maximum value

$\left(\text{i.e. } \dfrac{2}{\pi} \times \text{maximum value}\right)$

The **effective value** of an alternating current is that current which will produce the same heating effect as an equivalent direct current. The effective value is called the **root mean square (r.m.s.) value** and whenever an alternating quantity is given, it is assumed to be the rms value. For example, the domestic mains supply in Great Britain is 240 V and is assumed to mean '240 V rms'. The symbols used for r.m.s. values are I, V, E, etc. For a non-sinusoidal waveform as shown in Fig. 22.4 the r.m.s. value is given by:

$$I = \sqrt{\left(\frac{i_1^2 + i_2^2 + \cdots + i_n^2}{n}\right)}$$

where n is the number of intervals used.

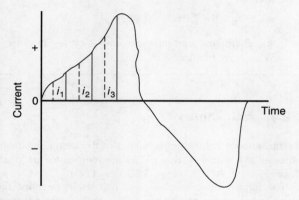

Fig. 22.4

> **For a sine wave:**
>
> **rms value = 0.707 × maximum value**
>
> $$\left(\text{i.e. } \frac{1}{\sqrt{2}} \times \text{maximum value}\right)$$

> $$\text{Form factor} = \frac{\text{r.m.s. value}}{\text{average value}}$$

For a sine wave, form factor = 1.11

> $$\text{Peak factor} = \frac{\text{maximum value}}{\text{r.m.s. value}}$$

For a sine wave, peak factor = 1.41.

The values of form and peak factors give an indication of the shape of waveforms.

Problem 4. For the periodic waveforms shown in Fig. 22.5 determine for each: (i) the frequency, (ii) the average value over half a cycle, (iii) the r.m.s. value, (iv) the form factor, and (v) the peak factor.

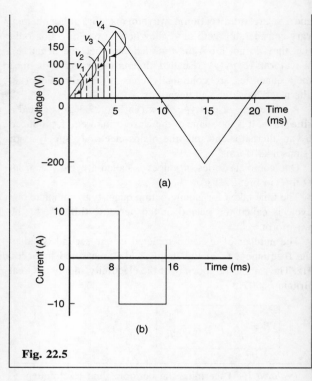

Fig. 22.5

(a) **Triangular waveform** (Fig. 22.5(a))

(i) Time for one complete cycle = 20 ms = periodic time, T

Hence frequency $f = \dfrac{1}{T} = \dfrac{1}{20 \times 10^{-3}}$

$$= \frac{1000}{20} = \textbf{50 Hz}$$

(ii) Area under the triangular waveform for a half cycle = $\frac{1}{2} \times$ base $\times$ height $= \frac{1}{2} \times (10 \times 10^{-3}) \times 200$

$$= 1 \text{ volt second}$$

Average value of waveform $= \dfrac{\text{area under curve}}{\text{length of base}}$

$$= \frac{1 \text{ volt second}}{10 \times 10^{-3} \text{ second}}$$

$$= \frac{1000}{10} = \textbf{100 V}$$

(iii) In Fig. 22.5(a), the first $\frac{1}{4}$ cycle is divided into 4 intervals

Thus rms value $= \sqrt{\left(\dfrac{v_1^2 + v_2^2 + v_3^2 + v_4^2}{4}\right)}$

$$= \sqrt{\left(\frac{25^2 + 75^2 + 125^2 + 175^2}{4}\right)}$$

$$= \textbf{114.6 V}$$

(Note that the greater the number of intervals chosen, the greater the accuracy of the result. For example, if twice the number of ordinates as that chosen above are used, the r.m.s. value is found to be 115.6 V)

(iv) Form factor $= \dfrac{\text{r.m.s. value}}{\text{average value}} = \dfrac{114.6}{100} = \mathbf{1.15}$

(v) Peak factor $= \dfrac{\text{maximum value}}{\text{r.m.s. value}} = \dfrac{200}{114.6} = \mathbf{1.75}$

(b) **Rectangular waveform** (Fig. 22.5(b))

(i) Time for one complete cycle $= 16\,\text{ms} =$ periodic time, T

Hence frequency, $f = \dfrac{1}{T} = \dfrac{1}{16 \times 10^{-3}} = \dfrac{1000}{16}$

$= \mathbf{62.5\,Hz}$

(ii) Average value over half a cycle

$= \dfrac{\text{area under curve}}{\text{length of base}} = \dfrac{10 \times (8 \times 10^{-3})}{8 \times 10^{-3}} = \mathbf{10\,A}$

(iii) The r.m.s. value $= \sqrt{\left(\dfrac{i_1^2 + i_2^2 + i_3^2 + i_4^2}{4}\right)} = \mathbf{10\,A}$

however many intervals are chosen, since the waveform is rectangular.

(iv) Form factor $= \dfrac{\text{r.m.s. value}}{\text{average value}} = \dfrac{10}{10} = \mathbf{1}$

(v) Peak factor $= \dfrac{\text{maximum value}}{\text{r.m.s. value}} = \dfrac{10}{10} = \mathbf{1}$

Problem 5. The following table gives the corresponding values of current and time for a half cycle of alternating current:

time t (ms)	0	0.5	1.0	1.5	2.0	2.5
current i (A)	0	7	14	23	40	56

time t (ms)	3.0	3.5	4.0	4.5	5.0
current i (A)	68	76	60	5	0

Assuming the negative half cycle is identical in shape to the positive half cycle, plot the waveform and find (a) the frequency of the supply, (b) the instantaneous values of current after 1.25 ms and 3.8 ms, (c) the peak or maximum value, (d) the mean or average value, and (e) the r.m.s. value of the waveform.

The half cycle of alternating current is shown plotted in Fig. 22.6.

(a) Time for a half cycle $= 5\,\text{ms}$; hence the time for one cycle, i.e. the periodic time, $T = 10\,\text{ms}$ or $0.01\,\text{s}$

Frequency, $f = \dfrac{1}{T} = \dfrac{1}{0.01} = \mathbf{100\,Hz}$

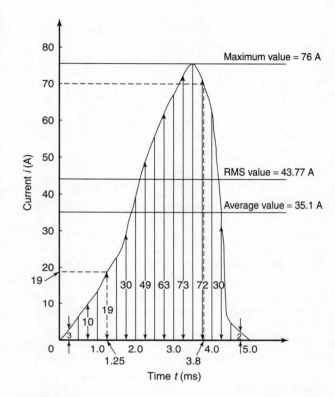

Fig. 22.6

(b) Instantaneous value of current after 1.25 ms is **19 A**, from Fig. 22.6
Instantaneous value of current after 3.8 ms is **70 A**, from Fig. 22.6

(c) Peak or maximum value $= \mathbf{76\,A}$

(d) Mean or average value $= \dfrac{\text{area under curve}}{\text{length of base}}$

Using the mid-ordinate rule with 10 intervals, each of width 0.5 ms gives: area under curve

$= (0.5 \times 10^{-3})[3 + 10 + 19 + 30 + 49 + 63$

$+\, 73 + 72 + 30 + 2]$ (see Fig. 22.6)

$= (0.5 \times 10^{-3})(351)$

Hence mean or average value $= \dfrac{(0.5 \times 10^{-3})(351)}{5 \times 10^{-3}}$

$= \mathbf{35.1\,A}$

(e) R.m.s. value $= \sqrt{\left(\dfrac{\begin{array}{c}3^2 + 10^2 + 19^2 + 30^2 + 49^2 + 63^2 \\ +\, 73^2 + 72^2 + 30^2 + 2^2\end{array}}{10}\right)}$

$= \sqrt{\left(\dfrac{19\,157}{10}\right)} = \mathbf{43.8\,A}$

Problem 6. Calculate the r.m.s. value of a sinusoidal current of maximum value 20 A

For a sine wave, r.m.s. value = 0.707 × maximum value

$$= 0.707 \times 20 = \mathbf{14.14\,A}$$

Problem 7. Determine the peak and mean values for a 240 V mains supply.

For a sine wave, r.m.s. value of voltage $V = 0.707 \times V_m$

A 240 V mains supply means that 240 V is the r.m.s. value,

hence $V_m = \dfrac{V}{0.707} = \dfrac{240}{0.707} = \mathbf{339.5\,V = peak\ value}$

Mean value $V_{AV} = 0.637 V_m = 0.637 \times 339.5 = \mathbf{216.3\,V}$

Problem 8. A supply voltage has a mean value of 150 V. Determine its maximum value and its r.m.s. value.

For a sine wave, mean value = 0.637 × maximum value

Hence **maximum value** $= \dfrac{\text{mean value}}{0.637} = \dfrac{150}{0.637} = \mathbf{235.5\,V}$

R.m.s. value = 0.707 × maximum value = 0.707 × 235.5

$$= \mathbf{166.5\,V}$$

Now try the following exercises

Exercise 97   Further problems on a.c. values of waveforms (Answers on page 306)

1. An alternating current varies with time over half a cycle as follows:

Current (A)	0	0.7	2.0	4.2	8.4	8.2
time (ms)	0	1	2	3	4	5

Current (A)	2.5	1.0	0.4	0.2	0
time (ms)	6	7	8	9	10

   The negative half cycle is similar. Plot the curve and determine:
   (a) the frequency
   (b) the instantaneous values at 3.4 ms and 5.8 ms
   (c) its mean value and
   (d) its r.m.s. value

2. For the waveforms shown in Fig. 22.7 determine for each (i) the frequency (ii) the average value over half a cycle (iii) the r.m.s. value (iv) the form factor (v) the peak factor.

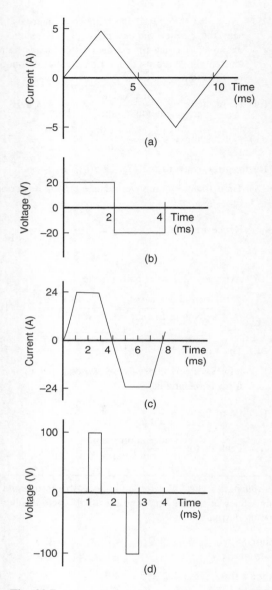

Fig. 22.7

3. An alternating voltage is triangular in shape, rising at a constant rate to a maximum of 300 V in 8 ms and then falling to zero at a constant rate in 4 ms. The negative half cycle is identical in shape to the positive half cycle. Calculate (a) the mean voltage over half a cycle, and (b) the r.m.s. voltage.

4. Calculate the r.m.s. value of a sinusoidal curve of maximum value 300 V.

5. Find the peak and mean values for a 200 V mains supply.

6. A sinusoidal voltage has a maximum value of 120 V. Calculate its r.m.s. and average values.

7. A sinusoidal current has a mean value of 15.0 A. Determine its maximum and r.m.s. values.

Exercise 98 Short answer questions on alternating voltages and currents

1. Briefly explain the principle of operation of the simple alternator

2. What is meant by (a) waveform (b) cycle

3. What is the difference between an alternating and a unidirectional waveform?

4. The time to complete one cycle of a waveform is called the ......

5. What is frequency? Name its unit

6. The mains supply voltage has a special shape of waveform called a ......

7. Define peak value

8. What is meant by the r.m.s. value?

9. What is the mean value of a sinusoidal alternating e.m.f. which has a maximum value of 100 V?

10. The effective value of a sinusoidal waveform is ...... × maximum value.

Exercise 99 Multi-choice questions on alternating voltages and currents (Answers on page 306)

1. The number of complete cycles of an alternating current occurring in one second is known as:

   (a) the maximum value of the alternating current
   (b) the frequency of the alternating current
   (c) the peak value of the alternating current
   (d) the r.m.s. or effective value

2. The value of an alternating current at any given instant is:
   (a) a maximum value   (b) a peak value
   (c) an instantaneous value (d) an r.m.s. value

3. An alternating current completes 100 cycles in 0.1 s. Its frequency is:

   (a) 20 Hz (b) 100 Hz (c) 0.002 Hz (d) 1 kHz

4. In Fig. 22.8, at the instant shown, the generated e.m.f. will be:

   (a) zero   (b) an r.m.s. value
   (c) an average value   (d) a maximum value

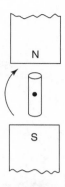

Fig. 22.8

5. The supply of electrical energy for a consumer is usually by a.c. because:
   (a) transmission and distribution are more easily effected
   (b) it is most suitable for variable speed motors
   (c) the volt drop in cables is minimal
   (d) cable power losses are negligible

6. Which of the following statements is false?

   (a) It is cheaper to use a.c. than d.c.
   (b) Distribution of a.c. is more convenient than with d.c. since voltages may be readily altered using transformers
   (c) An alternator is an a.c. generator
   (d) A rectifier changes d.c. to a.c.

7. An alternating voltage of maximum value 100 V is applied to a lamp. Which of the following direct voltages, if applied to the lamp, would cause the lamp to light with the same brilliance?

   (a) 100 V   (b) 63.7 V   (c) 70.7 V   (d) 141.4 V

8. The value normally stated when referring to alternating currents and voltages is the:

   (a) instantaneous value (b) r.m.s. value
   (c) average value   (d) peak value

9. State which of the following is false. For a sine wave:

   (a) the peak factor is 1.414
   (b) the r.m.s. value is 0.707 × peak value
   (c) the average value is 0.637 × r.m.s. value
   (d) the form factor is 1.11

10. An a.c. supply is 70.7 V, 50 Hz. Which of the following statements is false?

   (a) The periodic time is 20 ms
   (b) The peak value of the voltage is 70.7 V
   (c) The r.m.s. value of the voltage is 70.7 V
   (d) The peak value of the voltage is 100 V

23

Electrical measuring instruments and measurements

At the end of this chapter you should be able to:

- recognise the importance of testing and measurements in electric circuits

- appreciate the essential devices comprising an analogue instrument

- explain the operation of attraction and repulsion type moving-iron instruments

- explain the operation of a moving-coil rectifier instrument

- compare moving-coil, moving-iron and moving coil rectifier instruments

- calculate values of shunts for ammeters and multipliers for voltmeters

- understand the advantages of electronic instruments

- understand the operation of an ohmmeter/megger

- appreciate the operation of multimeters/Avometers

- understand the operation of a wattmeter

- appreciate instrument 'loading' effect

- understand the operation of a C.R.O. for d.c. and a.c. measurements

- calculate periodic time, frequency, peak to peak values from waveforms on a C.R.O.

- understand null methods of measurement for a Wheatstone bridge and d.c. potentiometer

- appreciate the most likely source of errors in measurements

- appreciate calibration accuracy of instruments

23.1 Introduction

Tests and measurements are important in designing, evaluating, maintaining and servicing electrical circuits and equipment. In order to detect electrical quantities such as current, voltage, resistance or power, it is necessary to transform an electrical quantity or condition into a visible indication. This is done with the aid of instruments (or meters) that indicate the magnitude of quantities either by the position of a pointer moving over a graduated scale (called an analogue instrument) or in the form of a decimal number (called a digital instrument).

23.2 Analogue instruments

All analogue electrical indicating instruments require three essential devices:

(a) A **deflecting or operating device**. A mechanical force is produced by the current or voltage which causes the pointer to deflect from its zero position.

(b) **A controlling device**. The controlling force acts in opposition to the deflecting force and ensures that the deflection shown on the meter is always the same for a given measured quantity. It also prevents the pointer always going to the maximum deflection. There are two main types of controlling device – spring control and gravity control.

(c) **A damping device**. The damping force ensures that the pointer comes to rest in its final position quickly and without undue oscillation. There are three main types of damping used – eddy-current damping, air-friction damping and fluid-friction damping.

There are basically **two types of scale** – linear and non-linear.

A **linear scale** is shown in Fig. 23.1(a), where the divisions or graduations are evenly spaced. The voltmeter shown has a range 0–100 V, i.e. a full-scale deflection (f.s.d.) of 100 V. A **non-linear scale** is shown in Fig. 23.1(b) where the scale is cramped at the beginning and the graduations are uneven throughout the range. The ammeter shown has a f.s.d. of 10 A.

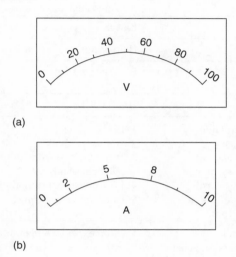

(a)

(b)

Fig. 23.1

23.3 Moving-iron instrument

(a) An **attraction type** of moving-iron instrument is shown diagrammatically in Fig. 23.2(a). When current flows in the solenoid, a pivoted soft-iron disc is attracted towards the solenoid and the movement causes a pointer to move across a scale.

(b) In the **repulsion type** moving-iron instrument shown diagrammatically in Fig. 23.2(b), two pieces of iron are placed inside the solenoid, one being fixed, and the other attached to the spindle carrying the pointer. When current passes through the solenoid, the two pieces of iron are magnetised in the same direction and therefore repel each other. The pointer thus moves across the scale. The force moving the pointer is, in each type, proportional to I^2 and because of this the direction of current does not matter. The moving-iron instrument can be used on d.c. or a.c.; the scale, however, is non-linear.

23.4 The moving-coil rectifier instrument

A moving-coil instrument, which measures only d.c., may be used in conjunction with a bridge rectifier circuit as shown

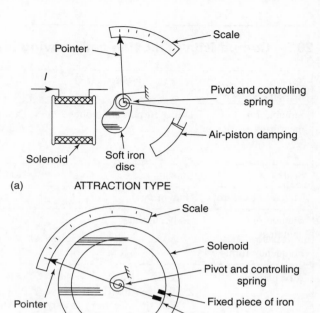

(a)   ATTRACTION TYPE

(b)   REPULSION TYPE

Fig. 23.2

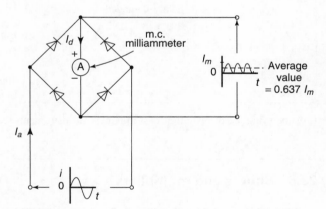

Fig. 23.3

in Fig. 23.3 to provide an indication of alternating currents and voltages (see Chapter 22). The average value of the full wave rectified current is $0.637 I_m$. However, a meter being used to measure a.c. is usually calibrated in r.m.s. values. For sinusoidal quantities the indication is $\dfrac{0.707 I_m}{0.637 I_m}$ i.e. 1.11 times the mean value.

Rectifier instruments have scales calibrated in r.m.s. quantities and it is assumed by the manufacturer that the a.c. is sinusoidal.

23.5   Comparison of moving-coil, moving-iron and moving-coil rectifier instruments

Type of instrument	Moving-coil	Moving-iron	Moving-coil rectifier
Suitable for measuring	Direct current and voltage	Direct and alternating currents and voltage (reading in r.m.s. value)	Alternating current and voltage (reads average value but scale is adjusted to give r.m.s. value for sinusoidal waveforms)
Scale	Linear	Non-linear	Linear
Method of control	Hairsprings	Hairsprings	Hairsprings
Method of damping	Eddy current	Air	Eddy current
Frequency limits	–	20–200 Hz	20–100 kHz
Advantages	1 Linear scale 2 High sensitivity 3 Well shielded from stray magnetic fields 4 Lower power consumption	1 Robust construction 2 Relatively cheap 3 Measures dc and ac 4 In frequency range 20–100 Hz reads rms correctly regardless of supply wave-form	1 Linear scale 2 High sensitivity 3 Well shielded from stray magnetic fields 4 Low power consumption 5 Good frequency range
Disadvantages	1 Only suitable for dc 2 More expensive than moving iron type 3 Easily damaged	1 Non-linear scale 2 Affected by stray magnetic fields 3 Hysteresis errors in dc circuits 4 Liable to temperature errors 5 Due to the inductance of the solenoid, readings can be affected by variation of frequency	1 More expensive than moving iron type 2 Errors caused when supply is non-sinusoidal

(For the principle of operation of a moving-coil instrument, see Chapter 20, page 162).

23.6   Shunts and multipliers

An **ammeter**, which measures current, has a low resistance (ideally zero) and must be connected in series with the circuit.

A **voltmeter**, which measures p.d., has a high resistance (ideally infinite) and must be connected in parallel with the part of the circuit whose p.d. is required.

There is no difference between the basic instrument used to measure current and voltage since both use a milliammeter as their basic part. This is a sensitive instrument which gives f.s.d. for currents of only a few milliamperes. When an ammeter is required to measure currents of larger magnitude, a proportion of the current is diverted through a low-value resistance connected in parallel with the meter. Such a diverting resistor is called a **shunt**.

From Fig. 23.4(a), $V_{PQ} = V_{RS}$. Hence $I_a r_a = I_S R_S$

Thus the value of the shunt,

$$R_S = \frac{I_a r_a}{I_S} \text{ ohms}$$

The milliammeter is converted into a voltmeter by connecting a high value resistance (called a **multiplier**) in series with it as shown in Fig. 23.4(b).

From Fig. 23.4(b),

$$V = V_a + V_M = I r_a + I R_M$$

Thus the value of the multiplier,

$$R_M = \frac{V - I r_a}{I} \text{ ohms}$$

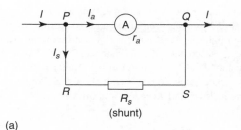

(a)

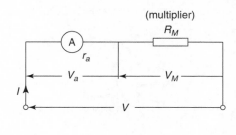

(b)

Fig. 23.4

Problem 1. A moving-coil instrument gives a f.s.d. when the current is 40 mA and its resistance is 25 Ω. Calculate the value of the shunt to be connected in parallel with the meter to enable it to be used as an ammeter for measuring currents up to 50 A.

The circuit diagram is shown in Fig. 23.5,

where r_a = resistance of instrument = 25 Ω,
R_s = resistance of shunt,
I_a = maximum permissible current flowing in instrument = 40 mA = 0.04 A,
I_s = current flowing in shunt and
I = total circuit current required to gives f.s.d. = 50 A

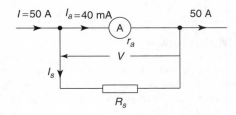

Fig. 23.5

Since $I = I_a + I_s$

then $I_s = I - I_a = 50 - 0.04 = 49.96$ A

$V = I_a r_a = I_s R_s$

Hence $R_s = \dfrac{I_a r_a}{I_s} = \dfrac{(0.04)(25)}{49.96} = 0.02002\ \Omega = \mathbf{20.02\,m\Omega}$

Thus for the moving-coil instrument to be used as an ammeter with a range 0–50 A, a resistance of value 20.02 mΩ needs to be connected in parallel with the instrument.

Problem 2. A moving-coil instrument having a resistance of 10 Ω, gives a f.s.d. when the current is 8 mA. Calculate the value of the multiplier to be connected in series with the instrument so that it can be used as a voltmeter for measuring p.d.s. up to 100 V.

The circuit diagram is shown in Fig. 23.6,

where r_a = resistance of instrument = 10 Ω,
R_M = resistance of multiplier,
I = total permissible instrument current = 8 mA
= 0.008 A,
V = total p.d. required to give f.s.d. = 100 V

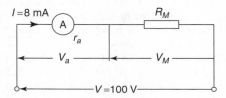

Fig. 23.6

$$V = V_a + V_M = I r_a + I R_M$$

i.e. $100 = (0.008)(10) + (0.008)R_M$

or $100 - 0.08 = 0.008 R_M$

thus $R_M = \dfrac{99.92}{0.008} = 12\,490\ \Omega = \mathbf{12.49\,k\Omega}$

Hence for the moving-coil instrument to be used as a voltmeter with a range 0–100 V, a resistance of value 12.49 kΩ needs to be connected in series with the instrument.

Now try the following exercise

Exercise 100   Further problems on shunts and multipliers (Answers on page 307)

1. A moving-coil instrument gives f.s.d. for a current of 10 mA. Neglecting the resistance of the instrument, calculate the approximate value of series resistance needed to enable the instrument to measure up to

(a) 20 V   (b) 100 V   (c) 250 V

2. A meter of resistance $50\,\Omega$ has a f.s.d. of $4\,mA$. Determine the value of shunt resistance required in order that f.s.d. should be
   (a) $15\,mA$ (b) $20\,A$ (c) $100\,A$

3. A moving-coil instrument having a resistance of $20\,\Omega$, gives a f.s.d. when the current is $5\,mA$. Calculate the value of the multiplier to be connected in series with the instrument so that it can be used as a voltmeter for measuring p.d.'s up to $200\,V$.

4. A moving-coil instrument has a f.s.d. of $20\,mA$ and a resistance of $25\,\Omega$. Calculate the values of resistance required to enable the instrument to be used (a) as a $0–10\,A$ ammeter, and (b) as a $0–100\,V$ voltmeter. State the mode of resistance connection in each case.

23.7 Electronic instruments

Electronic measuring instruments have advantages over instruments such as the moving-iron or moving-coil meters, in that they have a much higher input resistance (some as high as $1000\,M\Omega$) and can handle a much wider range of frequency (from d.c. up to MHz).

The **digital voltmeter (DVM)** is one which provides a digital display of the voltage being measured. Advantages of a DVM over analogue instruments include higher accuracy and resolution, no observational or parallex errors (see Section 23.16) and a very high input resistance, constant on all ranges. A digital multimeter is a DVM with additional circuitry which makes it capable of measuring a.c. voltage, d.c. and a.c. current and resistance.

Instruments for a.c. measurements are generally calibrated with a sinusoidal alternating waveform to indicate r.m.s. values when a sinusoidal signal is applied to the instrument. Some instruments, such as the moving-iron and electrodynamic instruments, give a true r.m.s. indication. With other instruments the indication is either scaled up from the mean value (such as with the rectified moving-coil instrument) or scaled down from the peak value. Sometimes quantities to be measured have complex waveforms, and whenever a quantity is non-sinusoidal, errors in instrument readings can occur if the instrument has been calibrated for sine waves only. Such waveform errors can be largely eliminated by using electronic instruments.

23.8 The ohmmeter

An **ohmmeter** is an instrument for measuring electrical resistance. A simple ohmmeter circuit is shown in Fig. 23.7(a). Unlike the ammeter or voltmeter, the ohmmeter circuit does not receive the energy necessary for its operation from the circuit under test. In the ohmmeter this energy is supplied by

a self-contained source of voltage, such as a battery. Initially, terminals XX are short-circuited and R adjusted to give f.s.d. on the milliammeter. If current I is at a maximum value and voltage E is constant, then resistance $R = E/I$ is at a minimum value. Thus f.s.d. on the milliammeter is made zero on the resistance scale. When terminals XX are open circuited no current flows and $R(= E/O)$ is infinity, ∞.

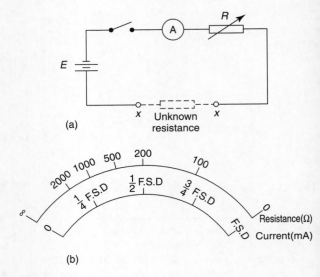

(a)

(b)

Fig. 23.7

The milliammeter can thus be calibrated directly in ohms. A cramped (non-linear) scale results and is 'back to front', as shown in Fig. 23.7(b). When calibrated, an unknown resistance is placed between terminals XX and its value determined from the position of the pointer on the scale. An ohmmeter designed for measuring low values of resistance is called a **continuity tester**. An ohmmeter designed for measuring high values of resistance (i.e. megohms) is called an **insulation resistance tester** (e.g. **'Megger'**).

23.9 Multimeters

Instruments are manufactured that combine a moving-coil meter with a number of shunts and series multipliers, to provide a range of readings on a single scale graduated to read current and voltage. If a battery is incorporated then resistance can also be measured. Such instruments are called **multimeters** or **universal instruments** or **multirange instruments**. An **'Avometer'** is a typical example. A particular range may be selected either by the use of separate terminals or by a selector switch. Only one measurement can be performed at a time. Often such instruments can be used in a.c. as well as d.c. circuits when a rectifier is incorporated in the instrument.

23.10 Wattmeters

A **wattmeter** is an instrument for measuring electrical power in a circuit. Fig. 23.8 shows typical connections of a wattmeter used for measuring power supplied to a load. The instrument has two coils:

(i) a current coil, which is connected in series with the load, like an ammeter, and

(ii) a voltage coil, which is connected in parallel with the load, like a voltmeter.

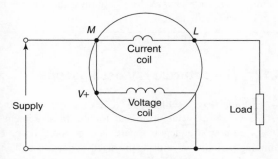

Fig. 23.8

23.11 Instrument 'loading' effect

Some measuring instruments depend for their operation on power taken from the circuit in which measurements are being made. Depending on the 'loading' effect of the instrument (i.e. the current taken to enable it to operate), the prevailing circuit conditions may change.

The resistance of voltmeters may be calculated since each have a stated sensitivity (or 'figure of merit'), often stated in 'kΩ per volt' of f.s.d. A voltmeter should have as high a resistance as possible (– ideally infinite). In a.c. circuits the impedance of the instrument varies with frequency and thus the loading effect of the instrument can change.

Problem 3. Calculate the power dissipated by the voltmeter and by resistor R in Fig. 23.9 when (a) $R = 250\,\Omega$ (b) $R = 2\,\mathrm{M}\Omega$. Assume that the voltmeter sensitivity (or figure of merit) is $10\,\mathrm{k}\Omega/\mathrm{V}$.

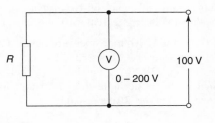

Fig. 23.9

(a) Resistance of voltmeter, R_v = sensitivity × f.s.d.

Hence, $R_v = (10\,\mathrm{k}\Omega/\mathrm{V}) \times (200\,\mathrm{V}) = 2000\,\mathrm{k}\Omega = 2\,\mathrm{M}\Omega$

Current flowing in voltmeter,

$$I_v = \frac{V}{R_v} = \frac{100}{2 \times 10^6} = 50 \times 10^{-6}\,\mathrm{A}$$

Power dissipated by

voltmeter $= VI_v = (100)(50 \times 10^{-6}) = \mathbf{5\,mW}$

When $R = 250\,\Omega$, current in resistor,

$$I_R = \frac{V}{R} = \frac{100}{250} = \mathbf{0.4\,A}$$

Power dissipated in load resistor R

$$= VI_R = (100)(0.4) = \mathbf{40\,W}$$

Thus the power dissipated in the voltmeter is insignificant in comparison with the power dissipated in the load.

(b) When $R = 2\,\mathrm{M}\Omega$, current in resistor,

$$I_R = \frac{V}{R} = \frac{100}{2 \times 10^6} = 50 \times 10^{-6}\,\mathrm{A}$$

Power dissipated in load resistor R

$$= VI_R = 100 \times 50 \times 10^{-6} = \mathbf{5\,mW}$$

In this case the higher load resistance reduced the power dissipated such that the voltmeter is using as much power as the load.

Problem 4. An ammeter has a f.s.d. of 100 mA and a resistance of 50 Ω. The ammeter is used to measure the current in a load of resistance 500 Ω when the supply voltage is 10 V. Calculate (a) the ammeter reading expected (neglecting its resistance), (b) the actual current in the circuit, (c) the power dissipated in the ammeter, and (d) the power dissipated in the load.

From Fig. 23.10,

(a) expected ammeter reading $= \dfrac{V}{R} = \dfrac{10}{500} = \mathbf{20\,mA}$

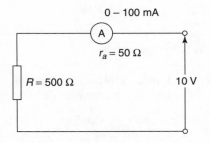

Fig. 23.10

(b) Actual ammeter reading $= \dfrac{V}{R + r_a} = \dfrac{10}{500 + 50}$

$$= \mathbf{18.18\,mA}$$

Thus the ammeter itself has caused the circuit conditions to change from 20 mA to 18.18 mA.

(c) Power dissipated in the ammeter

$$= I^2 r_a = (18.18 \times 10^{-3})^2 (50) = \mathbf{16.53\,mW}$$

(d) Power dissipated in the load resistor

$$= I^2 R = (18.18 \times 10^{-3})^2 (500) = \mathbf{165.3\,mW}$$

Problem 5. (a) A current of 20 A flows through a load having a resistance of 2 Ω. Determine the power dissipated in the load. (b) A wattmeter, whose current coil has a resistance of 0.01 Ω is connected as shown in Fig. 23.13. Determine the wattmeter reading.

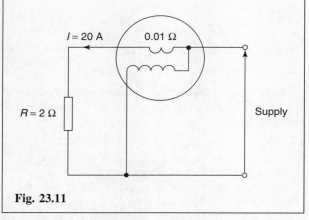

Fig. 23.11

(a) Power dissipated in the load,

$$P = I^2 R = (20)^2 (2) = \mathbf{800\,W}$$

(b) With the wattmeter connected in the circuit the total resistance R_T is $2 + 0.01 = 2.01\,\Omega$

The wattmeter reading is thus

$$I^2 R_T = (20)^2 (2.01) = \mathbf{804\,W}$$

Now try the following exercise

Exercise 101   Further problems on instrument 'load-ing' effects (Answers on page 307)

1. A 0–1 A ammeter having a resistance of 50 Ω is used to measure the current flowing in a 1 kΩ resistor when the supply voltage is 250 V. Calculate:

(a) the approximate value of current (neglecting the ammeter resistance),

(b) the actual current in the circuit,

(c) the power dissipated in the ammeter,

(d) the power dissipated in the 1 kΩ resistor.

2. (a) A current of 15 A flows through a load having a resistance of 4 Ω. Determine the power dissipated in the load. (b) A wattmeter, whose current coil has a resistance of 0.02 Ω is connected (as shown in Fig. 23.11) to measure the power in the load. Determine the wattmeter reading assuming the current in the load is still 15 A.

23.12   The cathode ray oscilloscope

The **cathode ray oscilloscope** (c.r.o.) may be used in the observation of waveforms and for the measurement of voltage, current, frequency, phase and periodic time. For examining periodic waveforms the electron beam is deflected horizontally (i.e. in the X direction) by a sawtooth generator acting as a timebase. The signal to be examined is applied to the vertical deflection system (Y direction) usually after amplification.

Oscilloscopes normally have a transparent grid of 10 mm by 10 mm squares in front of the screen, called a graticule. Among the timebase controls is a 'variable' switch which gives the sweep speed as time per centimetre. This may be in s/cm, ms/cm or μs/cm, a large number of switch positions being available. Also on the front panel of a c.r.o. is a Y amplifier switch marked in volts per centimetre, with a large number of available switch positions.

(i) With **direct voltage measurements**, only the Y amplifier 'volts/cm' switch on the c.r.o. is used. With no voltage applied to the Y plates the position of the spot trace on the screen is noted. When a direct voltage is applied to the Y plates the new position of the spot trace is an indication of the magnitude of the voltage. For example, in Fig. 23.12(a), with no voltage applied to the Y plates, the spot trace is in the centre of the screen (initial position) and then the spot trace moves 2.5 cm to the final position shown, on application of a d.c. voltage. With the 'volts/cm' switch on 10 volts/cm the magnitude of the direct voltage is 2.5 cm × 10 volts/cm, i.e. 25 volts

(ii) With **alternating voltage measurements**, let a sinusoidal waveform be displayed on a c.r.o. screen as shown in Fig. 23.12(b). If the time/cm switch is on, say, 5 ms/cm then the **periodic time** T of the sinewave is 5 ms/cm × 4 cm, i.e. **20 ms** or **0.02 s**

Since frequency $f = \dfrac{1}{T}$, **frequency** $= \dfrac{1}{0.02} = \mathbf{50\,Hz}$

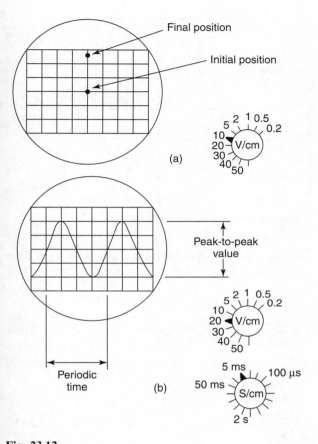

Fig. 23.12

If the 'volts/cm' switch is on, say, 20 volts/cm then the **amplitude** or **peak value** of the sinewave shown is 20 volts/cm × 2 cm, i.e. 40 V

Since r.m.s. voltage = $\dfrac{\text{peak voltage}}{\sqrt{2}}$

(see chapter 22),

r.m.s. voltage $= \dfrac{40}{\sqrt{2}} = \mathbf{28.28\ volts}$.

Double beam oscilloscopes are useful whenever two signals are to be compared simultaneously.

The c.r.o. demands reasonable skill in adjustment and use. However its greatest advantage is in observing the shape of a waveform – a feature not possessed by other measuring instruments.

Problem 6. Describe how a simple c.r.o. is adjusted to give (a) a spot trace, (b) a continuous horizontal trace on the screen, explaining the functions of the various controls.

(a) To obtain a spot trace on a typical c.r.o. screen:

(i) Switch on the c.r.o.

(ii) Switch the timebase control to off. This control is calibrated in time per centimetres – for example, 5 ms/cm or 100 μs/cm. Turning it to zero ensures no signal is applied to the X-plates. The Y-plate input is left open-circuited.

(iii) Set the intensity, X-shift and Y-shift controls to about the mid-range positions.

(iv) A spot trace should now be observed on the screen. If not, adjust either or both of the X and Y-shift controls. The X-shift control varies the position of the spot trace in a horizontal direction whilst the Y-shift control varies its vertical position.

(v) Use the X and Y-shift controls to bring the spot to the centre of the screen and use the focus control to focus the electron beam into a small circular spot.

(b) To obtain a continuous horizontal trace on the screen the same procedure as in (a) is initially adopted. Then the timebase control is switched to a suitable position, initially the millisecond timebase range, to ensure that the repetition rate of the sawtooth is sufficient for the persistence of the vision time of the screen phosphor to hold a given trace.

Problem 7. For the c.r.o. square voltage waveform shown in Fig. 23.13 determine (a) the periodic time, (b) the frequency and (c) the peak-to-peak voltage. The 'time/cm' (or timebase control) switch is on 100 μs/cm and the 'volts/cm' (or signal amplitude control) switch is on 20 V/cm.

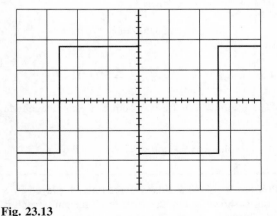

Fig. 23.13

(In Figures 23.13 to 23.19 assume that the squares shown are 1 cm by 1 cm)

(a) The width of one complete cycle is 5.2 cm

Hence the periodic time,

$$T = 5.2 \text{ cm} \times 100 \times 10^{-6} \text{ s/cm} = \mathbf{0.52\,ms}$$

(b) Frequency, $f = \dfrac{1}{T} = \dfrac{1}{0.52 \times 10^{-3}} = \mathbf{1.92\,Hz}$

(c) The peak-to-peak height of the display is 3.6 cm, hence the

$$\text{peak-to-peak voltage} = 3.6 \text{ cm} \times 20 \text{ V/cm} = \mathbf{72\,V}$$

Problem 8. For the c.r.o. display of a pulse waveform shown in Fig. 23.14 the 'time/cm' switch is on 50 ms/cm and the 'volts/cm' switch is on 0.2 V/cm. Determine (a) the periodic time, (b) the frequency, (c) the magnitude of the pulse voltage.

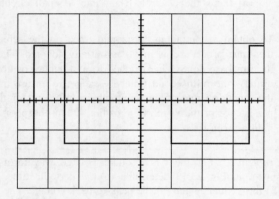

Fig. 23.14

(a) The width of one complete cycle is 3.5 cm

Hence the periodic time,

$$T = 3.5 \text{ cm} \times 50 \text{ ms/cm} = \mathbf{175\,ms}$$

(b) Frequency, $f = \dfrac{1}{T} = \dfrac{1}{0.175} = \mathbf{5.71\,Hz}$

(c) The height of a pulse is 3.4 cm hence the magnitude of the pulse

$$\text{voltage} = 3.4 \text{ cm} \times 0.2 \text{ V/cm} = \mathbf{0.68\,V}$$

Problem 9. A sinusoidal voltage trace displayed by a c.r.o. is shown in Fig. 23.15. If the 'time/cm' switch is on 500 μs/cm and the 'volts/cm' switch is on 5 V/cm, find, for the waveform, (a) the frequency, (b) the peak-to-peak voltage, (c) the amplitude, (d) the r.m.s. value.

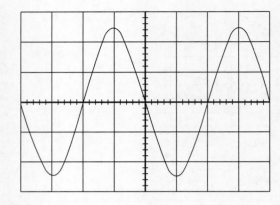

Fig. 23.15

(a) The width of one complete cycle is 4 cm. Hence the periodic time, T is 4 cm × 500 μs/cm, i.e. 2 ms

$$\text{Frequency,} \quad f = \dfrac{1}{T} = \dfrac{1}{2 \times 10^{-3}} = \mathbf{500\,Hz}$$

(b) The peak-to-peak height of the waveform is 5 cm. Hence the peak-to-peak

$$\text{voltage} = 5 \text{ cm} \times 5 \text{ V/cm} = \mathbf{25\,V}$$

(c) Amplitude $= \dfrac{1}{2} = 25 \text{ V} = \mathbf{12.5\,V}$

(d) The peak value of voltage is the amplitude, i.e. 12.5 V, and

$$\text{r.m.s. voltage} = \dfrac{\text{peak voltage}}{\sqrt{2}} = \dfrac{12.5}{\sqrt{2}} = \mathbf{8.84\,V}$$

Problem 10. For the double-beam oscilloscope displays shown in Fig. 23.16 determine (a) their frequency, (b) their r.m.s. values, (c) their phase difference. The 'time/cm' switch is on 100 μs/cm and the 'volts/cm' switch on 2 V/cm.

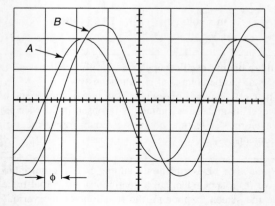

Fig. 23.16

(a) The width of each complete cycle is 5 cm for both waveforms.

Hence the periodic time, T, of each waveform is $5\,\text{cm} \times 100\,\mu\text{s/cm}$, i.e. 0.5 ms

Frequency of each waveform,

$$f = \frac{1}{T} = \frac{1}{0.5 \times 10^{-3}} = \textbf{2 kHz}$$

(b) The peak value of waveform A is $2\,\text{cm} \times 2\,\text{V/cm} = \textbf{4 V}$,

hence the r.m.s. value of waveform $A = \dfrac{4}{\sqrt{2}} = \textbf{2.83 V}$

The peak value of waveform B is $2.5\,\text{cm} \times 2\,\text{V/cm} = 5\,\text{V}$,

hence the r.m.s. value of waveform $B = \dfrac{5}{\sqrt{2}} = \textbf{3.54 V}$

(c) Since 5 cm represents 1 cycle, then 5 cm represents 360°,

i.e. 1 cm represents $\dfrac{360}{5} = 72°$

The phase angle $\phi = 0.5\,\text{cm} = 0.5\,\text{cm} \times 72°/\text{cm} = 36°$

Hence waveform A leads waveform B by 36°

Now try the following exercise

Exercise 102 Further problems on the cathode ray oscilloscope (Answers on page 307)

1. For the square voltage waveform displayed on a c.r.o. shown in Fig. 23.17, find (a) its frequency, (b) its peak-to-peak voltage

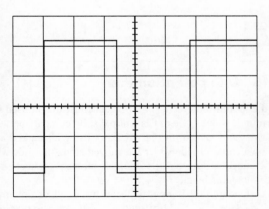

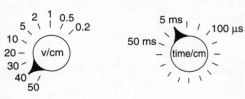

Fig. 23.17

2. For the pulse waveform shown in Fig. 23.18, find (a) its frequency, (b) the magnitude of the pulse voltage

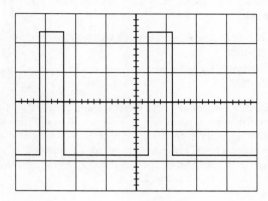

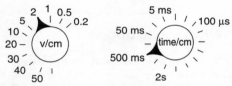

Fig. 23.18

3. For the sinusoidal waveform shown in Fig. 23.19, determine (a) its frequency, (b) the peak-to-peak voltage, (c) the r.m.s. voltage

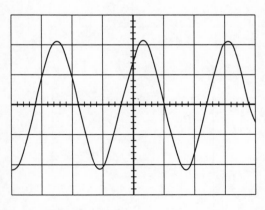

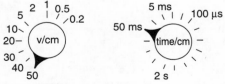

Fig. 23.19

23.13   Null method of measurement

A **null method of measurement** is a simple, accurate and widely used method which depends on an instrument reading being adjusted to read zero current only. The method assumes:

(i) if there is any deflection at all, then some current is flowing

(ii) if there is no deflection, then no current flows (i.e. a null condition)

Hence it is unnecessary for a meter sensing current flow to be calibrated when used in this way. A sensitive milliammeter or microammeter with centre zero position setting is called a **galvanometer**. Two examples where the method is used are in the Wheatstone bridge (see Section 23.14), in the d.c. potentiometer (see Section 23.15)

23.14   Wheatstone bridge

Figure 23.20 shows a **Wheatstone bridge** circuit which compares an unknown resistance R_x with others of known values, i.e. R_1 and R_2, which have fixed values, and R_3, which is variable. R_3 is varied until zero deflection is obtained on the galvanometer G. No current then flows through the meter, $V_A = V_B$, and the bridge is said to be 'balanced'.

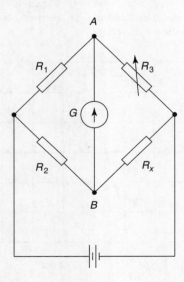

Fig. 23.20

At balance,

$$R_1 R_x = R_2 R_3 \text{ i.e. } \boxed{R_x = \frac{R_2 R_3}{R_1} \text{ ohms.}}$$

Problem 11. In a Wheatstone bridge $ABCD$, a galvanometer is connected between A and C, and a battery between B and D. A resistor of unknown value is connected between A and B. When the bridge is balanced, the resistance between B and C is $100\,\Omega$, that between C and D is $10\,\Omega$ and that between D and A is $400\,\Omega$. Calculate the value of the unknown resistance.

The Wheatstone bridge is shown in Fig. 23.21 where R_x is the unknown resistance. At balance, equating the products of opposite ratio arms, gives:

$$(R_x)(10) = (100)(400)$$

and

$$R_x = \frac{(100)(400)}{10} = 4000\,\Omega$$

Hence, the unknown resistance, $R_x = 4\,k\Omega$

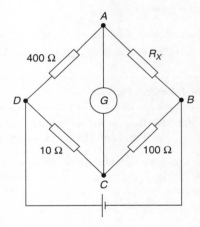

Fig. 23.21

23.15   D.C. potentiometer

The **d.c. potentiometer** is a null-balance instrument used for determining values of e.m.f.'s and p.d.s. by comparison with a known e.m.f. or p.d. In Fig. 23.22(a), using a standard cell of known e.m.f. E_1, the slider S is moved along the slide wire until balance is obtained (i.e. the galvanometer deflection is zero), shown as length l_1

The standard cell is now replaced by a cell of unknown e.m.f. E_2 (see Fig. 23.22(b)) and again balance is obtained (shown as l_2).

Since $E_1 \propto l_1$ and $E_2 \propto l_2$ then

$$\frac{E_1}{E_2} = \frac{l_1}{l_2} \text{ and } \boxed{E_2 = E_1 \left(\frac{l_2}{l_1}\right) \text{ volts}}$$

A potentiometer may be arranged as a resistive two-element potential divider in which the division ratio is adjustable

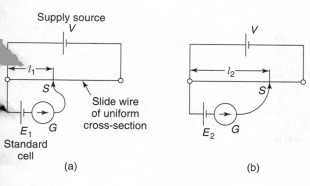

Supply source

Slide wire of uniform cross-section

E_1
Standard cell

G

E_2

G

(a)

(b)

Fig. 23.22

to give a simple variable d.c. supply. Such devices may be constructed in the form of a resistive element carrying a sliding contact which is adjusted by a rotary or linear movement of the control knob.

Problem 12. In a d.c. potentiometer, balance is obtained at a length of 400 mm when using a standard cell of 1.0186 volts. Determine the e.m.f. of a dry cell if balance is obtained with a length of 650 mm.

$E_1 = 1.0186$ V, $l_1 = 400$ mm and $l_2 = 650$ mm.

With reference to Fig. 23.22,

$$\frac{E_1}{E_2} = \frac{l_1}{l_2}$$

from which, $E_2 = E_1 \left(\frac{l_2}{l_1}\right) = (1.0186)\left(\frac{650}{400}\right)$

$$= \mathbf{1.655 \ volts.}$$

Now try the following exercise

Exercise 103   Further problems on the Wheatstone bridge and d.c. potentiometer (Answers on page 307)

1. In a Wheatstone bridge *PQRS*, a galvanometer is connected between *Q* and *S* and a voltage source between *P* and *R*. An unknown resistor R_x is connected between *P* and *Q*. When the bridge is balanced, the resistance between *Q* and *R* is 200 Ω, that between *R* and *S* is 10 Ω and that between *S* and *P* is 150 Ω. Calculate the value of R_x

2. Balance is obtained in a d.c. potentiometer at a length of 31.2 cm when using a standard cell of 1.0186 volts. Calculate the e.m.f. of a dry cell if balance is obtained with a length of 46.7 cm.

23.16   Measurement errors

Errors are always introduced when using instruments to measure electrical quantities. The errors most likely to occur in measurements are those due to:

(i) the limitations of the instrument

(ii) the operator

(iii) the instrument disturbing the circuit

(i) Errors in the limitations of the instrument

The **calibration accuracy** of an instrument depends on the precision with which it is constructed. Every instrument has a margin of error which is expressed as a percentage of the instruments full scale deflection. For example, industrial grade instruments have an accuracy of ±2% of f.s.d. Thus if a voltmeter has a f.s.d. of 100 V and it indicates 40 V say, then the actual voltage may be anywhere between 40 ± (2% of 100), or 40 ± 2, i.e. between 38 V and 42 V.

When an instrument is calibrated, it is compared against a standard instrument and a graph is drawn of 'error' against 'meter deflection'.

A typical graph is shown in Fig. 23.23 where it is seen that the accuracy varies over the scale length. Thus a meter with a ±2% f.s.d. accuracy would tend to have an accuracy which is much better than ±2% f.s.d. over much of the range.

(ii) Errors by the operator

It is easy for an operator to misread an instrument. With linear scales the values of the sub-divisions are reasonably easy to determine; non-linear scale graduations are more difficult to estimate. Also, scales differ from instrument to instrument and some meters have more than one scale (as with multimeters) and mistakes in reading indications are easily made. When reading a meter scale it should be viewed from an angle perpendicular to the surface of the scale at the location of the pointer; a meter scale should not be viewed 'at an angle'.

(iii) Errors due to the instrument disturbing the circuit

Any instrument connected into a circuit will affect that circuit to some extent. Meters require some power to operate, but provided this power is small compared with the power in the measured circuit, then little error will result. Incorrect positioning of instruments in a circuit can be a source of errors. For example, let a resistance be measured by the voltmeter-ammeter method as shown in Fig. 23.24. Assuming 'perfect' instruments, the resistance should be given by the voltmeter reading divided by the ammeter reading (i.e. $R = V/I$). However, in Fig. 23.24(a), $V/I = R + r_a$ and in Fig. 23.24(b) the current through the ammeter is that through

194 *Science for Engineering*

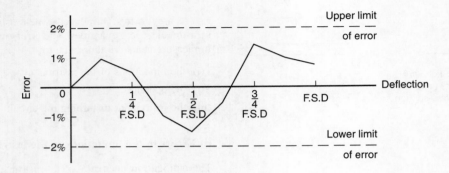

Fig. 23.23

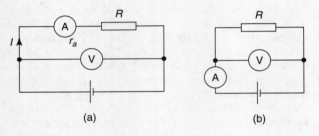

Fig. 23.24

the resistor plus that through the voltmeter. Hence the voltmeter reading divided by the ammeter reading will not give the true value of the resistance R for either method of connection.

Now try the following exercises

Exercise 104 Short answer questions on electrical measuring instruments and measurements

1. What is the main difference between an analogue and a digital type of measuring instrument?

2. Name the three essential devices for all analogue electrical indicating instruments

3. Complete the following statements:
   (a) An ammeter has a ... resistance and is connected ... with the circuit
   (b) A voltmeter has a ... resistance and is connected ... with the circuit

4. State two advantages and two disadvantages of a moving coil instrument

5. What effect does the connection of (a) a shunt (b) a multiplier have on a milliammeter?

6. State two advantages and two disadvantages of a moving coil instrument

7. Name two advantages of electronic measuring instruments compared with moving coil or moving iron instruments

8. Briefly explain the principle of operation of an ohmmeter

9. Name a type of ohmmeter used for measuring (a) low resistance values (b) high resistance values

10. What is a multimeter?

11. When may a rectifier instrument be used in preference to either a moving coil or moving iron instrument?

12. Name five quantities that a c.r.o. is capable of measuring

13. What is meant by a null method of measurement?

14. Sketch a Wheatstone bridge circuit used for measuring an unknown resistance in a d.c. circuit and state the balance condition

15. How may a d.c. potentiometer be used to measure p.d.'s

16. Define 'calibration accuracy' as applied to a measuring instrument

17. State three main areas where errors are most likely to occur in measurements

Exercise 105 Multi-choice questions on electrical measuring instruments and measurements (Answers on page 307)

1. Which of the following would apply to a moving coil instrument?
   (a) An uneven scale, measuring d.c.

Formulae for electrical applications

Formula	Formula Symbols	Units
Charge = current × time	$Q = It$	C
Resistance = $\dfrac{\text{potential difference}}{\text{current}}$	$R = \dfrac{V}{I}$	Ω
Electrical power = potential difference × current	$P = VI = I^2R = \dfrac{V^2}{R}$	W
Terminal p.d. = source e.m.f. − (current) (resistance)	$V = E - Ir$	V
Resistance = $\dfrac{\text{resistivity} \times \text{length of conductor}}{\text{cross sectional area}}$	$R = \dfrac{\rho l}{a}$	Ω
Total resistance of resistors in series	$R = R_1 + R_2 + ..$	Ω
Total resistance of resistors in parallel	$\dfrac{1}{R} = \dfrac{1}{R_1} + \dfrac{1}{R_2} + ..$	
Magnetic flux density = $\dfrac{\text{magnetic flux}}{\text{area}}$	$B = \dfrac{\Phi}{A}$	T
Force on conductor = flux density × current × length of conductor	$F = Bil$	N
Force on a charge = charge × velocity × flux density	$F = QvB$	N
Induced e.m.f. = flux density × current × conductor velocity	$E = Blv$	V
Induced e.m.f. = number of coil turns × rate of change of flux	$E = -N\dfrac{\text{d}\Phi}{\text{d}t}$	V
Induced e.m.f. = inductance × rate of change of current	$E = -L\dfrac{\text{d}I}{\text{d}t}$	V
Inductance = $\dfrac{\text{number of coil turns} \times \text{flux}}{\text{current}}$	$L = \dfrac{N\Phi}{I}$	H
Mutually induced e.m.f.	$E_2 = -M\dfrac{\text{d}I_1}{\text{d}t}$	V
Ideal transformer	$\dfrac{V_1}{V_2} = \dfrac{N_1}{N_2} = \dfrac{I_2}{I_1}$	

Shunt resistor $\qquad R_S = \dfrac{I_a r_a}{I_S} \qquad \Omega$

Multiplier resistor $\qquad R_M = \dfrac{V - I r_a}{I} \qquad \Omega$

Wheatstone bridge $\qquad R_x = \dfrac{R_2 R_3}{R_1} \qquad \Omega$

Potentiometer $\qquad E_2 = E_1 \left(\dfrac{l_2}{l_1} \right) \qquad V$

Periodic time $\qquad T = \dfrac{1}{f} \qquad s$

R.m.a. current $\qquad I = \sqrt{\dfrac{i_1^2 + i_2^2 + \ldots + i_n}{n}} \qquad A$

For a sine wave:

Average or mean value $\qquad I_{AV} = \dfrac{2}{\pi} I_m$

R.m.s value $\qquad I = \dfrac{1}{\sqrt{2}} I_m$

Form factor $= \dfrac{\text{rms}}{\text{average}}$

Peak factor $= \dfrac{\text{maximum}}{\text{rms}}$

Section 4

Mechanical Applications

24

Speed and velocity

At the end of this chapter you should be able to:

- define speed
- calculate speed given distance and time
- plot a distance/time graph from given data
- determine the average speed from a distance/time graph
- determine the distance travelled from a speed/time graph
- define velocity

24.1 Speed

Speed is the rate of covering distance and is given by:

$$\text{speed} = \frac{\text{distance travelled}}{\text{time taken}}$$

The usual units for speed are metres per second, (m/s or $m\,s^{-1}$), or kilometres per hour, (km/h or $km\,h^{-1}$). Thus if a person walks 5 kilometres in 1 hour, the speed of the person is $\frac{5}{1}$, that is, 5 kilometres per hour.

The symbol for the SI unit of speed (and velocity) is written as $m\,s^{-1}$, called the 'index notation'. However, engineers usually use the symbol m/s, called the 'oblique notation', and it is this notation which is largely used in this chapter and other chapters on mechanics. One of the exceptions is when labelling the axes of graphs, when two obliques occur, and in this case the index notation is used. Thus for speed or velocity, the axis markings are speed/$m\,s^{-1}$ or velocity/$m\,s^{-1}$

Problem 1. A man walks 600 metres in 5 minutes. Determine his speed in (a) metres per second and (b) kilometres per hour.

(a) $\text{Speed} = \dfrac{\text{distance travelled}}{\text{time taken}} = \dfrac{600\,\text{m}}{5\,\text{min}}$

$= \dfrac{600\,\text{m}}{5\,\text{min}} \times \dfrac{1\,\text{min}}{60\,\text{s}} = \mathbf{2\,m/s}$

(b) $2\,\text{m/s} = \dfrac{2\,\text{m}}{1\,\text{s}} \times \dfrac{1\,\text{km}}{1000\,\text{m}} \times \dfrac{3600\,\text{s}}{1\,\text{h}} = 2 \times 3.6 = \mathbf{7.2\,km/h}$

(Note: to change from m/s to km/h, multiply by 3.6)

Problem 2. A car travels at 50 kilometres per hour for 24 minutes. Find the distance travelled in this time.

Since $\text{speed} = \dfrac{\text{distance travelled}}{\text{time taken}}$ then,

$\text{distance travelled} = \text{speed} \times \text{time taken}$

$\text{Time} = 24\,\text{minutes} = \dfrac{24}{60}\,\text{hours, hence}$

$\text{distance travelled} = 50\dfrac{\text{km}}{\text{h}} \times \dfrac{24}{60}\,\text{h} = \mathbf{20\,km}.$

Problem 3. A train is travelling at a constant speed of 25 metres per second for 16 kilometres. Find the time taken to cover this distance.

Since $\text{speed} = \dfrac{\text{distance travelled}}{\text{time taken}}$ then,

$\text{time taken} = \dfrac{\text{distance travelled}}{\text{speed}}$

16 kilometres = 16 000 metres, hence,

$\text{time taken} = \dfrac{16\,000}{\dfrac{25\,\text{m}}{1\,\text{s}}} = 16\,000\,\text{m} \times \dfrac{1\,\text{s}}{25\,\text{m}} = 640\,\text{s}$

and $640\,s = 640\,s \times \dfrac{1\,min}{60\,s} = \mathbf{10}\,\dfrac{2}{3}$ **min** or **10 min 40 s**

Now try the following exercise

> **Exercise 106   Further problems on speed (Answers on page 307)**
>
> 1. A train covers a distance of 96 km in 1 h 20 min. Determine the average speed of the train (a) in km/h and (b) in m/s.
> 2. A horse trots at an average speed of 12 km/h for 18 minutes; determine the distance covered by the horse in this time.
> 3. A ship covers a distance of 1365 km at an average speed of 15 km/h. How long does it take to cover this distance?

24.2   Distance/time graph

One way of giving data on the motion of an object is graphically. A graph of distance travelled (the scale on the vertical axis of the graph) against time (the scale on the horizontal axis of the graph) is called a **distance/time graph**. Thus if an aeroplane travels 500 kilometres in its first hour of flight and 750 kilometres in its second hour of flight, then after 2 hours, the total distance travelled is (500 + 750) kilometres, that is, 1250 kilometres. The distance/time graph for this flight is shown in Fig. 24.1.

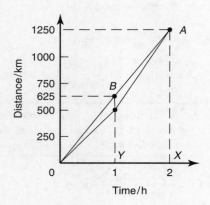

Fig. 24.1

The **average speed** is given by:

$$\frac{\text{total distance travelled}}{\text{total time taken}}$$

Thus, the average speed of the aeroplane is

$$\frac{(500 + 750)\,km}{(1+1)\,h} = \frac{1250}{2} = 625\,km/h.$$

If points O and A are joined in Fig. 24.1, the slope of line OA is defined as

$$\frac{\text{change in distance (vertical)}}{\text{change in time (horizontal)}}$$

for any two points on line OA.

For point A, the change in distance is AX, that is, 1250 kilometres, and the change in time is OX, that is, 2 hours. Hence the average speed is $\dfrac{1250}{2}$, i.e. 625 kilometres per hour.

Alternatively, for point B on line OA, the change in distance is BY, that is, 625 kilometres, and the change in time is OY, that is 1 hour, hence the average speed is $\dfrac{625}{1}$, i.e. 625 kilometres per hour.

In general, the average speed of an object travelling between points M and N is given by the slope of line MN on the distance/time graph.

> **Problem 4.**   A person travels from point O to A, then from A to B and finally from B to C. The distances of A, B and C from O and the times, measured from the start to reach points A, B and C are as shown:
>
> | | A | B | C |
> |---------------|-----|-----|-----|
> | Distance (m) | 100 | 200 | 250 |
> | Time (s) | 40 | 60 | 100 |
>
> Plot the distance/time graph and determine the speed of travel for each of the three parts of the journey

The vertical scale of the graph is distance travelled and the scale is selected to span 0 to 250 m, the total distance travelled from the start. The horizontal scale is time and spans 0 to 100 seconds, the total time taken to cover the whole journey. Co-ordinates corresponding to A, B and C are plotted and OA, AB and BC are joined by straight lines. The resulting distance/time graph is shown in Fig. 24.2.

The speed is given by the slope of the distance/time graph. Speed for part OA of the journey = slope of OA

$$= \frac{AX}{OX} = \frac{100\,m}{40\,s} = \mathbf{2.5\,m/s}$$

Speed for part AB of the journey = slope of AB

$$= \frac{Bm}{Am} = \frac{(200-100)\,m}{(60-40)\,s} = \mathbf{5\,m/s}$$

Speed for part BC of the journey = slope of BC

$$= \frac{Cn}{Bn} = \frac{(250-200)\,m}{(100-60)\,s} = \mathbf{1.25\,m/s}$$

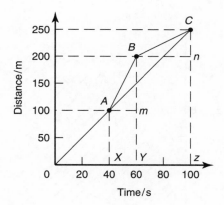

Fig. 24.2

Problem 5. Determine the average speed (both in m/s and km/h) for the whole journey for the information given in Problem 4.

$$\text{Average speed} = \frac{\text{total distance travelled}}{\text{total time taken}}$$

$$= \text{slope of line } OC$$

From Fig. 24.2, slope of line OC

$$= \frac{Cz}{Oz} = \frac{250\,\text{m}}{100\,\text{s}} = \mathbf{2.5\,m/s}$$

$$2.5\,\text{m/s} = \frac{2.5\,\text{m}}{1\,\text{s}} \times \frac{1\,\text{km}}{1000\,\text{m}} \times \frac{3600\,\text{s}}{1\,\text{h}}$$

$$= (2.5 \times 3.6)\,\text{km/h} = \mathbf{9\,km/h}$$

Hence the average speed is 2.5 m/s or 9 km/h

Problem 6. A coach travels from town A to town B, a distance of 40 kilometres at an average speed of 55 kilometres per hour. It then travels from town B to town C, a distance of 25 kilometres in 35 minutes. Finally, it travels from town C to town D at an average speed of 60 kilometres per hour in 45 minutes. Determine

(a) the time taken to travel from A to B

(b) the average speed of the coach from B to C

(c) the distance from C to D

(d) the average speed of the whole journey from A to D

(a) From town A to town B:

Since speed $= \dfrac{\text{distance travelled}}{\text{time taken}}$ then,

$$\text{time taken} = \frac{\text{distance travelled}}{\text{speed}}$$

$$= \frac{40\,\text{km}}{\dfrac{55\,\text{km}}{1\,\text{h}}} = 40\,\text{km} \times \frac{1\,\text{h}}{55\,\text{km}}$$

$$= 0.727\,\text{h or } \mathbf{43.64\,minutes}$$

(b) From town B to town C:

Since speed $= \dfrac{\text{distance travelled}}{\text{time taken}}$

and $35\,\text{min} = \dfrac{35}{60}\,\text{h}$, then,

$$\text{speed} = \frac{25\,\text{km}}{\dfrac{35}{60}\,\text{h}} = \frac{25 \times 60}{35}\,\text{km/h} = \mathbf{42.86\,km/h}$$

(c) From town C to town D:

Since speed $= \dfrac{\text{distance travelled}}{\text{time taken}}$

then, distance travelled $=$ speed $\times$ time taken, and $45\,\text{min} = \frac{3}{4}\,\text{h}$, hence,

$$\text{distance travelled} = 60\,\frac{\text{km}}{\text{h}} \times \frac{3}{4}\,\text{h} = \mathbf{45\,km}$$

(d) From town A to town D:

$$\text{average speed} = \frac{\text{total distance travelled}}{\text{total time taken}}$$

$$= \frac{(40 + 25 + 45)\,\text{km}}{\left(\dfrac{43.64}{60} + \dfrac{35}{60} + \dfrac{45}{60}\right)\,\text{h}} = \frac{110\,\text{km}}{\dfrac{123.64}{60}\,\text{h}}$$

$$= \frac{110 \times 60}{123.64}\,\text{km/h} = \mathbf{53.38\,km/h}$$

Now try the following exercise

Exercise 107 Further problems on distance/time graphs (Answers on page 307)

1. Using the information given in the distance/time graph shown in Fig. 24.3, determine the average speed when travelling from O to A, A to B, B to C, O to C and A to C.

2. The distances travelled by an object from point O and the corresponding times taken to reach A, B, C and D, respectively, from the start are as shown:

Points	Start	A	B	C	D
Distance (m)	0	20	40	60	80
Time (s)	0	5	12	18	25

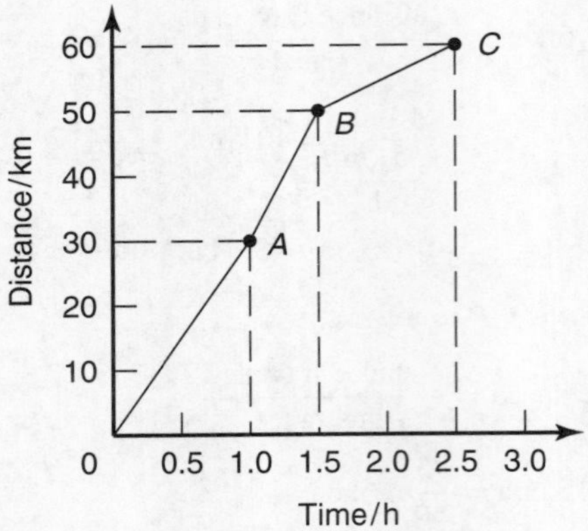

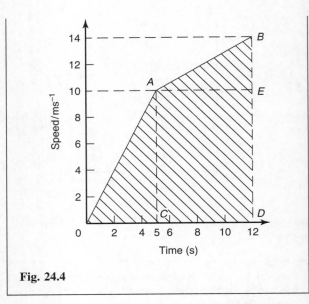

Fig. 24.3

Draw the distance/time graph and hence determine the average speeds from O to A, A to B, B to C, C to D and O to D.

3. A train leaves station A and travels via stations B and C to station D. The times the train passes the various stations are as shown:

Station	A	B	C	D
Times	10.55 am	11.40 am	12.15 pm	12.50 pm

The average speeds are:

A to B, 56 km/h, B to C, 72 km/h, and C to D, 60 km/h

Calculate the total distance from A to D

4. A gun is fired 5 km north of an observer and the sound takes 15 s to reach him. Determine the average velocity of sound waves in air at this place.

5. The light from a star takes 2.5 years to reach an observer. If the velocity of light is 330×10^6 m/s, determine the distance of the star from the observer in kilometres, based on a 365 day year

24.3   Speed/time graph

If a graph is plotted of speed against time, the area under the graph gives the distance travelled. This is demonstrated in Problem 7.

Problem 7. The motion of an object is described by the speed/time graph given in Fig. 24.4. Determine the distance covered by the object when moving from O to B.

Fig. 24.4

The distance travelled is given by the area beneath the speed/time graph, shown shaded in Fig. 24.4. Area of

triangle $OAC = \dfrac{1}{2} \times$ base $\times$ perpendicular height

$$= \frac{1}{2} \times 5\,\text{s} \times 10\frac{\text{m}}{\text{s}} = 25\,\text{m}$$

Area of rectangle $AEDC$

$$= \text{base} \times \text{height}$$
$$= (12 - 5)\,\text{s} \times (10 - 0)\frac{\text{m}}{\text{s}} = 70\,\text{m}$$

Area of triangle ABE

$$= \frac{1}{2} \times \text{base} \times \text{perpendicular height}$$
$$= \frac{1}{2} \times (12 - 5)\,\text{s} \times (14 - 10)\frac{\text{m}}{\text{s}}$$
$$= \frac{1}{2} \times 7\,\text{s} \times 4\frac{\text{m}}{\text{s}} = 14\,\text{m}$$

Hence **the distance covered by the object moving from O to B** is:

$$(25 + 70 + 14)\,\text{m} = \mathbf{109\,m}$$

Now try the following exercise

Exercise 108   Further problems on speed/time graphs (Answers on page 307)

1. The speed/time graph for a car journey is shown in Fig. 24.5. Determine the distance travelled by the car.

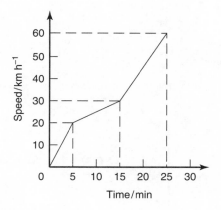

Fig. 24.5

2. The motion of an object is as follows:

   A to *B*, distance 122 m, time 64 s,

   B to *C*, distance 80 m at an average speed of 20 m/s,

   C to *D*, time 7 s at an average speed of 14 m/s

Determine the overall average speed of the object when travelling from *A* to *D*.

24.4   Velocity

The **velocity** of an object is the speed of the object **in a specified direction**. Thus, if a plane is flying due south at 500 kilometres per hour, its speed is 500 kilometres per hour, but its velocity is 500 kilometres per hour due south. It follows that if the plane had flown in a circular path for one hour at a speed of 500 kilometres per hour, so that one hour after taking off it is again over the airport, its average velocity in the first hour of flight is zero.

The average velocity is given by

$$\frac{\text{distance travelled in a specified direction}}{\text{time taken}}$$

If a plane flies from place *O* to place *A*, a distance of 300 kilometres in one hour, A being due north of *O*, then *OA* in Fig. 23.6 represents the first hour of flight. It then flies from *A* to *B*, a distance of 400 kilometres during the second hour of flight, *B* being due east of *A*, thus *AB* in Fig. 24.6 represents its second hour of flight.

Its average velocity for the two hour flight is:

$$\frac{\text{distance } OB}{2 \text{ hours}} = \frac{500 \text{ km}}{2 \text{ h}} = 250 \text{ km/h in direction } OB$$

A graph of velocity (scale on the vertical axis) against time (scale on the horizontal axis) is called a velocity/time graph. The graph shown in Fig. 24.7 represents a plane flying for 3 hours at a constant speed of 600 kilometres per hour in

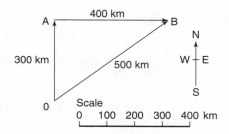

Fig. 24.6

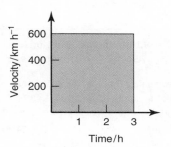

Fig. 24.7

a specified direction. The shaded area represents velocity (vertically) multiplied by time (horizontally), and has units of

$$\frac{\text{kilometres}}{\text{hours}} \times \text{hours},$$

i.e. kilometres, and represents the distance travelled in a specific direction. In this case,

$$\text{distance} = 600 \frac{\text{km}}{\text{h}} \times 3 \text{ h} = 1800 \text{ km}.$$

Another method of determining the distance travelled is from:

distance travelled = average velocity × time

Thus if a plane travels due south at 600 kilometres per hour for 20 minutes, the distance covered is

$$\frac{600 \text{ km}}{1 \text{ h}} \times \frac{20}{60} \text{ h}, \text{ i.e. } 200 \text{ km}.$$

Now try the following exercises

Exercise 109   Short answer questions on speed and velocity

1. Speed is defined as ....

2. Speed is given by $\dfrac{\cdots\cdots\cdots\cdots}{\cdots\cdots\cdots\cdots}$

3. The usual units for speed are .... or ....

4. Average speed is given by $\dfrac{\cdots\cdots\cdots\cdots}{\cdots\cdots\cdots\cdots}$

5. The velocity of an object is ....

6. Average velocity is given by $\dfrac{\cdots\cdots\cdots}{\cdots\cdots\cdots}$

7. The area beneath a velocity/time graph represents the ....

8. Distance travelled = ... × ....

9. The slope of a distance/time graph gives the ....

10. The average speed can be determined from a distance/time graph from ....

Exercise 110   Multi-choice questions on speed and velocity (Answers on page 307)

An object travels for 3 s at an average speed of 10 m/s and then for 5 s at an average speed of 15 m/s. In questions 1 to 3, select the correct answers from those given below:

   (a) 105 m/s   (b) 3 m   (c) 30 m

   (d) 13.125 m/s   (e) 3.33 m   (f) 0.3 m

   (g) 75 m   (h) $\frac{1}{3}$ m   (i) 12.5 m/s

1. The distance travelled in the first 3 s

2. The distance travelled in the latter 5 s

3. The average speed over the 8 s period

4. Which of the following statements is false?

   (a) Speed is the rate of covering distance

   (b) Speed and velocity are both measured in m/s units

   (c) Speed is the velocity of travel in a specified direction

   (d) The area beneath the velocity/time graph gives distance travelled

In questions 5 to 7, use the table to obtain the quantities stated, selecting the correct answer from (a) to (i) of those given below.

Distance	Time	Speed
20 m	30 s	X
5 km	Y	20 km/h
Z	3 min	10 m/min

   (a) 30 m   (b) $\frac{1}{4}$ h   (c) 600 m/s

   (d) $3\frac{1}{3}$ m   (e) $\frac{2}{3}$ m/s   (f) $\frac{3}{10}$ m

   (g) 4 h   (h) $1\frac{1}{4}$ m/s   (i) 100 h

5. Quantity X

6. Quantity Y

7. Quantity Z

Questions 8 to 10, refer to the distance/time graph shown in Fig. 24.8.

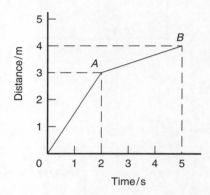

Fig. 24.8

8. The average speed when travelling from O to A is:

   (a) 3 m/s   (b) 1.5 m/s   (c) 0.67 m/s   (d) 0.66 m/s

9. The average speed when travelling from A to B is:

   (a) 3 m/s   (b) 1.5 m/s   (c) 0.67 m/s   (d) 0.33 m/s

10. The average overall speed when travelling from O to B is:

   (a) 0.8 m/s   (b) 1.2 m/s   (c) 1.5 m/s   (d) 20 m/s

11. A car travels at 60 km/h for 25 minutes. The distance travelled in that time is:

   (a) 25 km   (b) 1500 km   (c) 2.4 km   (d) 416.7 km

An object travels for 4 s at an average speed of 16 m/s, and for 6 s at an average speed of 24 m/s. Use this data to answer questions 12 to 14.

12. The distance travelled in the first 4 s is:

   (a) 0.25 km   (b) 64 m   (c) 4 m   (d) 0.64 km

13. The distance travelled in the latter 6 s is:

   (a) 4 m   (b) 0.25 km   (c) 144 m   (d) 14.4 km

14. The average speed over the 10 s period is:

   (a) 20 m/s   (b) 50 m/s   (c) 40 m/s   (d) 20.8 m/s

25

Acceleration

At the end of this chapter you should be able to:

- define acceleration and state its units
- draw a velocity/time graph
- determine acceleration from a velocity/time graph
- appreciate that 'free fall' has a constant acceleration of $9.8 \, \text{m/s}^2$
- use the equation of motion $v = u + at$ in calculations

25.1 Introduction to acceleration

Acceleration is the rate of change of velocity with time. The average acceleration, a, is given by:

$$a = \frac{\text{change in velocity}}{\text{time taken}}$$

The usual units are metres per second squared (m/s^2 or m s^{-2}). If u is the initial velocity of an object in metres per second, v is the final velocity in metres per second and t is the time in seconds elapsing between the velocities of u and v, then:

$$\boxed{\text{average acceleration, } a = \frac{v - u}{t} \text{ m/s}^2}$$

25.2 Velocity/time graph

A graph of velocity (scale on the vertical axis) against time (scale on the horizontal axis) is called a velocity/time graph, as introduced in Chapter 24. For the velocity/time graph shown in Fig. 25.1, the slope of line OA is given by $\dfrac{AX}{OX}$.

AX is the change in velocity from an initial velocity, u, of zero to a final velocity, v, of 4 metres per second. OX is the time taken for this change in velocity, thus

$$\frac{AX}{OX} = \frac{\text{change in velocity}}{\text{time taken}}$$

$$= \text{the acceleration in the first two seconds}$$

From the graph: $\dfrac{AX}{OX} = \dfrac{4 \, \text{m/s}}{2 \, \text{s}} = 2 \, \text{m/s}^2$ i.e. the acceleration is $2 \, \text{m/s}^2$

Similarly, the slope of line AB in Fig. 25.1 is given by $\dfrac{BY}{AY}$ i.e. the acceleration between 2 s and 5 s is

$$\frac{8 - 4}{5 - 2} = \frac{4}{3} = 1\frac{1}{3} \, \text{m/s}^2$$

In general, the slope of a line on a velocity/time graph gives the acceleration.

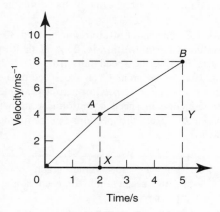

Fig. 25.1

The words 'velocity' and 'speed' are commonly interchanged in everyday language. Acceleration is a vector quantity and is correctly defined as the rate of change of velocity with respect to time. However, acceleration is also the rate of change of speed with respect to time in a certain specified direction.

Problem 1. The speed of a car travelling along a straight road changes uniformly from zero to 50 km/h in 20 s. It then maintains this speed for 30 s and finally reduces speed uniformly to rest in 10 s. Draw the speed/time graph for this journey.

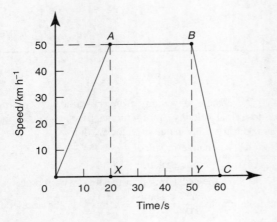

Fig. 25.2

The vertical scale of the speed/time graph is speed (km h^{-1}) and the horizontal scale is time (s). Since the car is initially at rest, then at time 0 seconds, the speed is 0 km/h. After 20 s, the speed is 50 km/h, which corresponds to point A on the speed/time graph shown in Fig. 25.2. Since the change in speed is uniform, a straight line is drawn joining points 0 and A.

The speed is constant at 50 km/h for the next 30 s, hence, horizontal line AB is drawn in Fig. 25.2 for the time period 20 s to 50 s. Finally, the speed falls from 50 km/h at 50 s to zero in 10 s, hence point C on the speed/time graph in Fig. 25.2 corresponds to a speed of zero and a time of 60 s. Since the reduction in speed is uniform, a straight line is drawn joining BC. Thus, the speed/time graph for the journey is as shown in Fig. 25.2.

Problem 2. For the speed/time graph shown in Fig. 25.2, find the acceleration for each of the three stages of the journey.

From above, the slope of line OA gives the uniform acceleration for the first 20 s of the journey.

$$\text{Slope of } OA = \frac{AX}{OX} = \frac{(50-0)\,\text{km/h}}{(20-0)\,\text{s}} = \frac{50\,\text{km/h}}{20\,\text{s}}$$

Expressing 50 km/h in metre-second units gives:

$$50\frac{\text{km}}{\text{h}} = \frac{50\,\text{km}}{1\,\text{h}} \times \frac{1000\,\text{m}}{1\,\text{km}} \times \frac{1\,\text{h}}{3600\,\text{s}} = \frac{50}{3.6}\,\text{m/s}$$

(Note: to change from km/h to m/s, divide by 3.6)

Thus, $\dfrac{50\,\text{km/h}}{20\,\text{s}} = \dfrac{\frac{50}{3.6}\,\text{m/s}}{20\,\text{s}} = 0.694\,\text{m/s}^2$

i.e. **the acceleration during the first 20 s is 0.694 m/s^2**
    Acceleration is defined as

$$\frac{\text{change of velocity}}{\text{time taken}} \quad \text{or} \quad \frac{\text{change of speed}}{\text{time taken}}$$

since the car is travelling along a straight road.
    Since there is no change in speed for the next 30 s (line AB in Fig. 25.2 is horizontal), **then the acceleration for this period is zero**.
    The slope of line BC gives the uniform deceleration for the final 10 s of the journey.

$$\text{Slope of } BC = \frac{BY}{YC} = \frac{50\,\text{km/h}}{10\,\text{s}}$$

$$= \frac{\frac{50}{3.6}\,\text{m/s}}{10\,\text{s}}$$

$$= 1.39\,\text{m/s}^2$$

i.e. **the deceleration during the final 10 s is 1.39 m/s^2**
Alternatively, **the acceleration is -1.39 m/s^2**

Now try the following exercise

Exercise 111　Further problems on velocity/time graphs and acceleration (Answers on page 307)

1. A coach increases velocity from 4 km/h to 40 km/h at an average acceleration of 0.2 m/s^2. Find the time taken for this increase in velocity.

2. A ship changes velocity from 15 km/h to 20 km/h in 25 min. Determine the average acceleration in m/s^2 of the ship during this time.

3. A cyclist travelling at 15 km/h changes velocity uniformly to 20 km/h in 1 min, maintains this velocity for 5 min and then comes to rest uniformly during the next 15 s. Draw a velocity/time graph and hence determine the accelerations in m/s^2 (a) during the first minute, (b) for the next 5 minutes, and (c) for the last 10 s.

4. Assuming uniform accelerations between points, draw the velocity/time graph for the data given below, and

hence determine the accelerations from A to B, B to C and C to D:

Point	A	B	C	D
Speed (m/s)	25	5	30	15
Time (s)	15	25	35	45

25.3 Free-fall and equation of motion

f a dense object such as a stone is dropped from a height, called **free-fall**, it has a constant acceleration of approximately $9.8\,\text{m/s}^2$. In a vacuum, all objects have this same constant acceleration, vertically downwards, that is, a feather has the same acceleration as a stone. However, if free-fall takes place in air, dense objects have the constant acceleration of $9.8\,\text{m/s}^2$ over short distances, but objects which have a low density, such as feathers, have little or no acceleration.

For bodies moving with a constant acceleration, the average acceleration is the constant value of the acceleration, and since from Section 25.1:

$$a = \frac{v - u}{t}$$

then

$$a \times t = v - u \text{ from which } \boldsymbol{v = u + at}$$

where u is the initial velocity in m/s,
   v is the final velocity in m/s,
   a is the constant acceleration in m/s^2 and
   t is the time in s.

When symbol 'a' has a negative value, it is called **deceleration** or **retardation**. The equation $v = u + at$ is called an **equation of motion**.

Problem 3. A stone is dropped from an aeroplane. Determine (a) its velocity after 2 s and (b) the increase in velocity during the third second, in the absence of all forces except that due to gravity.

The stone is free-falling and thus has an acceleration, a, of approximately $9.8\,\text{m/s}^2$ (taking downward motion as positive). From above:

final velocity, $v = u + at$

a) The initial downward velocity of the stone, u, is zero. The acceleration a, is $9.8\,\text{m/s}^2$ downwards and the time during which the stone is accelerating is 2 s.
Hence, final velocity, $v = u + at = 0 + 9.8 \times 2 = 19.6\,\text{m/s}$, i.e. **the velocity of the stone after 2 s is approximately 19.6 m/s.**

b) From part (a) above, the velocity after two seconds, u, is 19.6 m/s. The velocity after 3 s, applying $v = u + at$,

is $v = 19.6 + 9.8 \times 3 = 49\,\text{m/s}$ Thus, **the change in velocity during the third second is** $(49 - 19.6)\,\text{m/s}$, that is, **29.4 m/s.**
(Since the value $a = 9.8\,\text{m/s}^2$ is only an approximate value, then the answer is an approximate value.)

Problem 4. Determine how long it takes an object, which is free-falling, to change its speed from 100 km/h to 150 km/h, assuming all other forces, except that due to gravity, are neglected.

The initial velocity, u is 100 km/h, i.e. $\dfrac{100}{3.6}$ m/s (see Problem 2). The final velocity, v, is 150 km/h, i.e. $\dfrac{150}{3.6}$ m/s. Since the object is free-falling, the acceleration, a, is approximately $9.8\,\text{m/s}^2$ downwards (i.e. in a positive direction).

From above, $v = u + at$, i.e. $\dfrac{150}{3.6} = \dfrac{100}{3.6} + 9.8 \times t$

Transposing, gives $\quad 9.8 \times t = \dfrac{150 - 100}{3.6} = \dfrac{50}{3.6}$

Hence, $\qquad\qquad\qquad t = \dfrac{50}{3.6 \times 9.8} = 1.42\,\text{s}$

Since the value of a is only approximate, and rounding-off errors have occurred in calculations, then **the approximate time for the velocity to change from 100 km/h to 150 km/h is 1.42 s**

Problem 5. A train travelling at 30 km/h accelerates uniformly to 50 km/h in 2 minutes. Determine the acceleration.

$$30\,\text{km/h} = \frac{30}{3.6}\,\text{m/s (see Problem 2)},$$

$$50\,\text{km/h} = \frac{50}{3.6}\,\text{m/s}$$

and $\quad 2\,\text{min} = 2 \times 60 = 120\,\text{s}.$

From above,

$$v = u + at, \quad \text{i.e.} \quad \frac{50}{3.6} = \frac{30}{3.6} + a \times 120$$

Transposing, gives $\quad 120 \times a = \dfrac{50 - 30}{3.6}$

and $\qquad\qquad\qquad\qquad a = \dfrac{20}{3.6 \times 120}$

$$= 0.0463\,\text{m/s}^2$$

i.e. **the uniform acceleration of the train is 0.0463 m/s^2**

Problem 6. A car travelling at 50 km/h applies its brakes for 6 s and decelerates uniformly at 0.5 m/s². Determine its velocity in km/h after the 6 s braking period.

The initial velocity, $u = 50$ km/h $= \dfrac{50}{3.6}$ m/s (see Problem 2). From above, $v = u + at$. Since the car is decelerating, i.e. it has a negative acceleration, then $a = -0.5$ m/s² and t is 6 s

Thus, final velocity, $v = \dfrac{50}{3.6} + (-0.5)(6)$

$$= 13.89 - 3 = 10.89 \text{ m/s}$$

$$= 10.89 \times 3.6 = 39.2 \text{ km/h}$$

(Note: to convert m/s to km/h, multiply by 3.6)
Thus, the velocity after braking is 39.2 km/h.

Problem 7. A cyclist accelerates uniformly at 0.3 m/s² for 10 s, and his speed after accelerating is 20 km/h. Find his initial speed.

The final speed, $v = \dfrac{20}{3.6}$ m/s, time, $t = 10$ s, and acceleration, $a = 0.3$ m/s²
From above, $v = u + at$, where u is the initial speed.

Hence, $\dfrac{20}{3.6} = u + 0.3 \times 10$

from which, $u = \dfrac{20}{3.6} - 3 = 2.56$ m/s

$$= 2.56 \times 3.6 \text{ km/h (see Problem 6)}$$

$$= 9.2 \text{ km/h}$$

i.e. **the initial speed of the cyclist is 9.2 km/h.**

Now try the following exercises

Exercise 112   Further problems on free-fall and equation of motion (Answers on page 307)

1. An object is dropped from the third floor of a building. Find its approximate velocity 1.25 s later if all forces except that of gravity are neglected.

2. During free fall, a ball is dropped from point A and is travelling at 100 m/s when it passes point B. Calculate the time for the ball to travel from A to B if all forces except that of gravity are neglected.

3. A piston moves at 10 m/s at the centre of its motion and decelerates uniformly at 0.8 m/s². Determine its velocity 3 s after passing the centre of its motion.

4. The final velocity of a train after applying its brakes for 1.2 min is 24 km/h. If its uniform retardation is 0.06 m/s², find its velocity before the brakes are applied.

5. A plane in level flight at 400 km/h starts to descend at a uniform acceleration of 0.6 m/s². It levels off when its velocity is 670 km/h. Calculate the time during which it is losing height.

6. A lift accelerates from rest uniformly at 0.9 m/s² for 1.5 s, travels at constant velocity for 7 s and then comes to rest in 3 s. Determine its velocity when travelling at constant speed and its acceleration during the final 3 s of its travel.

Exercise 113   Short answer questions on acceleration (Answers on page 307)

1. Acceleration is defined as ....

2. Acceleration is given by $\dfrac{\cdots}{\cdots}$

3. The usual units for acceleration are ......

4. The slope of a velocity/time graph gives the ......

5. The value of free-fall acceleration for a dense object is approximately ......

6. The relationship between initial velocity, u, final velocity, v, acceleration, a, and time, t, is ......

7. A negative acceleration is called a ...... or a ......

Exercise 114   Multi-choice questions on acceleration (Answers on page 307)

1. If a car accelerates from rest, the acceleration is defined as the rate of change of:

   (a) energy   (b) velocity
   (c) mass   (d) displacement

2. An engine is travelling along a straight, level track at 15 m/s. The driver switches off the engine and applies the brakes to bring the engine to rest with uniform retardation in 5 s. The retardation of the engine is:

   (a) 3 m/s²   (b) 4 m/s²   (c) $\dfrac{1}{3}$ m/s²   (d) 75 m/s²

   Six statements, (a) to (f), are given below, some of the statements being true and the remainder false.

   (a) Acceleration is the rate of change of velocity with distance.

   (b) Average acceleration $= \dfrac{\text{(change of velocity)}}{\text{(time taken)}}$.

(c) Average acceleration $= (u - v)/t$, where u is the initial velocity, v is the final velocity and t is the time.

(d) The slope of a velocity/time graph gives the acceleration.

(e) The acceleration of a dense object during free-fall is approximately $9.8\,\text{m/s}^2$ in the absence of all other forces except gravity.

(f) When the initial and final velocities are u and v, respectively, a is the acceleration and t the time, then $u = v + at$.

In problems 3 and 4, select the statements required from those given.

3. (b), (c), (d), (e) Which statement is false?

4. (a), (c), (e), (f) Which statement is true?

A car accelerates uniformly from 5 m/s to 15 m/s in 20 s. It stays at the velocity attained at 20 s for 2 min. Finally, the brakes are applied to give a uniform deceleration and it comes to rest in 10 s. Use this data in Problems 5

to 9, selecting the correct answer from (a)–(l) given below.

(a) $-1.5\,\text{m/s}^2$ (b) $\dfrac{2}{15}\,\text{m/s}^2$ (c) 0

(d) $0.5\,\text{m/s}^2$ (e) $1.389\,\text{km/h}$ (f) $7.5\,\text{m/s}^2$

(g) $54\,\text{km/h}$ (h) $2\,\text{m/s}^2$ (i) $18\,\text{km/h}$

(j) $-\dfrac{1}{10}\,\text{m/s}^2$ (k) $1.467\,\text{km/h}$ (l) $-\dfrac{2}{3}\,\text{m/s}^2$

5. The initial speed of the car in km/h

6. The speed of the car after 20 s in km/h

7. The acceleration during the first 20 s period

8. The acceleration during the 2 min period

9. The acceleration during the final 10 s

10. A cutting tool accelerates from 50 mm/s to 150 mm/s in 0.2 seconds. The average acceleration of the tool is:

(a) $500\,\text{m/s}^2$ (b) $1\,\text{m/s}^2$

(c) $20\,\text{m/s}^2$ (d) $0.5\,\text{m/s}^2$

26

Force, mass and acceleration

At the end of this chapter you should be able to:

- define force and state its unit
- appreciate 'gravitational force'
- state Newton's three laws of motion
- perform calculations involving force $F = ma$
- define 'centripetal acceleration'
- perform calculations involving centripetal force = mv^2/r

26.1 Introduction

When an object is pushed or pulled, a **force** is applied to the object. This force is measured in **newtons (N)**. The effects of pushing or pulling an object are:

 (i) to cause a change in the motion of the object, and

(ii) to cause a change in the shape of the object.

If a change occurs in the motion of the object, that is, its velocity changes from u to v, then the object accelerates. Thus, it follows that acceleration results from a force being applied to an object. If a force is applied to an object and it does not move, then the object changes shape, that is, deformation of the object takes place. Usually the change in shape is so small that it cannot be detected by just watching the object. However, when very sensitive measuring instruments are used, very small changes in dimensions can be detected.

A force of attraction exists between all objects. The factors governing the size of this force F are the masses of the objects and the distances between their centres:

$$F \propto \frac{m_1 m_2}{d^2}$$

Thus, if a person is taken as one object and the earth as a second object, a force of attraction exists between the person and the earth. This force is called the **gravitational force** and is the force which gives a person a certain weight when standing on the earth's surface. It is also this force which gives freely falling objects a constant acceleration in the absence of other forces.

26.2 Newton's laws of motion

To make a stationary object move or to change the direction in which the object is moving requires a force to be applied externally to the object. This concept is known as **Newton's first law of motion** and may be stated as:

An object remains in a state of rest, or continues in a state of uniform motion in a straight line, unless it is acted on by an externally applied force.

Since a force is necessary to produce a change of motion, an object must have some resistance to a change in its motion. The force necessary to give a stationary pram a given acceleration is far less than the force necessary to give a stationary car the same acceleration. The resistance to a change in motion is called the **inertia** of an object and the amount of inertia depends on the mass of the object. Since a car has a much larger mass than a pram, the inertia of a car is much larger than that of a pram.

Newton's second law of motion may be stated as:

The acceleration of an object acted upon by an external force is proportional to the force and is in the same direction as the force.

Thus, force α acceleration, or force = a constant $\times$ acceleration, this constant of proportionality being the mass of the object, i.e.

force = mass × acceleration

The unit of force is the newton (N) and is defined in terms of mass and acceleration. One newton is the force required to give a mass of 1 kilogram an acceleration of 1 metre per second squared. Thus

$$F = ma$$

where F is the force in newtons (N), m is the mass in kilograms (kg) and a is the acceleration in metres per second squared (m/s^2), i.e.

$$1\,N = \frac{1\,kg\,m}{s^2}$$

It follows that $1\,m/s^2 = 1\,N/kg$. Hence a gravitational acceleration of $9.8\,m/s^2$ is the same as a gravitational field of $9.8\,N/kg$.

Newton's third law of motion may be stated as:

For every force, there is an equal and opposite reacting force

Thus, an object on, say, a table, exerts a downward force on the table and the table exerts an equal upward force on the object, known as a **reaction force** or just a **reaction**.

Problem 1. Calculate the force needed to accelerate a boat of mass 20 tonne uniformly from rest to a speed of 21.6 km/h in 10 minutes.

The mass of the boat, m, is 20 t, that is 20 000 kg.
The law of motion, $v = u + at$ can be used to determine the acceleration a.
The initial velocity, u, is zero, the final velocity,

$$v = 21.6\,km/h = \frac{21.6}{3.6} = 6\,m/s,$$

and the time, $t = 10\,min = 600\,s$. Thus $v = u + at$, i.e. $6 = 0 + a \times 600$, from which,

$$a = \frac{6}{600} = 0.01\,m/s^2$$

From Newton's second law, $F = ma$, i.e.

force = 20 000 × 0.01 N = **200 N**

Problem 2. The moving head of a machine tool requires a force of 1.2 N to bring it to rest in 0.8 s from a cutting speed of 30 m/min. Find the mass of the moving head.

From Newton's second law, $F = ma$, thus $m = F/a$, where force is given as 1.2 N. The law of motion $v = u + at$ can be

used to find acceleration a, where $v = 0$, $u = 30\,m/min = (30/60)\,m/s = 0.5\,m/s$, and $t = 0.8\,s$. Thus,

$$0 = 0.5 + a \times 0.8$$

from which, $a = -\dfrac{0.5}{0.8} = -0.625\,m/s^2$

or a retardation of $0.625\,m/s^2$

Thus the **mass**, $m = \dfrac{F}{a} = \dfrac{1.2}{0.625} = \mathbf{1.92\,kg}$

Problem 3. A lorry of mass 1350 kg accelerates uniformly from 9 km/h to reach a velocity of 45 km/h in 18 s. Determine (a) the acceleration of the lorry, (b) the uniform force needed to accelerate the lorry.

(a) The law of motion $v = u + at$ can be used to determine the acceleration, where final velocity $v = (45/3.6)\,m/s$, initial velocity $u = (9/3.6)\,m/s$ and time $t = 18\,s$

Thus $\dfrac{45}{3.6} = \dfrac{9}{3.6} + a \times 18$

from which, $a = \dfrac{1}{18}\left(\dfrac{45}{3.6} - \dfrac{9}{3.6}\right) = \dfrac{1}{18}\left(\dfrac{36}{3.6}\right)$

$$= \dfrac{10}{18} = \dfrac{5}{9}\,m/s^2 \quad or \quad \mathbf{0.556\,m/s^2}$$

(b) From Newton's second law of motion,

force, $F = ma = 1350 \times \dfrac{5}{9} = \mathbf{750\,N}$

Problem 4. Find the weight of an object of mass 1.6 kg at a point on the earth's surface where the gravitational field is 9.81 N/kg.

The weight of an object is the force acting vertically downwards due to the force of gravity acting on the object. Thus:

weight = force acting vertically downwards
= mass × gravitational field
= 1.6 × 9.81 = **15.696 N**

Problem 5. A bucket of cement of mass 40 kg is tied to the end of a rope connected to a hoist. Calculate the tension in the rope when the bucket is suspended but stationary. Take the gravitational field, g, as 9.81 N/kg.

The **tension** in the rope is the same as the force acting in the rope. The force acting vertically downwards due to the weight of the bucket must be equal to the force acting upwards in the rope, i.e. the tension.

Weight of bucket of cement, $F = mg = 40 \times 9.81 = 392.4\,N$.
Thus, **the tension in the rope = 392.4 N**

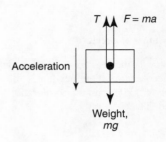

Fig. 26.2

Problem 6. The bucket of cement in Problem 5 is now hoisted vertically upwards with a uniform acceleration of $0.4 \, \text{m/s}^2$. Calculate the tension in the rope during the period of acceleration.

With reference to Fig. 26.1, the forces acting on the bucket are:

(i) a tension (or force) of T acting in the rope

(ii) a force of mg acting vertically downwards, i.e. the weight of the bucket and cement.

The resultant force $F = T - mg$; hence, $ma = T - mg$

i.e. $\qquad 40 \times 0.4 = T - 40 \times 9.81$

from which, **tension, $T = 408.4 \, \text{N}$**

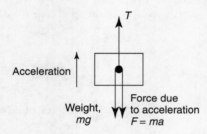

Fig. 26.1

By comparing this result with that of Problem 5, it can be seen that there is an increase in the tension in the rope when an object is accelerating upwards.

Problem 7. The bucket of cement in Problem 5 is now lowered vertically downwards with a uniform acceleration of $1.4 \, \text{m/s}^2$. Calculate the tension in the rope during the period of acceleration.

With reference to Fig. 26.2, the forces acting on the bucket are:

(i) a tension (or force) of T acting vertically upwards

(ii) a force of mg acting vertically downwards, i.e. the weight of the bucket and cement.

The resultant force, $\quad F = mg - T$

Hence, $\qquad\qquad ma = mg - T$

from which, **tension, $T = m(g - a) = 40(9.81 - 1.4)$**

$$= \mathbf{336.4 \, N}$$

By comparing this result with that of Problem 5, it can be seen that there is a decrease in the tension in the rope when an object is accelerating downwards.

Now try the following exercise

Exercise 115   Further problems on Newton's laws of motion (Answers on page 307)

(Take g as $9.81 \, \text{m/s}^2$, and express answers to three significant figure accuracy)

1. A car initially at rest accelerates uniformly to a speed of 55 km/h in 14 s. Determine the accelerating force required if the mass of the car is 800 kg.

2. The brakes are applied on the car in question 1 when travelling at 55 km/h and it comes to rest uniformly in a distance of 50 m. Calculate the braking force and the time for the car to come to rest.

3. The tension in a rope lifting a crate vertically upwards is 2.8 kN. Determine its acceleration if the mass of the crate is 270 kg.

4. A ship is travelling at 18 km/h when it stops its engines. It drifts for a distance of 0.6 km and its speed is then 14 km/h. Determine the value of the forces opposing the motion of the ship, assuming the reduction in speed is uniform and the mass of the ship is 2000 t.

5. A cage having a mass of 2 t is being lowered down a mine shaft. It moves from rest with an acceleration of $4 \, \text{m/s}^2$, until it is travelling at 15 m/s. It then travels at constant speed for 700 m and finally comes to rest in 6 s. Calculate the tension in the cable supporting the cage during:

   (a) the initial period of acceleration,
   (b) the period of constant speed travel,
   (c) the final retardation period.

26.3   Centripetal acceleration

When an object moves in a circular path at constant speed, its direction of motion is continually changing and hence its velocity (which depends on both magnitude and direction) is also continually changing. Since acceleration is the (change in velocity)/(time taken) the object has an acceleration.

Let the object be moving with a constant angular velocity of ω and a tangential velocity of magnitude v and let the change of velocity for a small change of angle of $\theta(=\omega t)$ be V (see Fig. 26.3(a)). Then, $v_2 - v_1 = V$.

The vector diagram is shown in Fig. 26.3(b) and since the magnitudes of v_1 and v_2 are the same, i.e. v, the vector diagram is also an isosceles triangle.

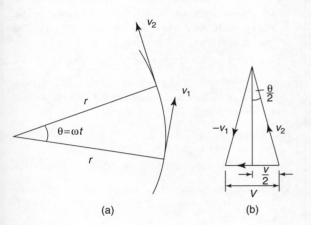

(a)   (b)

Fig. 26.3

Bisecting the angle between v_2 and v_1 gives:

$$\sin\frac{\theta}{2} = \frac{V/2}{v_2} = \frac{V}{2v}$$

i.e. $$V = 2v\sin\frac{\theta}{2} \qquad (1)$$

Since $\theta = \omega t$, then $t = \dfrac{\theta}{\omega}$ (2)

Dividing (1) by (2) gives: $\dfrac{V}{t} = \dfrac{2v\sin\frac{\theta}{2}}{\frac{\theta}{\omega}} = \dfrac{v\omega\sin\frac{\theta}{2}}{\frac{\theta}{2}}$

For small angles, $\dfrac{\sin\frac{\theta}{2}}{\frac{\theta}{2}}$ is very nearly equal to unity,

hence, $\dfrac{V}{t} = \dfrac{\text{change of velocity}}{\text{change of time}} = \text{acceleration}, a = v\omega$

But, $\omega = v/r$, thus $v\omega = v \times \dfrac{v}{r} = \dfrac{v^2}{r}$

That is, **the acceleration a is** $\dfrac{v^2}{r}$ and is towards the centre of the circle of motion (along V). It is called the **centripetal acceleration**. If the mass of the rotating object is m, then by

Newton's second law, the **centripetal force** is $\dfrac{mv^2}{r}$, and its direction is towards the centre of the circle of motion.

Problem 8. A vehicle of mass 750 kg travels round a bend of radius 150 m, at 50.4 km/h. Determine the centripetal force acting on the vehicle.

The centripetal force is given by $\dfrac{mv^2}{r}$ and its direction is towards the centre of the circle.

$$m = 750\,\text{kg}, v = 50.4\,\text{km/h}$$
$$= \frac{50.4}{3.6}\,\text{m/s} = 14\,\text{m/s} \quad\text{and}\quad r = 150\,\text{m}$$

Thus, **centripetal force** $= \dfrac{750 \times 14^2}{150} = \textbf{980 N}$

Problem 9. An object is suspended by a thread 250 mm long and both object and thread move in a horizontal circle with a constant angular velocity of 2.0 rad/s. If the tension in the thread is 12.5 N, determine the mass of the object.

Centripetal force (i.e. tension in thread) $= \dfrac{mv^2}{r} = 12.5\,\text{N}$.

The angular velocity, $\omega = 2.0\,\text{rad/s}$

and radius, $r = 250\,\text{mm} = 0.25\,\text{m}$

Since linear velocity $v = \omega r, v = 2.0 \times 0.25 = 0.5\,\text{m/s}$,

and since $F = \dfrac{mv^2}{r}$, then $m = \dfrac{Fr}{v^2}$

i.e. **mass of object**, $m = \dfrac{12.5 \times 0.25}{0.5^2} = \textbf{12.5 kg}$

Problem 10. An aircraft is turning at constant altitude, the turn following the arc of a circle of radius 1.5 km. If the maximum allowable acceleration of the aircraft is 2.5 g, determine the maximum speed of the turn in km/h. Take g as 9.8 m/s².

The acceleration of an object turning in a circle is $\dfrac{v^2}{r}$. Thus, to determine the maximum speed of turn $\dfrac{v^2}{r} = 2.5\,g$. Hence,

speed of turn, $v = \sqrt{2.5\,gr} = \sqrt{2.5 \times 9.8 \times 1500}$
$$= \sqrt{36750} = 191.7\,\text{m/s}$$
$$= 191.7 \times 3.6\,\text{km/h} = \textbf{690 km/h}$$

Now try the following exercises

Exercise 116 Further problems on centripetal acceleration (Answers on page 308)

1. Calculate the centripetal force acting on a vehicle of mass 1 tonne when travelling round a bend of radius 125 m at 40 km/h. If this force should not exceed 750 N, determine the reduction in speed of the vehicle to meet this requirement.

2. A speed-boat negotiates an S-bend consisting of two circular arcs of radii 100 m and 150 m. If the speed of the boat is constant at 34 km/h, determine the change in acceleration when leaving one arc and entering the other.

3. An object is suspended by a thread 400 mm long and both object and thread move in a horizontal circle with a constant angular velocity of 3.0 rad/s. If the tension in the thread is 36 N, determine the mass of the object.

Exercise 117 Short answer questions on force, mass and acceleration

1. Force is measured in . . . .
2. The two effects of pushing or pulling an object are . . . or . . . .
3. A gravitational force gives free-falling objects a . . . in the absence of all other forces.
4. State Newton's first law of motion.
5. Describe what is meant by the inertia of an object.
6. State Newton's second law of motion.
7. Define the newton.
8. State Newton's third law of motion.
9. Explain why an object moving round a circle at a constant angular velocity has an acceleration.
10. Define centripetal acceleration in symbols.
11. Define centripetal force in symbols.

Exercise 118 Multi-choice questions on force, mass and acceleration (Answers on page 308)

1. The unit of force is the:
   (a) watt   (b) kelvin   (c) newton   (d) joule

2. If a = acceleration and F = force, then mass m is given by:
   (a) $m = a - F$ (b) $m = \dfrac{F}{a}$
   (c) $m = F - a$ (d) $m = \dfrac{a}{F}$

3. The weight of an object of mass 2 kg at a point on the earth's surface when the gravitational field is 10 N/kg is:
   (a) 20 N   (b) 0.2 N   (c) 20 kg   (d) 5 N

4. The force required to accelerate a loaded barrow of 80 kg mass up to 0.2 m/s² on friction-less bearings is:
   (a) 400 N   (b) 3.2 N   (c) 0.0025 N   (d) 16 N

5. A bucket of cement of mass 30 kg is tied to the end of a rope connected to a hoist. If the gravitational field $g = 10$ N/kg, the tension in the rope when the bucket is suspended but stationary is:
   (a) 300 N   (b) 3 N   (c) 300 kg   (d) 0.67 N

A man of mass 75 kg is standing in a lift of mass 500 kg. Use this data to determine the answers to questions 6 to 9. Take g as 10 m/s².

6. The tension in a cable when the lift is moving at a constant speed vertically upward is:
   (a) 4250 N   (b) 5750 N   (c) 4600 N   (d) 6900 N

7. The tension in the cable supporting the lift when the lift is moving at a constant speed vertically downwards is:
   (a) 4250 N   (b) 5750 N   (c) 4600 N   (d) 6900 N

8. The reaction force between the man and the floor of the lift when the lift is travelling at a constant speed vertically upwards is:
   (a) 750 N   (b) 900 N   (c) 600 N   (d) 475 N

9. The reaction force between the man and the floor of the lift when the lift is travelling at a constant speed vertically downwards is:
   (a) 750 N   (b) 900 N   (c) 600 N   (d) 475 N

A ball of mass 0.5 kg is tied to a thread and rotated at a constant angular velocity of 10 rad/s in a circle of radius 1 m. Use this data to determine the answers to questions 10 and 11.

10. The centripetal acceleration is:
    (a) 50 m/s²   (b) $\dfrac{100}{2\pi}$ m/s²
    (c) $\dfrac{50}{2\pi}$ m/s²   (d) 100 m/s²

11. The tension in the thread is:
    (a) 25 N   (b) $\dfrac{50}{2\pi}$ N   (c) $\dfrac{25}{2\pi}$ N   (d) 50 N

12. Which of the following statements is false?
    (a) An externally applied force is needed to change the direction of a moving object.

(b) For every force, there is an equal and opposite reaction force.

(c) A body travelling at a constant velocity in a circle has no acceleration.

(d) Centripetal acceleration acts towards the centre of the circle of motion.

Assignment 9

This assignment covers the material contained in chapters 24 to 26. The marks for each question are shown in brackets at the end of each question.

1. A vehicle is travelling at a constant speed of 15 metres per second for 20 kilometres. Find the time taken to cover this distance. (4)

2. A train travels from station P to station Q, a distance of 50 kilometres at an average speed of 80 kilometres per hour. It then travels from station Q to station R, a distance of 30 kilometres in 30 minutes. Finally, it travels from station R to station S at an average speed of 72 kilometres per hour in 45 minutes. Determine (a) the time taken to travel from P to Q, (b) the average speed of the train from Q to R, (c) the distance from R to S, and (d) the average speed of the whole journey from P to S. (12)

3. The motion of a car is described by the speed/time graph given in Fig. A9.1. Determine the distance covered by the car when moving from O to Y. (6)

4. For the speed/time graph of a vehicle shown in Fig. A9.2, find the acceleration for each of the three stages of the journey. (8)

5. A lorry travelling at 40 km/h accelerates uniformly to 76 km/h in 1 minute 40 seconds. Determine the acceleration. (6)

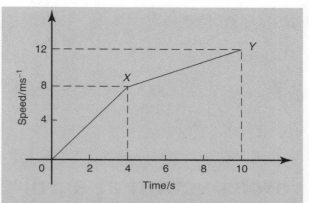

Fig. A9.1

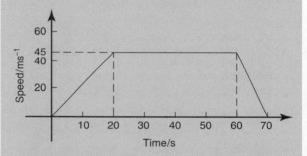

Fig. A9.2

6. Determine the mass of the moving head of a machine tool if it requires a force of 1.5 N to bring it to rest in 0.75 s from a cutting speed of 25 m/min. (5)

7. Find the weight of an object of mass 2.5 kg at a point on the earth's surface where the gravitational field is 9.8 N/kg. (4)

8. A van of mass 1200 kg travels round a bend of radius 120 m, at 54 km/h. Determine the centripetal force acting on the vehicle. (5)

27

Forces acting at a point

At the end of this chapter you should be able to:

- distinguish between scalar and vector quantities
- define 'centre of gravity' of an object
- define 'equilibrium' of an object
- understand the terms 'coplanar' and 'concurrent'
- determine the resultant of two coplanar forces using

  (a) the triangle of forces method
  (b) the parallelogram of forces method

- calculate the resultant of two coplanar forces using

  (a) the cosine and sine rules
  (b) resolution of forces

- determine the resultant of more than two coplanar forces using

  (a) the polygon of forces method
  (b) calculation by resolution of forces

- determine unknown forces when three or more coplanar forces are in equilibrium

27.1 Scalar and vector quantities

Quantities used in engineering and science can be divided into two groups:

(a) **Scalar quantities** have a size (or magnitude) only and need no other information to specify them. Thus, 10 centimetres, 50 seconds, 7 litres and 3 kilograms are all examples of scalar quantities.

(b) **Vector quantities** have both a size or magnitude and a direction, called the line of action of the quantity. Thus, a velocity of 50 kilometres per hour due east, an

acceleration of 9.81 metres per second squared vertically downwards and a force of 15 newtons at an angle of 30 degrees are all examples of vector quantities.

27.2 Centre of gravity and equilibrium

The **centre of gravity** of an object is a point where the resultant gravitational force acting on the body may be taken to act. For objects of uniform thickness lying in a horizontal plane, the centre of gravity is vertically in line with the point of balance of the object. For a thin uniform rod the point of balance and hence the centre of gravity is halfway along the rod as shown in Fig. 27.1(a).

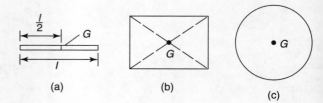

(a) (b) (c)

Fig. 27.1

A thin flat sheet of a material of uniform thickness is called a **lamina** and the centre of gravity of a rectangular lamina lies at the point of intersection of its diagonals, as shown in Fig. 27.1(b). The centre of gravity of a circular lamina is at the centre of the circle, as shown in Fig. 27.1(c).

An object is in **equilibrium** when the forces acting on the object are such that there is no tendency for the object to move. The state of equilibrium of an object can be divided into three groups.

(i) If an object is in **stable equilibrium** and it is slightly disturbed by pushing or pulling (i.e. a disturbing force is applied), the centre of gravity is raised and when the disturbing force is removed, the object returns to its original position. Thus a ball bearing in a hemispherical cup is in stable equilibrium, as shown in Fig. 27.2(a).

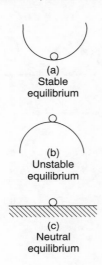

(a)
Stable
equilibrium

(b)
Unstable
equilibrium

(c)
Neutral
equilibrium

Fig. 27.2

(ii) An object is in **unstable equilibrium** if, when a disturbing force is applied, the centre of gravity is lowered and the object moves away from its original position. Thus, a ball bearing balanced on top of a hemispherical cup is in unstable equilibrium, as shown in Fig. 27.2(b).

(iii) When an object in **neutral equilibrium** has a disturbing force applied, the centre of gravity remains at the same height and the object does not move when the disturbing force is removed. Thus, a ball bearing on a flat horizontal surface is in neutral equilibrium, as shown in Fig. 27.2(c).

27.3 Forces

When forces are all acting in the same plane, they are called **coplanar**. When forces act at the same time and at the same point, they are called **concurrent forces**.

Force is a **vector quantity** and thus has both a magnitude and a direction. A vector can be represented graphically by a line drawn to scale in the direction of the line of action of the force.

To distinguish between vector and scalar quantities, various ways are used. These include:

(i) **bold print**

(ii) two capital letters with an arrow above them to denote the sense of direction, e.g. $\overrightarrow{AB}$, where A is the starting point and B the end point of the vector

(iii) a line over the top of letters, e.g. $\overline{AB}$ or $\bar{a}$

(iv) letters with an arrow above, e.g. $\vec{a}$, $\vec{A}$

(v) underlined letters, e.g. $\underline{a}$

(vi) $xi + jy$, where i and j are axes at right-angles to each other; for example, $3i + 4j$ means 3 units in the i direction and 4 units in the j direction, as shown in Fig. 27.3

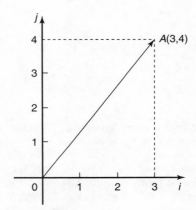

Fig. 27.3

(vii) a column matrix $\begin{pmatrix} a \\ b \end{pmatrix}$; for example, the vector $\boldsymbol{OA}$ shown in Fig. 27.3 could be represented by $\begin{pmatrix} 3 \\ 4 \end{pmatrix}$

Thus, in Fig. 27.3, $\boldsymbol{OA} \equiv \overrightarrow{OA} \equiv \overline{OA} \equiv 3i + 4j \equiv \begin{pmatrix} 3 \\ 4 \end{pmatrix}$

The method adopted in this text is to denote vector quantities in **bold print**. Thus, *ab* in Fig. 27.4 represents a force of 5 newtons acting in a direction due east.

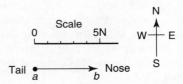

Fig. 27.4

27.4 The resultant of two coplanar forces

For two forces acting at a point, there are three possibilities.

(a) For forces acting in the same direction and having the same line of action, the single force having the same effect as both of the forces, called the **resultant force** or just the **resultant**, is the arithmetic sum of the separate forces. Forces of F_1 and F_2 acting at point P, as shown in Fig. 27.5(a), have exactly the same effect on point P as force F shown in Fig. 27.5(b), where $F = F_1 + F_2$ and acts in the same direction as F_1 and F_2. Thus F is the resultant of F_1 and F_2

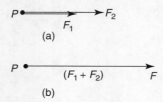

(a)

(b)

Fig. 27.5

(b) For forces acting in opposite directions along the same line of action, the resultant force is the arithmetic difference between the two forces. Forces of F_1 and F_2 acting at point P as shown in Fig. 27.6(a) have exactly the same effect on point P as force F shown in Fig. 27.6(b), where $F = F_2 - F_1$ and acts in the direction of F_2, since F_2 is greater than F_1. Thus F is the resultant of F_1 and F_2

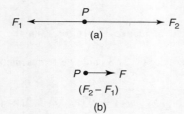

(a)

(b)

Fig. 27.6

(c) When two forces do not have the same line of action, the magnitude and direction of the resultant force may be found by a procedure called vector addition of forces. There are two graphical methods of performing **vector addition**, known as the **triangle of forces** method (see Section 27.5) and the **parallelogram of forces** method (see Section 27.6).

Problem 1. Determine the resultant force of two forces of 5 kN and 8 kN,

(a) acting in the same direction and having the same line of action,

(b) acting in opposite directions but having the same line of action.

(a) The vector diagram of the two forces acting in the same direction is shown in Fig. 27.7(a), which assumes that

the line of action is horizontal, although since it is not specified, could be in any direction. From above, the resultant force F is given by: $F = F_1 + F_2$, i.e. $F = (5 + 8)\,\text{kN} = \mathbf{13\,kN}$ in the direction of the original forces.

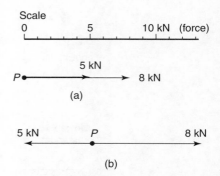

(a)

(b)

Fig. 27.7

(b) The vector diagram of the two forces acting in opposite directions is shown in Fig. 27.7(b), again assuming that the line of action is in a horizontal direction. From above, the resultant force F is given by: $F = F_2 - F_1$, i.e. $F = (8 - 5)\,\text{kN} = \mathbf{3\,kN}$ in the direction of the 8 kN force.

27.5 Triangle of forces method

A simple procedure for the triangle of forces method of vector addition is as follows:

(i) Draw a vector representing one of the forces, using an appropriate scale and in the direction of its line of action.

(ii) From the **nose** of this vector and using the same scale, draw a vector representing the second force in the direction of its line of action.

(iii) The resultant vector is represented in both magnitude and direction by the vector drawn from the tail of the first vector to the nose of the second vector.

Problem 2. Determine the magnitude and direction of the resultant of a force of 15 N acting horizontally to the right and a force of 20 N, inclined at an angle of 60° to the 15 N force. Use the triangle of forces method.

Using the procedure given above and with reference to Fig. 27.8.

(i) *ab* is drawn 15 units long horizontally.

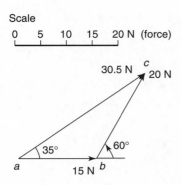

Fig. 27.8

(ii) From b, **bc** is drawn 20 units long, inclined at an angle of 60° to *ab*.
(Note, in angular measure, an angle of 60° from *ab* means 60° in an anticlockwise direction)

(iii) By measurement, the resultant **ac** is 30.5 units long inclined at an angle of 35° to *ab*. That is, the resultant force is **30.5 N**, inclined at an angle of **35°** to the 15 N force.

Problem 3. Find the magnitude and direction of the two forces given, using the triangle of forces method.

First force: 1.5 kN acting at an angle of 30°
Second force: 3.7 kN acting at an angle of −45°

From the above procedure and with reference to Fig. 27.9:

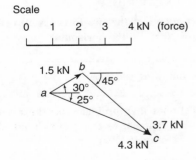

Fig. 27.9

(i) **ab** is drawn at an angle of 30° and 1.5 units in length.

(ii) From b, **bc** is drawn at an angle of −45° and 3.7 units in length. (Note, an angle of −45° means a clockwise rotation of 45° from a line drawn horizontally to the right.)

(iii) By measurement, the resultant **ac** is 4.3 units long at an angle of −25°. That is, the resultant force is **4.3 kN** at an angle of **−25°**

Now try the following exercise

Exercise 119   Further problems on the triangle of forces method (Answers on page 308)

In questions 1 to 5, use the triangle of forces method to determine the magnitude and direction of the resultant of the forces given.

1. 1.3 kN and 2.7 kN, having the same line of action and acting in the same direction.

2. 470 N and 538 N having the same line of action but acting in opposite directions.

3. 13 N at 0° and 25 N at 30°

4. 5 N at 60° and 8 N at 90°

5. 1.3 kN at 45° and 2.8 kN at −30°

27.6 The parallelogram of forces method

A simple procedure for the parallelogram of forces method of vector addition is as follows:

(i) Draw a vector representing one of the forces, using an appropriate scale and in the direction of its line of action.

(ii) From the **tail** of this vector and using the same scale draw a vector representing the second force in the direction of its line of action.

(iii) Complete the parallelogram using the two vectors drawn in (i) and (ii) as two sides of the parallelogram.

(iv) The resultant force is represented in both magnitude and direction by the vector corresponding to the diagonal of the parallelogram drawn from the tail of the vectors in (i) and (ii).

Problem 4. Use the parallelogram of forces method to find the magnitude and direction of the resultant of a force of 250 N acting at an angle of 135° and a force of 400 N acting at an angle of −120°

From the procedure given above and with reference to Fig. 27.10:

(i) **ab** is drawn at an angle of 135° and 250 units in length.

(ii) **ac** is drawn at an angle of −120° and 400 units in length.

(iii) **bd** and **cd** are drawn to complete the parallelogram.

(iv) **ad** is drawn. By measurement, **ad** is 413 units long at an angle of −156°. That is, the resultant force is **413 N** at an angle of **−156°**

Scale

0 100 200 300 400 500 N (force)

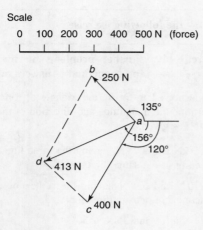

Fig. 27.10

Now try the following exercise

Exercise 120   Further problems on the parallelogram of forces method (Answers on page 308)

In questions 1 to 5, use the parallelogram of forces method to determine the magnitude and direction of the resultant of the forces given.

1. 1.7 N at 45° and 2.4 N at −60°
2. 9 N at 126° and 14 N at 223°
3. 23.8 N at −50° and 14.4 N at 215°
4. 0.7 kN at 147° and 1.3 kN at −71°
5. 47 N at 79° and 58 N at 247°

27.7 Resultant of coplanar forces by calculation

An alternative to the graphical methods of determining the resultant of two coplanar forces is by **calculation**. This can be achieved by **trigonometry** using the **cosine rule** and the **sine rule**, as shown in Problem 5 following, or by **resolution of forces** (see Section 27.10). Refer to Chapter 10 for more on trigonometry.

Problem 5. Use the cosine and sine rules to determine the magnitude and direction of the resultant of a force of 8 kN acting at an angle of 50° to the horizontal and a force of 5 kN acting at an angle of −30° to the horizontal.

The space diagram is shown in Fig. 27.11(a). A sketch is made of the vector diagram, *oa* representing the 8 kN force in magnitude and direction and *ab* representing the 5 kN force in magnitude and direction. The resultant is given by length

ob. By the cosine rule (see page 58),

$$ob^2 = oa^2 + ab^2 - 2(oa)(ab)\cos \angle oab$$
$$= 8^2 + 5^2 - 2(8)(5)\cos 100°$$
$$\text{(since } \angle oab = 180° - 50° - 30° = 100°)$$
$$= 64 + 25 - (-13.892) = 102.892$$

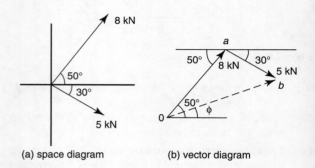

(a) space diagram   (b) vector diagram

Fig. 27.11

Hence $ob = \sqrt{102.892} = 10.14\,\text{kN}$

By the sine rule, $\dfrac{5}{\sin \angle aob} = \dfrac{10.14}{\sin 100°}$

from which, $\sin \angle aob = \dfrac{5 \sin 100°}{10.14} = 0.4856$

Hence $\angle aob = \sin^{-1}(0.4856) = 29.05°$. Thus angle ϕ in Fig. 27.11(b) is $50° - 29.05° = 20.95°$

Hence the resultant of the two forces is 10.14 kN acting at an angle of 20.95° to the horizontal.

Now try the following exercise

Exercise 121   Further problems on the resultant of coplanar forces by calculation (Answers on page 308)

1. Forces of 7.6 kN at 32° and 11.8 kN at 143° act at a point. Use the cosine and sine rules to calculate the magnitude and direction of their resultant.

In questions 2 to 5, calculate the resultant of the given forces by using the cosine and sine rules.

2. 13 N at 0° and 25 N at 30°
3. 1.3 kN at 45° and 2.8 kN at −30°
4. 9 N at 126° and 14 N at 223°
5. 0.7 kN at 147° and 1.3 kN at −71°

27.8 Resultant of more than two coplanar forces

For the three coplanar forces F_1, F_2 and F_3 acting at a point as shown in Fig. 27.12, the vector diagram is drawn using the nose-to-tail method of Section 27.5. The procedure is:

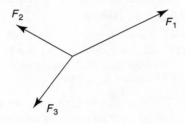

Fig. 27.12

(i) Draw *oa* to scale to represent force F_1 in both magnitude and direction (see Fig. 27.13)

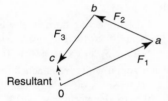

Fig. 27.13

(ii) From the nose of *oa*, draw *ab* to represent force F_2

(iii) From the nose of *ab*, draw *bc* to represent force F_3

(iv) The resultant vector is given by length *oc* in Fig. 27.13. The direction of resultant *oc* is from where we started, i.e. point 0, to where we finished, i.e. point *c*. When acting by itself, the resultant force, given by *oc*, has the same effect on the point as forces F_1, F_2 and F_3 have when acting together. The resulting vector diagram of Fig. 27.13 is called the **polygon of forces**.

Problem 6. Determine graphically the magnitude and direction of the resultant of these three coplanar forces, which may be considered as acting at a point. Force *A*, 12 N acting horizontally to the right; force *B*, 7 N inclined at 60° to force *A*; force *C*, 15 N inclined at 150° to force *A*.

The space diagram is shown in Fig. 27.14. The vector diagram shown in Fig. 27.15, is produced as follows:

(i) *oa* represents the 12 N force in magnitude and direction.

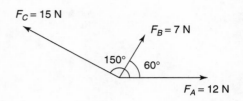

Fig. 27.14

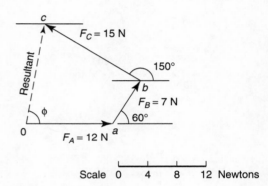

Fig. 27.15

(ii) From the nose of *oa*, *ab* is drawn inclined at 60° to *oa* and 7 units long.

(iii) From the nose of *ab*, *bc* is drawn 15 units long inclined at 150° to *oa* (i.e. 150° to the horizontal).

(iv) *oc* represents the resultant; by measurement, the resultant is 13.8 N inclined at $\phi = 80°$ to the horizontal.

Thus the resultant of the three forces, F_A, F_B and F_C is a force of 13.8 N at 80° to the horizontal.

Problem 7. The following coplanar forces are acting at a point, the given angles being measured from the horizontal: 100 N at 30°, 200 N at 80°, 40 N at −150°, 120 N at −100° and 70 N at −60°. Determine graphically the magnitude and direction of the resultant of the five forces.

The five forces are shown in the space diagram of Fig. 27.16. Since the 200 N and 120 N forces have the same line of action but are in opposite sense, they can be represented by a single force of 200 − 120, i.e. 80 N acting at 80° to the horizontal. Similarly, the 100 N and 40 N forces can be represented by a force of 100 − 40, i.e. 60 N acting at 30° to the horizontal. Hence the space diagram of Fig. 27.16 may be represented by the space diagram of Fig. 27.17. Such a simplification of the vectors is not essential but it is easier to construct the vector diagram from a space diagram having three forces, than one with five.

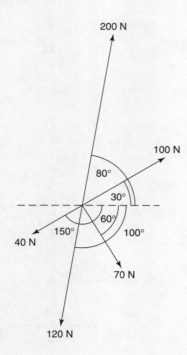

Fig. 27.16

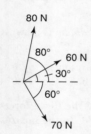

Fig. 27.17

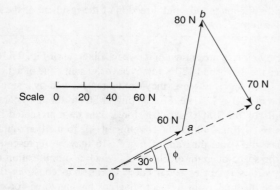

Fig. 27.18

The vector diagram is shown in Fig. 27.18, *oa* representing the 60 N force, *ab* representing the 80 N force and *bc* the 70 N force. The resultant, *oc*, is found by measurement to represent a force of 112 N and angle ϕ is 25°

Thus the five forces shown in Fig. 27.16 may be represented by a single force of 112 N at 25° to the horizontal.

Now try the following exercise

Exercise 122   Further problems on the resultant of more than two coplanar forces (Answers on page 308)

In questions 1 to 3, determine graphically the magnitude and direction of the resultant of the coplanar forces given which are acting at a point.

1. Force A, 12 N acting horizontally to the right, force B, 20 N acting at 140° to force A; force C, 16 N acting at 290° to force A.

2. Force 1, 23 kN acting at 80° to the horizontal; force 2, 30 kN acting at 37° to force 1; force 3, 15 kN acting at 70° to force 2.

3. Force P, 50 kN acting horizontally to the right; force Q, 20 kN at 70° to force P; force R, 40 kN at 170° to force P; force S, 80 kN at 300° to force P.

4. Four horizontal wires are attached to a telephone pole and exert tensions of 30 N to the south, 20 N to the east, 50 N to the north-east and 40 N to the north-west. Determine the resultant force on the pole and its direction.

27.9   Coplanar forces in equilibrium

When three or more coplanar forces are acting at a point and the vector diagram closes, there is no resultant. The forces acting at the point are in **equilibrium**.

Problem 8.   A load of 200 N is lifted by two ropes connected to the same point on the load, making angles of 40° and 35° with the vertical. Determine graphically the tensions in each rope when the system is in equilibrium.

The space diagram is shown in Fig. 27.19. Since the system is in equilibrium, the vector diagram must close. The vector diagram, shown in Fig. 27.20, is drawn as follows:

(i)   The load of 200 N is drawn vertically as shown by *oa*.

(ii)   The direction only of force F_1 is known, so from point *a*, *ad* is drawn at 40° to the vertical.

(iii)   The direction only of force F_2 is known, so from point *o*, *oc* is drawn at 35° to the vertical.

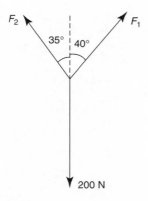

Fig. 27.19

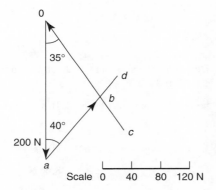

Fig. 27.20

(iv) Lines **ad** and **oc** cross at point *b*; hence the vector diagram is given by triangle *oab*. By measurement, **ab** is 119 N and **ob** is 133 N.

Thus the tensions in the ropes are $F_1 = 119$ N and $F_2 = 133$ N.

Problem 9. Five coplanar forces are acting on a body and the body is in equilibrium. The forces are: 12 kN acting horizontally to the right, 18 kN acting at an angle of 75°, 7 kN acting at an angle of 165°, 16 kN acting from the nose of the 7 kN force, and 15 kN acting from the nose of the 16 kN force. Determine the directions of the 16 kN and 15 kN forces relative to the 12 kN force.

With reference to Fig. 27.21, **oa** is drawn 12 units long horizontally to the right. From point *a*, **ab** is drawn 18 units long at an angle of 75°. From *b*, **bc** is drawn 7 units long at an angle of 165°. The direction of the 16 kN force is not known, thus arc *pq* is drawn with a compass, with centre at *c*, radius 16 units. Since the forces are at equilibrium, the polygon of forces must close. Using a compass with centre at 0, arc *rs* is drawn having a radius 15 units. The point where the arcs intersect is at *d*.

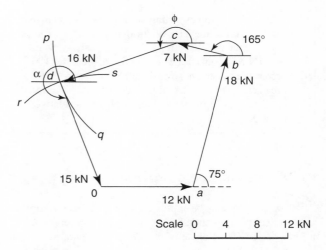

Fig. 27.21

By measurement, angle $\phi = 198°$ and $\alpha = 291°$

Thus the 16 kN force acts at an angle of 198° (or −162°) to the 12 kN force, and the 15 kN force acts at an angle of 291° (or −69°) to the 12 kN force.

Now try the following exercise

Exercise 123   Further problems on coplanar forces in equilibrium (Answers on page 308)

1. A load of 12.5 N is lifted by two strings connected to the same point on the load, making angles of 22° and 31° on opposite sides of the vertical. Determine the tensions in the strings.

2. A two-legged sling and hoist chain used for lifting machine parts is shown in Fig. 27.22. Determine the forces in each leg of the sling if parts exerting a downward force of 15 kN are lifted.

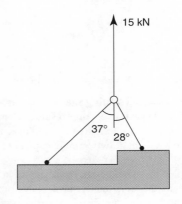

Fig. 27.22

3. Four coplanar forces acting on a body are such that it is in equilibrium. The vector diagram for the forces is such that the 60 N force acts vertically upwards, the 40 N force acts at 65° to the 60 N force, the 100 N force acts from the nose of the 60 N force and the 90 N force acts from the nose of the 100 N force. Determine the direction of the 100 N and 90 N forces relative to the 60 N force.

27.10   Resolution of forces

A vector quantity may be expressed in terms of its **horizontal** and **vertical components**. For example, a vector representing a force of 10 N at an angle of 60° to the horizontal is shown in Fig. 27.23. If the horizontal line *oa* and the vertical line *ab* are constructed as shown, then *oa* is called the horizontal component of the 10 N force, and *ab* the vertical component of the 10 N force.

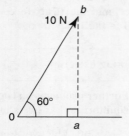

Fig. 27.23

By trigonometry,

$$\cos 60° = \frac{oa}{ob}$$

hence the horizontal component,

$$oa = 10 \cos 60°$$

Also, $\sin 60° = \dfrac{ab}{ob}$

hence the vertical component, $ab = 10 \sin 60°$

This process is called **finding the horizontal and vertical components of a vector** or **the resolution of a vector**, and can be used as an alternative to graphical methods for calculating the resultant of two or more coplanar forces acting at a point.

For example, to calculate the resultant of a 10 N force acting at 60° to the horizontal and a 20 N force acting at −30° to the horizontal (see Fig. 27.24) the procedure is as follows:

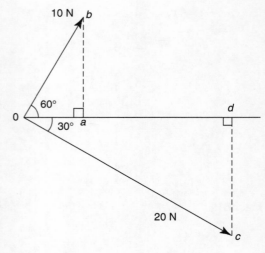

Fig. 27.24

(i) Determine the horizontal and vertical components of the 10 N force, i.e.

horizontal component, $oa = 10 \cos 60°$

$$= 5.0 \text{ N, and}$$

vertical component, $ab = 10 \sin 60°$

$$= 8.66 \text{ N}$$

(ii) Determine the horizontal and vertical components of the 20 N force, i.e.

horizontal component, $od = 20 \cos(-30°)$

$$= 17.32 \text{ N, and}$$

vertical component, $cd = 20 \sin(-30°)$

$$= -10.0 \text{ N}$$

(iii) Determine the total horizontal component, i.e.

$$oa + od = 5.0 + 17.32 = 22.32 \text{ N}$$

(iv) Determine the total vertical component, i.e.

$$ab + cd = 8.66 + (-10.0) = -1.34 \text{ N}$$

(v) Sketch the total horizontal and vertical components as shown in Fig. 27.25. The resultant of the two components is given by length *or* and, by

Pythagoras' theorem, **or** $= \sqrt{22.32^2 + 1.34^2}$

$$= 22.36 \text{ N}$$

and using trigonometry, angle

$$\phi = \tan^{-1}\frac{1.34}{22.32}$$

$$= 3.44°$$

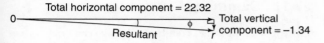

Fig. 27.25

Hence the resultant of the 10 N and 20 N forces shown in Fig. 27.24 is **22.36 N at an angle of −3.44° to the horizontal**.

Problem 10. Forces of 5.0 N at 25° and 8.0 N at 112° act at a point. By resolving these forces into horizontal and vertical components, determine their resultant.

The space diagram is shown in Fig. 27.26.

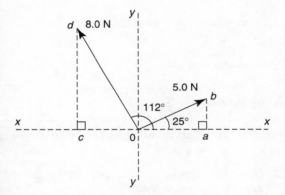

Fig. 27.26

(i) The horizontal component of the 5.0 N force,
$oa = 5.0 \cos 25° = 4.532$, and the vertical component of the 5.0 N force, $ab = 5.0 \sin 25° = 2.113$

(ii) The horizontal component of the 8.0 N force,
$oc = 8.0 \cos 112° = -2.997$. The vertical component of the 8.0 N force, $cd = 8.0 \sin 112° = 7.417$

(iii) Total horizontal component $= oa + oc$
$= 4.532 + (-2.997) = +1.535$

(iv) Total vertical component $= ab + cd$
$= 2.113 + 7.417 = +9.530$

(v) The components are shown sketched in Fig. 27.27

By Pythagoras' theorem,

$$r = \sqrt{1.535^2 + 9.530^2}$$

$$= 9.653$$

and by trigonometry, angle

$$\phi = \tan^{-1} \frac{9.530}{1.535} = 80.85°$$

Fig. 27.27

Hence the resultant of the two forces shown in Fig. 27.26 is a force of 9.653 N acting at 80.85° to the horizontal.

Problems 9 and 10 demonstrates the use of resolution of forces for calculating the resultant of two coplanar forces acting at a point. However the method may be used for more than two forces acting at a point, as shown in Problem 11.

Problem 11. Determine by resolution of forces the resultant of the following three coplanar forces acting at a point: 200 N acting at 20° to the horizontal; 400 N acting at 165° to the horizontal; 500 N acting at 250° to the horizontal.

A tabular approach using a calculator may be made as shown below.

Horizontal component

Force 1	$200 \cos 20° =$	187.94
Force 2	$400 \cos 165° =$	−386.37
Force 3	$500 \cos 250° =$	−171.01
Total horizontal component		$= -369.44$

Vertical component

Force 1	$200 \sin 20° =$	68.40
Force 2	$400 \sin 165° =$	103.53
Force 3	$500 \sin 250° =$	−469.85
Total vertical component		$= -297.92$

The total horizontal and vertical components are shown in Fig. 27.28.

$$\text{Resultant } r = \sqrt{369.44^2 + 297.92^2}$$

$$= 474.60$$

and angle

$$\phi = \tan^{-1} \frac{297.92}{369.44} = 38.88°$$

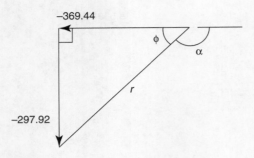

Fig. 27.28

from which,

$$\alpha = 180° - 38.88° = 141.12°$$

Thus the resultant of the three forces given is 474.6 N acting at an angle of −141.12° (or +218.88°) to the horizontal.

Now try the following exercise

Exercise 124   Further problems on resolution of forces (Answers on page 308)

1. Resolve a force of 23.0 N at an angle of 64° into its horizontal and vertical components.

2. Forces of 5 N at 21° and 9 N at 126° act at a point. By resolving these forces into horizontal and vertical components, determine their resultant

    In questions 3 and 4, determine the magnitude and direction of the resultant of the coplanar forces given which are acting at a point, by resolution of forces.

3. Force *A*, 12 N acting horizontally to the right, force *B*, 20 N acting at 140° to force *A*; force *C*, 16 N acting at 290° to force *A*.

4. Force 1, 23 kN acting at 80° to the horizontal; force 2, 30 kN acting at 37° to force 1; force 3, 15 kN acting at 70° to force 2.

5. Determine, by resolution of forces, the resultant of the following three coplanar forces acting at a point: 10 kN acting at 32° to the horizontal; 15 kN acting at 170° to the horizontal; 20 kN acting at 240° to the horizontal.

6. The following coplanar forces act at a point: force *A*, 15 N acting horizontally to the right; force *B*, 23 N at 81° to the horizontal; force *C*, 7 N at 210° to the horizontal; force *D*, 9 N at 265° to the horizontal; force *E*, 28 N at 324° to the horizontal. Determine the resultant of the five forces by resolution of the forces.

27.11   Summary

(a) To determine the **resultant of two coplanar forces** acting at a point, four methods are commonly used. They are:

    by drawing:

    (1) triangle of forces method, and

    (2) parallelogram of forces method, and

    by calculation:

    (3) use of cosine and sine rules, and

    (4) resolution of forces

(b) To determine the **resultant of more than two coplanar forces** acting at a point, two methods are commonly used. They are:

    by drawing:

    (1) polygon of forces method, and

    by calculation:

    (2) resolution of forces

Now try the following exercises

Exercise 125   Short answer questions on forces acting at a point

1. Give one example of a scalar quantity and one example of a vector quantity.

2. Explain the difference between a scalar and a vector quantity.

3. What is meant by the centre of gravity of an object?

4. Where is the centre of gravity of a rectangular lamina?

5. What is meant by neutral equilibrium?

6. State the meaning of the term 'coplanar'.

7. What is a concurrent force?

8. State what is meant by a triangle of forces.

9. State what is meant by a parallelogram of forces.

10. State what is meant by a polygon of forces.

11. When a vector diagram is drawn representing coplanar forces acting at a point, and there is no resultant, the forces are in ...

12. Two forces of 6 N and 9 N act horizontally to the right. The resultant is ... N acting ...

13. A force of 10 N acts at an angle of 50° and another force of 20 N acts at an angle of 230°. The resultant is a force ... N acting at an angle of ...°

14. What is meant by 'resolution of forces'?

15. A coplanar force system comprises a 20 kN force acting horizontally to the right, 30 kN at 45°, 20 kN at 180° and 25 kN at 225°. The resultant is a force of ... N acting at an angle of ...° to the horizontal.

Exercise 126 Multi-choice questions on forces acting at a point (Answers on page 308)

1. A physical quantity which has direction as well as magnitude is known as a:

   (a) force (b) vector (c) scalar (d) weight

2. Which of the following is not a scalar quantity?

   (a) velocity (b) potential energy

   (c) work (d) kinetic energy

3. Which of the following is not a vector quantity?

   (a) displacement (b) density

   (c) velocity (d) acceleration

4. Which of the following statements is false?

   (a) Scalar quantities have size or magnitude only

   (b) Vector quantities have both magnitude and direction

   (c) Mass, length and time are all scalar quantities

   (d) Distance, velocity and acceleration are all vector quantities

5. If the centre of gravity of an object which is slightly disturbed is raised and the object returns to its original position when the disturbing force is removed, the object is said to be in

   (a) neutral equilibrium (b) stable equilibrium

   (c) static equilibrium (d) unstable equilibrium

6. Which of the following statements is false?

   (a) The centre of gravity of a lamina is at its point of balance

   (b) The centre of gravity of a circular lamina is at its centre

   (c) The centre of gravity of a rectangular lamina is at the point of intersection of its two sides

   (d) The centre of gravity of a thin uniform rod is halfway along the rod

7. The magnitude of the resultant of the vectors shown in Fig. 27.29 is:

   (a) 2 N (b) 12 N (c) 35 N (d) −2 N

Fig. 27.29

8. The magnitude of the resultant of the vectors shown in Fig. 27.30 is:

   (a) 7 N (b) 5 N (c) 1 N (d) 12 N

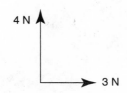

Fig. 27.30

9. Which of the following statements is false?

   (a) There is always a resultant vector required to close a vector diagram representing a system of coplanar forces acting at a point, which are not in equilibrium

   (b) A vector quantity has both magnitude and direction

   (c) A vector diagram representing a system of coplanar forces acting at a point when in equilibrium does not close

   (d) Concurrent forces are those which act at the same time at the same point

10. Which of the following statements is false?

    (a) The resultant of coplanar forces of 1 N, 2 N and 3 N acting at a point can be 4 N

    (b) The resultant of forces of 6 N and 3 N acting in the same line of action but opposite in sense is 3 N

    (c) The resultant of forces of 6 N and 3 N acting in the same sense and having the same line of action is 9 N

    (d) The resultant of coplanar forces of 4 N at 0°, 3 N at 90° and 8 N at 180° is 15 N

11. A space diagram of a force system is shown in Fig. 27.31. Which of the vector diagrams in Fig. 27.32 does not represent this force system?

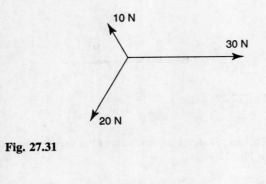

Fig. 27.31

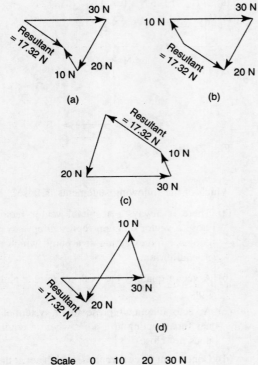

Scale   0   10   20   30 N

Fig. 27.32

12. With reference to Fig. 27.33, which of the following statements is false?

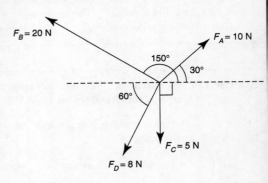

Fig. 27.33

(a) The horizontal component of F_A is 8.66 N
(b) The vertical component of F_B is 10 N
(c) The horizontal component of F_C is 0
(d) The vertical component of F_D is 4 N

13. The resultant of two forces of 3 N and 4 N can never be equal to:

(a) 2.5 N   (b) 4.5 N   (c) 6.5 N   (d) 7.5 N

14. The magnitude of the resultant of the vectors shown in Fig. 27.34 is:

(a) 5 N   (b) 13 N   (c) 1 N   (d) 63 N

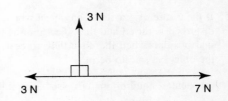

Fig. 27.34

28

Simply supported beams

At the end of this chapter you should be able to:

- define a 'moment' of a force and state its unit
- calculate the moment of a force from $M = F \times d$
- understand the conditions for equilibrium of a beam
- state the principle of moments
- perform calculations involving the principle of moments
- recognise typical practical applications of simply supported beams with point loadings
- perform calculations on simply supported beams having point loads

28.1 The moment of a force

When using a spanner to tighten a nut, a force tends to turn the nut in a clockwise direction. This turning effect of a force is called the **moment of a force** or more briefly, just a **moment**. The size of the moment acting on the nut depends on two factors:

(a) the size of the force acting at right angles to the shank of the spanner, and

(b) the perpendicular distance between the point of application of the force and the centre of the nut.

In general, with reference to Fig. 28.1, the moment M of a force acting about a point P is force × perpendicular distance between the line of action of the force and P, i.e.

$$M = F \times d$$

The unit of a moment is the **newton metre (Nm)**. Thus, if force F in Fig. 28.1 is 7 N and distance d is 3 m, then the moment M is $7\,\text{N} \times 3\,\text{m}$, i.e. 21 Nm.

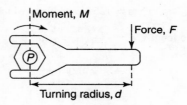

Fig. 28.1

Problem 1. A force of 15 N is applied to a spanner at an effective length of 140 mm from the centre of a nut. Calculate (a) the moment of the force applied to the nut, (b) the magnitude of the force required to produce the same moment if the effective length is reduced to 100 mm.

From above, $M = F \times d$, where M is the turning moment, F is the force applied at right angles to the spanner and d is the effective length between the force and the centre of the nut. Thus, with reference to Fig. 28.2(a):

(a) Turning moment, $M = 15\,\text{N} \times 140\,\text{mm} = 2100\,\text{Nmm}$

$$= 2100\,\text{Nmm} \times \frac{1\,\text{m}}{1000\,\text{mm}}$$

$$= \mathbf{2.1\,Nm}$$

(b) Turning moment, M is 2100 Nmm and the effective length d becomes 100 mm (see Fig. 28.2(b)).

Applying $\qquad M = F \times d$

gives: $\qquad 2100\,\text{Nmm} = F \times 100\,\text{mm}$

from which, **force,** $F = \dfrac{2100\,\text{Nmm}}{100\,\text{mm}} = \mathbf{21\,N}$

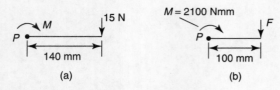

(a)   (b)

Fig. 28.2

> *Problem 2.* A moment of 25 Nm is required to operate a lifting jack. Determine the effective length of the handle of the jack if the force applied to it is (a) 125 N (b) 0.4 kN.

From above, moment $M = F \times d$, where F is the force applied at right angles to the handle and d is the effective length of the handle. Thus:

(a) $\qquad\qquad$ 25 Nm $= 125$ N $\times d$, from which

$$\textbf{effective length, } d = \frac{25\,\text{Nm}}{125\,\text{N}} = \frac{1}{5}\,\text{m}$$

$$= \frac{1}{5} \times 1000\,\text{mm}$$

$$= \textbf{200 mm}$$

(b) Turning moment M is 25 Nm and the force F becomes 0.4 kN, i.e. 400 N.

Since $M = F \times d$, then 25 Nm $= 400$ N $\times d$

Thus $\qquad$ **effective length,** $d = \dfrac{25\,\text{Nm}}{400\,\text{N}} = \dfrac{1}{16}\text{m}$

$$= \frac{1}{16} \times 1000\,\text{mm}$$

$$= \textbf{62.5 mm}$$

Now try the following exercise

Exercise 127   Further problems on the moment of a force (Answers on page 308)

1. Determine the moment of a force of 25 N applied to a spanner at an effective length of 180 mm from the centre of a nut.

2. A moment of 7.5 Nm is required to turn a wheel. If a force of 37.5 N is applied to the rim of the wheel, calculate the effective distance from the rim to the hub of the wheel.

3. Calculate the force required to produce a moment of 27 Nm on a shaft, when the effective distance from the centre of the shaft to the point of application of the force is 180 mm.

28.2 Equilibrium and the principle of moments

If more than one force is acting on an object and the forces do not act at a single point, then the turning effect of the forces, that is, the moment of the forces, must be considered.

Figure 28.3 shows a beam with its support (known as its pivot or fulcrum) at P, acting vertically upwards, and forces F_1 and F_2 acting vertically downwards at distances a and b, respectively, from the fulcrum.

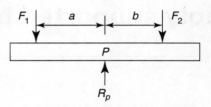

Fig. 28.3

A beam is said to be in **equilibrium** when there is no tendency for it to move. There are two conditions for equilibrium:

(i) The sum of the forces acting vertically downwards must be equal to the sum of the forces acting vertically upwards, i.e. for Fig. 28.3,

$$R_p = F_1 + F_2$$

(ii) The total moment of the forces acting on a beam must be zero; for the total moment to be zero:

> *the sum of the clockwise moments about any point must be equal to the sum of the anticlockwise moments about that point.*

This statement is known as the **principle of moments**.

Hence, taking moments about P in Fig. 28.3,

$$F_2 \times b = \text{the clockwise moment, and}$$

$$F_1 \times a = \text{the anticlockwise moment}$$

Thus for equilibrium: $\boxed{F_1 a = F_2 b}$

> *Problem 3.* A system of forces is as shown in Fig. 28.4.
>
> (a) If the system is in equilibrium find the distance d.
>
> (b) If the point of application of the 5 N force is moved to point P, distance 200 mm from the support, find the new value of F to replace the 5 N force for the system to be in equilibrium.

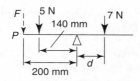

Fig. 28.4

(a) From above, the clockwise moment M_1 is due to a force of 7 N acting at a distance d from the support, called the **fulcrum**, i.e. $M_1 = 7\,\text{N} \times d$

The anticlockwise moment M_2 is due to a force of 5 N acting at a distance of 140 mm from the fulcrum, i.e. $M_2 = 5\,\text{N} \times 140\,\text{mm}$.

Applying the principle of moments, for the system to be in equilibrium about the fulcrum:

clockwise moment = anticlockwise moment

i.e. $\qquad\qquad 7\,\text{N} \times d = 5 \times 140\,\text{Nmm}$

Hence, $\qquad$ **distance, $d = \dfrac{5 \times 140\,\text{Nmm}}{7\,\text{N}} = 100\,\text{mm}$**

(b) When the 5 N force is replaced by force F at a distance of 200 mm from the fulcrum, the new value of the anticlockwise moment is $F \times 200$. For the system to be in equilibrium:

clockwise moment = anticlockwise moment

i.e. $\qquad\qquad (7 \times 100)\,\text{Nmm} = F \times 200\,\text{mm}$

Hence, **new value of force, $F = \dfrac{700\,\text{Nmm}}{200\,\text{mm}} = 3.5\,\text{N}$**

Problem 4. A beam is supported at its centre on a fulcrum and forces act as shown in Fig. 28.5. Calculate (a) force F for the beam to be in equilibrium, (b) the new position of the 23 N force when F is decreased to 21 N, if equilibrium is to be maintained.

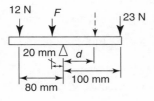

Fig. 28.5

(a) The clockwise moment, M_1, is due to the 23 N force acting at a distance of 100 mm from the fulcrum, i.e.

$$M_1 = 23 \times 100 = 2300\,\text{Nmm}$$

There are two forces giving the anticlockwise moment M_2. One is the force F acting at a distance of 20 mm

from the fulcrum and the other a force of 12 N acting at a distance of 80 mm. Thus,

$$M_2 = (F \times 20) + (12 \times 80)\,\text{Nmm}$$

Applying the principle of moments about the fulcrum:

clockwise moment = anticlockwise moments

i.e. $\qquad\qquad 2300 = (F \times 20) + (12 \times 80)$

Hence $\qquad\qquad F \times 20 = 2300 - 960$

i.e. $\qquad\qquad$ **force, $F = \dfrac{1340}{20} = 67\,\text{N}$**

(b) The clockwise moment is now due to a force of 23 N acting at a distance of, say, d from the fulcrum. Since the value of F is decreased to 21 N, the anticlockwise moment is $(21 \times 20) + (12 \times 80)\,\text{Nmm}$.

Applying the principle of moments,

$$23 \times d = (21 \times 20) + (12 \times 80)$$

i.e. **distance, $d = \dfrac{420 + 960}{23} = \dfrac{1380}{23} = 60\,\text{mm}$**

Problem 5. For the centrally supported uniform beam shown in Fig. 28.6, determine the values of forces F_1 and F_2 when the beam is in equilibrium.

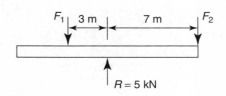

Fig. 28.6

At equilibrium: (i) $R = F_1 + F_2$ i.e. $5 = F_1 + F_2$ $\qquad$ (1)

and $\qquad$ (ii) $F_1 \times 3 = F_2 \times 7$ $\qquad\qquad\qquad\qquad$ (2)

From equation (1), $F_2 = 5 - F_1$

Substituting for F_2 in equation (2) gives:

$$F_1 \times 3 = (5 - F_1) \times 7$$

i.e. $\qquad\qquad 3F_1 = 35 - 7F_1$

$$10F_1 = 35$$

from which, $\qquad\qquad F_1 = 3.5\,\text{kN}$

Since $F_2 = 5 - F_1$, $\qquad F_2 = 1.5\,\text{kN}$

Thus at equilibrium, force $F_1 = 3.5\,\text{kN}$ and force $F_2 = 1.5\,\text{kN}$

Now try the following exercise

Exercise 128   Further problems on equilibrium and the principle of moments (Answers on page 308)

1. Determine distance d and the force acting at the support A for the force system shown in Fig. 28.7, when the system is in equilibrium.

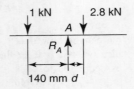

Fig. 28.7

2. If the 1 kN force shown in Fig. 28.7 is replaced by a force F at a distance of 250 mm to the left of R_A, find the value of F for the system to be in equilibrium.

3. Determine the values of the forces acting at A and B for the force system shown in Fig. 28.8.

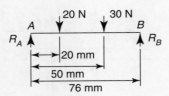

Fig. 28.8

4. The forces acting on a beam are as shown in Fig. 28.9. Neglecting the mass of the beam, find the value of R_A and distance d when the beam is in equilibrium.

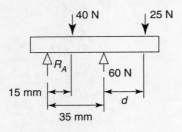

Fig. 28.9

28.3   Simply supported beams having point loads

A **simply supported beam** is one which rests on two supports and is free to move horizontally.

Two typical simply supported beams having loads acting at given points on the beam, called **point loading**, are shown in Fig. 28.10.

A man whose mass exerts a force F vertically downwards, standing on a wooden plank which is simply supported at its ends, may, for example, be represented by the beam diagram of Fig. 28.10(a) if the mass of the plank is neglected. The forces exerted by the supports on the plank, R_A and R_B, act vertically upwards, and are called **reactions**.

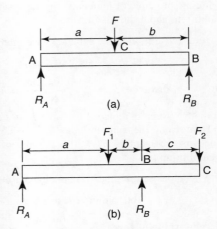

Fig. 28.10

When the forces acting are all in one plane, the algebraic sum of the moments can be taken about **any** point.

For the beam in Fig. 28.10(a) at equilibrium:

(i) $R_A + R_B = F$, and

(ii) taking moments about A, $Fa = R_B(a + b)$

(Alternatively, taking moments about C, $R_A a = R_B b$)

For the beam in Fig. 28.10(b), at equilibrium:

(i) $R_A + R_B = F_1 + F_2$, and

(ii) taking moments about B, $R_A(a + b) + F_2 c = F_1 b$

Typical **practical applications** of simply supported beams with point loadings include bridges, beams in buildings, and beds of machine tools.

Problem 6.   A beam is loaded as shown in Fig. 28.11. Determine (a) the force acting on the beam support at B, (b) the force acting on the beam support at A, neglecting the mass of the beam.

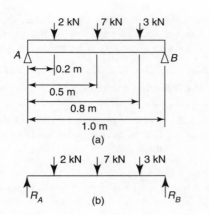

Fig. 28.11

A beam supported as shown in Fig. 28.11 is called a simply supported beam.

(a) Taking moments about point A and applying the principle of moments gives:

clockwise moments = anticlockwise moments

$(2 \times 0.2) + (7 \times 0.5) + (3 \times 0.8)\,kNm = R_B \times 1.0\,m,$

where R_B is the force supporting the beam at B, as shown in Fig. 28.11(b).

Thus $(0.4 + 3.5 + 2.4)\,kNm = R_B \times 1.0\,m$

i.e. $$R_B = \frac{6.3\,kNm}{1.0\,m} = \mathbf{6.3\,kN}$$

(b) For the beam to be in equilibrium, the forces acting upwards must be equal to the forces acting downwards, thus,

$R_A + R_B = (2 + 7 + 3)\,kN, \ R_B = 6.3\,kN,$

thus $R_A = 12 - 6.3 = \mathbf{5.7\,kN}$

Problem 7. For the beam shown in Fig. 28.12 calculate (a) the force acting on support A, (b) distance d, neglecting any forces arising from the mass of the beam.

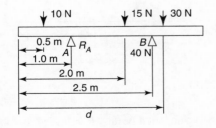

Fig. 28.12

(a) From section 28.2,

(the forces acting in an upward direction)

= (the forces acting in a downward direction)

Hence $(R_A + 40)\,N = (10 + 15 + 30)\,N$

$R_A = 10 + 15 + 30 - 40 = \mathbf{15\,N}$

(b) Taking moments about the left-hand end of the beam and applying the principle of moments gives:

clockwise moments = anticlockwise moments

$(10 \times 0.5) + (15 \times 2.0)\,Nm + 30\,N \times d$

$= (15 \times 1.0) + (40 \times 2.5)\,Nm$

i.e. $35\,Nm + 30\,N \times d = 115\,Nm$

from which, $$\textbf{distance, } d = \frac{(115 - 35)\,Nm}{30\,N}$$

$$= \mathbf{2.67\,m}$$

Problem 8. A metal bar AB is 4.0 m long and is supported at each end in a horizontal position. It carries loads of 2.5 kN and 5.5 kN at distances of 2.0 m and 3.0 m, respectively, from A. Neglecting the mass of the beam, determine the reactions of the supports when the beam is in equilibrium.

The beam and its loads are shown in Fig. 28.13. At equilibrium,

$$R_A + R_B = 2.5 + 5.5 = 8.0\,kN \qquad (1)$$

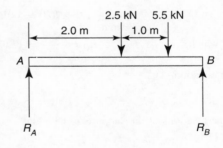

Fig. 28.13

Taking moments about A,

clockwise moments = anticlockwise moment.

i.e. $(2.5 \times 2.0) + (5.5 \times 3.0) = 4.0\,R_B$

or $5.0 + 16.5 = 4.0\,R_B$

from which, $$R_B = \frac{21.5}{4.0} = 5.375\,kN$$

From equation (1), $R_A = 8.0 - 5.375 = 2.625\,kN$

Thus the reactions at the supports at equilibrium are 2.625 kN at *A* and 5.375 kN at *B*

Problem 9. A beam *PQ* is 5.0 m long and is supported at its ends in a horizontal position as shown in Fig. 28.14. Its mass is equivalent to a force of 400 N acting at its centre as shown. Point loads of 12 kN and 20 kN act on the beam in the positions shown. When the beam is in equilibrium, determine (a) the reactions of the supports, R_P and R_Q, and (b) the position to which the 12 kN load must be moved for the force on the supports to be equal.

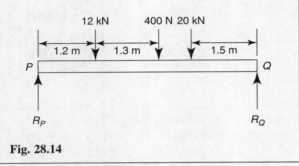

Fig. 28.14

(a) At equilibrium,

$$R_P + R_Q = 12 + 0.4 + 20 = 32.4 \text{ kN} \quad (1)$$

Taking moments about *P*:

clockwise moments = anticlockwise moments

i.e. $(12 \times 1.2) + (0.4 \times 2.5) + (20 \times 3.5) = (R_Q \times 5.0)$

$$14.4 + 1.0 + 70.0 = 5.0 R_Q$$

from which, $R_Q = \dfrac{85.4}{5.0} = \mathbf{17.08 \text{ kN}}$

From equation (1),

$$R_P = 32.4 - R_Q = 32.4 - 17.08 = \mathbf{15.32 \text{ kN}}$$

(b) For the reactions of the supports to be equal,

$$R_P = R_Q = \frac{32.4}{2} = 16.2 \text{ kN}$$

Let the 12 kN load be at a distance *d* metres from *P* (instead of at 1.2 m from *P*). Taking moments about point *P* gives:

$$12d + (0.4 \times 2.5) + (20 \times 3.5) = 5.0 R_Q$$

i.e. $12d + 1.0 + 70.0 = 5.0 \times 16.2$

and $12d = 81.0 - 71.0$

from which, $d = \dfrac{10.0}{12} = 0.833 \text{ m}$

Hence the 12 kN load needs to be moved to a position 833 mm from *P* for the reactions of the supports to be equal (i.e. 367 mm to the left of its original position).

Now try the following exercises

Exercise 129   Further problems on simply supported beams having point loads (Answers on page 308)

1. Calculate the force R_A and distance *d* for the beam shown in Fig. 28.15. The mass of the beam should be neglected and equilibrium conditions assumed.

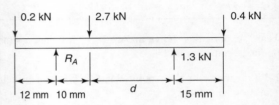

Fig. 28.15

2. For the force system shown in Fig. 28.16, find the values of *F* and *d* for the system to be in equilibrium.

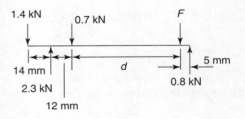

Fig. 28.16

3. For the force system shown in Fig. 28.17, determine distance *d* for the forces R_A and R_B to be equal, assuming equilibrium conditions.

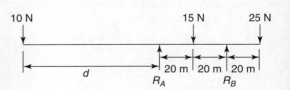

Fig. 28.17

4. A simply supported beam *AB* is loaded as shown in Fig. 28.18. Determine the load *F* in order that the reaction at *A* is zero.

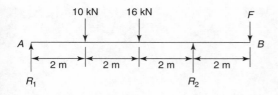

Fig. 28.18

5. A uniform wooden beam, 4.8 m long, is supported at its left-hand end and also at 3.2 m from the left-hand end. The mass of the beam is equivalent to 200 N acting vertically downwards at its centre. Determine the reactions at the supports.

6. For the simply supported beam PQ shown in Fig. 28.19, determine (a) the reaction at each support, (b) the maximum force which can be applied at Q without losing equilibrium.

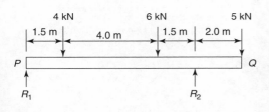

Fig. 28.19

Exercise 130   Short answer questions on simply supported beams

1. The moment of a force is the product of... and...
2. When a beam has no tendency to move it is in...
3. State the two conditions for equilibrium of a beam.
4. State the principle of moments.
5. What is meant by a simply supported beam?
6. State two practical applications of simply supported beams.

Exercise 13i   Multi-choice questions on simply supported beams (Answers on page 309)

1. A force of 10 N is applied at right angles to the handle of a spanner, 0.5 m from the centre of a nut. The moment on the nut is:
   (a) 5 Nm   (b) 2 N/m   (c) 0.5 m/N   (d) 15 Nm

2. The distance d in Fig. 28.20 when the beam is in equilibrium is:
   (a) 0.5 m   (b) 1.0 m   (c) 4.0 m   (d) 15 m

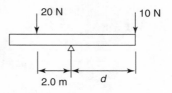

Fig. 28.20

3. With reference to Fig. 28.21, the clockwise moment about A is:
   (a) 70 Nm   (b) 10 Nm   (c) 60 Nm   (d) $5 \times R_B$ Nm

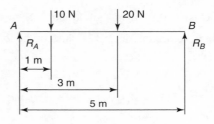

Fig. 28.21

4. The force acting at B (i.e. R_B) in Fig. 28.21 is:
   (a) 16 N   (b) 20 N   (c) 5 N   (d) 14 N

5. The force acting at A (i.e. R_A) in Fig. 28.21 is:
   (a) 16 N   (b) 10 N   (c) 15 N   (d) 14 N

6. Which of the following statements is false for the beam shown in Fig. 28.22 if the beam is in equilibrium?
   (a) The anticlockwise moment is 27 N.
   (b) The force F is 9 N.
   (c) The reaction at the support R is 18 N.
   (d) The beam cannot be in equilibrium for the given conditions.

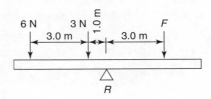

Fig. 28.22

7. With reference to Fig. 28.23, the reaction R_A is:
   (a) 10 N   (b) 30 N   (c) 20 N   (d) 40 N

8. With reference to Fig. 28.23, when moments are taken about R_A, the sum of the anticlockwise moments is:
   (a) 25 Nm   (b) 20 Nm   (c) 35 Nm   (d) 30 Nm

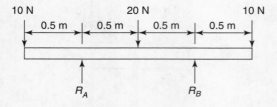

Fig. 28.23

9. With reference to Fig. 28.23, when moments are taken about the right-hand end, the sum of the clockwise moments is:

(a) 10 Nm   (b) 20 Nm   (c) 30 Nm   (d) 40 Nm

10. With reference to Fig. 28.23, which of the following statements is false?

(a) $(5 + R_B) = 25$ Nm   (b) $R_A = R_B$

(c) $(10 \times 0.5) = (10 \times 1) + (10 \times 1.5) + R_A$

(d) $R_A + R_B = 40$ N

Assignment 10

This assignment covers the material contained in Chapters 27 and 28. The marks for each question are shown in brackets at the end of each question.

1. A force of 25 N acts horizontally to the right and a force of 25 N is inclined at an angle of 30° to the 25 N force. Determine the magnitude and direction of the resultant of the two forces using (a) the triangle of forces method, (b) the parallelogram of forces method, and (c) by calculation. (13)

2. Determine graphically the magnitude and direction of the resultant of the following three coplanar forces, which may be considered as acting at a point.

Force P, 15 N acting horizontally to the right

force Q, 8 N inclined at 45° to force P

force R, 20 N inclined at 120° to force P (7)

3. Determine by resolution of forces the resultant of the following three coplanar forces acting at a point: 120 N acting at 40° to the horizontal; 250 N acting at 145° to the horizontal; 300 N acting at 260° to the horizontal. (8)

4. A moment of 18 Nm is required to operate a lifting jack. Determine the effective length of the handle of the jack (in millimetres) if the force applied to it is (a) 90 N (b) 0.36 kN (6)

5. For the centrally supported uniform beam shown in Fig. A10.1, determine the values of forces F_1 and F_2 when the beam is in equilibrium. (8)

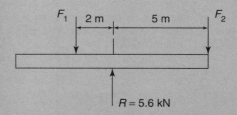

Fig. A10.1

6. For the beam shown in Fig. A10.2 calculate (a) the force acting on support Q, (b) distance d, neglecting any forces arising from the mass of the beam. (8)

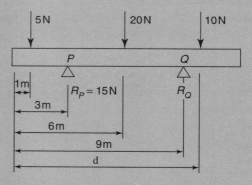

Fig. A10.2

29

Linear and angular motion

At the end of this chapter you should be able to:

- appreciate that 2π radians corresponds to 360°

- define linear and angular velocity

- perform calculations on linear and angular velocity using $\omega = 2\pi n$ and $v = \omega r$

- define linear and angular acceleration

- perform calculations on linear and angular acceleration using $\omega_2 = \omega_1 + \alpha t$ and $a = r\alpha$

- select appropriate equations of motion when performing simple calculations

- appreciate the difference between scalar and vector quantities

- use vectors to determine relative velocities, by drawing and by calculation

29.1 The radian

The unit of angular displacement is the radian, where one radian is the angle subtended at the centre of a circle by an arc equal in length to the radius, as shown in Fig. 29.1.

The relationship between angle in radians (θ), arc length (s) and radius of a circle (r) is:

$$s = r\theta \qquad (1)$$

Since the arc length of a complete circle is $2\pi r$ and the angle subtended at the centre is 360°, then from equation (1), for a complete circle,

$$2\pi r = r\theta \quad \text{or} \quad \theta = 2\pi \text{ radians}$$

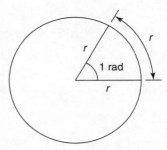

Fig. 29.1

Thus,

$$\boxed{2\pi \text{ radians corresponds to } 360°} \qquad (2)$$

29.2 Linear and angular velocity

Linear velocity

Linear velocity v is defined as the rate of change of linear displacement s with respect to time t, and for motion in a straight line:

$$\text{Linear velocity} = \frac{\text{change of displacement}}{\text{change of time}}$$

i.e.

$$\boxed{v = \frac{s}{t}} \qquad (3)$$

The unit of linear velocity is metres per second (m/s)

Angular velocity

The speed of revolution of a wheel or a shaft is usually measured in revolutions per minute or revolutions per second but these units do not form part of a coherent system of units.

The basis used in SI units is the angle turned through in one second.

Angular velocity is defined as the rate of change of angular displacement θ, with respect to time t, and for an object rotating about a fixed axis at a constant speed:

$$\text{angular velocity} = \frac{\text{angle turned through}}{\text{time taken}}$$

i.e.

$$\boxed{\omega = \frac{\theta}{t}} \qquad (4)$$

The unit of angular velocity is radians per second (rad/s).

An object rotating at a constant speed of n revolutions per second subtends an angle of $2\pi n$ radians in one second, that is, its angular velocity,

$$\boxed{\omega = 2\pi n \text{ rad/s}} \qquad (5)$$

From equation (1), $s = r\theta$, and from equation (4), $\theta = \omega t$,

hence $\quad s = r\omega t \quad$ or $\quad \dfrac{s}{t} = \omega r$

However, from equation (3),

$$v = \frac{s}{t}$$

hence $\quad \boxed{v = \omega r} \qquad (6)$

Equation (6) gives the relationship between linear velocity, v, and angular velocity, ω

Problem 1. A wheel of diameter 540 mm is rotating at $(1500/\pi)$ rev/min. Calculate the angular velocity of the wheel and the linear velocity of a point on the rim of the wheel.

From equation (5), angular velocity $\omega = 2\pi n$, where n is the speed of revolution in revolutions per second, i.e.

$$n = \frac{1500}{60\pi} \text{ revolutions per second.}$$

Thus, **angular velocity,** $\omega = 2\pi\left(\dfrac{1500}{60\pi}\right) = \mathbf{50\,rad/s}$

The linear velocity of a point on the rim, $v = \omega r$, where r is the radius of the wheel, i.e. 0.54/2 or 0.27 m.

Thus, **linear velocity,** $v = 50 \times 0.27 = \mathbf{13.5\,m/s}$

Problem 2. A car is travelling at 64.8 km/h and has wheels of diameter 600 mm.

(a) Find the angular velocity of the wheels in both rad/s and rev/min.

(b) If the speed remains constant for 1.44 km, determine the number of revolutions made by a wheel, assuming no slipping occurs.

(a) $64.8\,\text{km/h} = 64.8\dfrac{\text{km}}{\text{h}} \times 1000\dfrac{\text{m}}{\text{km}} \times \dfrac{1}{3600}\dfrac{\text{h}}{\text{s}}$

$\qquad = \dfrac{64.8}{3.6}\,\text{m/s} = 18\,\text{m/s}$

i.e. the linear velocity, v, is 18 m/s

The radius of a wheel is $(600/2)\,\text{mm} = 0.3\,\text{m}$

From equation (6), $v = \omega r$, hence $\omega = v/r$

i.e. the **angular velocity,** $\omega = \dfrac{18}{0.3} = \mathbf{60\,rad/s}$

From equation (5), angular velocity, $\omega = 2\pi n$, where n is in revolutions per second. Hence $n = \omega/2\pi$ and angular speed of a wheel in revolutions per minute is $60\omega/2\pi$; but $\omega = 60$ rad/s,

hence **angular speed** $= \dfrac{60 \times 60}{2\pi}$

$\qquad = \mathbf{573\text{ revolutions per minute}}$

(b) From equation (3), time taken to travel 1.44 km at a constant speed of 18 m/s is $\dfrac{1440\,\text{m}}{18\,\text{m/s}} = 80\,\text{s}$

Since a wheel is rotating at 573 revolutions per minute, then in 80/60 minutes it makes $\dfrac{573 \times 80}{60} = \mathbf{764\text{ revo-}}$
lutions.

Now try the following exercise

Exercise 132   Further problems on linear and angular velocity (Answers on page 309)

1. A pulley driving a belt has a diameter of 360 mm and is turning at $2700/\pi$ revolutions per minute. Find the angular velocity of the pulley and the linear velocity of the belt assuming that no slip occurs.

2. A bicycle is travelling at 36 km/h and the diameter of the wheels of the bicycle is 500 mm. Determine the angular velocity of the wheels of the bicycle and the linear velocity of a point on the rim of one of the wheels.

29.3   Linear and angular acceleration

Linear acceleration, a, is defined as the rate of change of linear velocity with respect to time (as introduced in

Chapter 25). For an object whose linear velocity is increasing uniformly:

$$\text{linear acceleration} = \frac{\text{change of linear velocity}}{\text{time taken}}$$

i.e.
$$\boxed{a = \frac{v_2 - v_1}{t}} \tag{7}$$

The unit of linear acceleration is metres per second squared (m/s^2). Rewriting equation (7) with v_2 as the subject of the formula gives:

$$\boxed{v_2 = v_1 + at} \tag{8}$$

Angular acceleration, α, is defined as the rate of change of angular velocity with respect to time. For an object whose angular velocity is increasing uniformly:

$$\text{Angular acceleration} = \frac{\text{change of angular velocity}}{\text{time taken}}$$

i.e.
$$\boxed{\alpha = \frac{\omega_2 - \omega_1}{t}} \tag{9}$$

The unit of angular acceleration is radians per second squared (rad/s^2). Rewriting equation (9) with ω_2 as the subject of the formula gives:

$$\boxed{\omega_2 = \omega_1 + \alpha t} \tag{10}$$

From equation (6), $v = \omega r$. For motion in a circle having a constant radius r, $v_2 = \omega_2 r$ and $v_1 = \omega_1 r$, hence equation (7) can be rewritten as

$$a = \frac{\omega_2 r - \omega_1 r}{t} = \frac{r(\omega_2 - \omega_1)}{t}$$

But from equation (9),

$$\frac{\omega_2 - \omega_1}{t} = \alpha$$

Hence
$$\boxed{a = r\alpha} \tag{11}$$

Problem 3. The speed of a shaft increases uniformly from 300 revolutions per minute to 800 revolutions per minute in 10 s. Find the angular acceleration, correct to 3 significant figures.

From equation (9), $\alpha = \dfrac{\omega_2 - \omega_1}{t}$

initial angular velocity,

$$\omega_1 = 300 \text{ rev/min} = 300/60 \text{ rev/s} = \frac{300 \times 2\pi}{60} \text{ rad/s}$$

final angular velocity,

$$\omega_2 = \frac{800 \times 2\pi}{60} \text{ rad/s} \quad \text{and} \quad \text{time, } t = 10 \text{ s}$$

Hence, **angular acceleration,**

$$\alpha = \frac{\dfrac{800 \times 2\pi}{60} - \dfrac{300 \times 2\pi}{60}}{10} \text{ rad/s}^2$$

$$= \frac{500 \times 2\pi}{60 \times 10} = \mathbf{5.24 \, rad/s^2}$$

Problem 4. If the diameter of the shaft in Problem 3 is 50 mm, determine the linear acceleration of the shaft, correct to 3 significant figures.

From equation (11), $a = r\alpha$

The shaft radius is $\dfrac{50}{2}$ mm $= 25$ mm $= 0.025$ m, and the angular acceleration, $\alpha = 5.24$ rad/s^2, thus the **linear acceleration, $a = r\alpha = 0.025 \times 5.24 = \mathbf{0.131 \, m/s^2}$**

Now try the following exercise

Exercise 133   Further problems on linear and angular acceleration (Answers on page 309)

1. A flywheel rotating with an angular velocity of 200 rad/s is uniformly accelerated at a rate of 5 rad/s^2 for 15 s. Find the angular velocity of the flywheel both in rad/s and revolutions per minute.

2. A disc accelerates uniformly from 300 revolutions per minute to 600 revolutions per minute in 25 s. Determine its angular acceleration and the linear acceleration of a point on the rim of the disc, if the radius of the disc is 250 mm.

29.4   Further equations of motion

From equation (3), $s = vt$, and if the linear velocity is changing uniformly from v_1 and v_2, then $s = $ mean linear velocity $\times$ time

i.e.
$$\boxed{s = \left(\frac{v_1 + v_2}{2}\right)t} \tag{12}$$

From equation (4), $\theta = \omega t$, and if the angular velocity is changing uniformly from ω_1 to ω_2, then $\theta = $ mean angular velocity $\times$ time

i.e.
$$\boxed{\theta = \left(\frac{\omega_1 + \omega_2}{2}\right)t} \tag{13}$$

Two further equations of linear motion may be derived from equations (8) and (11):

$$s = v_1 t + \frac{1}{2} a t^2 \qquad (14)$$

and

$$v_2^2 = v_1^2 + 2as \qquad (15)$$

Two further equations of angular motion may be derived from equations (10) and (12):

$$\theta = \omega_1 t + \frac{1}{2} \alpha t^2 \qquad (16)$$

and

$$\omega_2^2 = \omega_1^2 + 2\alpha\theta \qquad (17)$$

Table 29.1 summarises the principal equations of linear and angular motion for uniform changes in velocities and constant accelerations and also gives the relationships between linear and angular quantities.

Problem 5. The speed of a shaft increases uniformly from 300 rev/min to 800 rev/min in 10 s. Find the number of revolutions made by the shaft during the 10 s it is accelerating.

From equation (13), angle turned through,

$$\theta = \left(\frac{\omega_1 + \omega_2}{2} \right) t = \left(\frac{\dfrac{300 \times 2\pi}{60} + \dfrac{800 \times 2\pi}{60}}{2} \right) t \quad (10) \text{ rad}$$

However, there are 2π radians in 1 revolution, hence,

$$\text{number of revolutions} = \left(\frac{\dfrac{300 \times 2\pi}{60} + \dfrac{800 \times 2\pi}{60}}{2} \right) \left(\frac{10}{2\pi} \right)$$

$$= \frac{1}{2} \left(\frac{1100}{60} \right) (10) = \frac{1100}{12}$$

$$= 91\frac{2}{3} \textbf{ revolutions}.$$

Problem 6. The shaft of an electric motor, initially at rest, accelerates uniformly for 0.4 s at 15 rad/s². Determine the angle (in radians) turned through by the shaft in this time.

From equation (16),

$$\theta = \omega_1 t + \frac{1}{2} \alpha t^2$$

Table 29.1

s = arc length (m)	r = radius of circle (m)
t = time (s)	θ = angle (rad)
v = linear velocity (m/s)	ω = angular velocity (rad/s)
v_1 = initial linear velocity (m/s)	ω_1 = initial angular velocity (rad/s)
v_2 = final linear velocity (m/s)	ω_2 = final angular velocity (rad/s)
a = linear acceleration (m/s²)	α = angular acceleration (rad/s²)
n = speed of revolution (rev/s)	

Equation number	Linear motion	Angular motion
(1)	$s = r\theta$ m	
(2)	2π rad = 360°	
(3) and (4)	$v = \dfrac{s}{t}$	$\omega = \dfrac{\theta}{t}$ rad/s
(5)	$\omega = 2\pi n$ rad/s	
(6)	$v = \omega r$ m/s²	
(8) and (10)	$v_2 = (v_1 + at)$ m/s	$\omega_2 = (\omega_1 + \alpha t)$ rad/s
(11)	$a = r\alpha$ m/s²	
(12) and (13)	$s = \left(\dfrac{v_1 + v_2}{2} \right) t$	$\theta = \left(\dfrac{\omega_1 + \omega_2}{2} \right) t$
(14) and (16)	$s = v_1 t + \dfrac{1}{2} a t^2$	$\theta = \omega_1 t + \dfrac{1}{2} \alpha t^2$
(15) and (17)	$v_2^2 = v_1^2 + 2as$	$\omega_2^2 = \omega_1^2 + 2\alpha\theta$

Since the shaft is initially at rest,

$$\omega_1 = 0 \quad \text{and} \quad \theta = \frac{1}{2}\alpha t^2$$

the angular acceleration,

$$\alpha = 15 \,\text{rad/s}^2 \quad \text{and} \quad \text{time } t = 0.4 \,\text{s}$$

Hence, **angle turned through**,

$$\theta = \frac{1}{2} \times 15 \times 0.4^2 = \mathbf{1.2\,rad}.$$

Problem 7. A flywheel accelerates uniformly at $2.05 \,\text{rad/s}^2$ until it is rotating at 1500 rev/min. If it completes 5 revolutions during the time it is accelerating, determine its initial angular velocity in rad/s, correct to 4 significant figures.

Since the final angular velocity is 1500 rev/min,

$$\omega_2 = 1500 \,\frac{\text{rev}}{\text{min}} \times \frac{1\,\text{min}}{60\,\text{s}} \times \frac{2\pi\,\text{rad}}{1\,\text{rev}} = 50\pi \,\text{rad/s}$$

5 revolutions $= 5\,\text{rev} \times \dfrac{2\pi\,\text{rad}}{1\,\text{rev}} = 10\pi \,\text{rad}$

From equation (17), $\omega_2^2 = \omega_1^2 + 2\alpha\theta$

i.e. $\quad (50\pi)^2 = \omega_1^2 + (2 \times 2.05 \times 10\pi)$

from which, $\quad \omega_1^2 = (50\pi)^2 - (2 \times 2.05 \times 10\pi)$

$$= (50\pi)^2 - 41\pi = 24\,545$$

i.e. $\quad \omega_1 = \sqrt{24\,545} = 156.7 \,\text{rad/s}.$

Thus the initial angular velocity is 156.7 rad/s, correct to 4 significant figures.

Now try the following exercise

Exercise 134   Further problems on equations of motion (Answers on page 309)

1. A grinding wheel makes 300 revolutions when slowing down uniformly from 1000 rad/s to 400 rad/s. Find the time for this reduction in speed.

2. Find the angular retardation for the grinding wheel in question 1.

3. A disc accelerates uniformly from 300 revolutions per minute to 600 revolutions per minute in 25 s. Calculate the number of revolutions the disc makes during this accelerating period.

4. A pulley is accelerated uniformly from rest at a rate of $8 \,\text{rad/s}^2$. After 20 s the acceleration stops and the pulley runs at constant speed for 2 min, and then the pulley comes uniformly to rest after a further 40 s. Calculate: (a) the angular velocity after the period of acceleration, (b) the deceleration, (c) the total number of revolutions made by the pulley.

29.5   Relative velocity

As stated in Chapter 27, quantities used in engineering and science can be divided into two groups:

(a) **Scalar quantities** have a size or magnitude only and need no other information to specify them. Thus 20 centimetres, 5 seconds, 3 litres and 4 kilograms are all examples of scalar quantities.

(b) **Vector quantities** have both a size (or magnitude), and a direction, called the line of action of the quantity. Thus, a velocity of 30 km/h due west, and an acceleration of $7 \,\text{m/s}^2$ acting vertically downwards, are both vector quantities.

A vector quantity is represented by a straight line lying along the line of action of the quantity and having a length which is proportional to the size of the quantity, as shown in Chapter 27. Thus *ab* in Fig. 29.2 represents a velocity of 20 m/s, whose line of action is due west. The bold letters, *ab*, indicate a vector quantity and the order of the letters indicate that the time of action is from *a* to *b*.

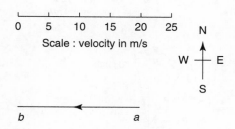

Fig. 29.2

Consider two aircraft A and B flying at a constant altitude, A travelling due north at 200 m/s and B travelling 30° east of north, written N 30° E, at 300 m/s, as shown in Fig. 29.3.

Relative to a fixed point 0, *oa* represents the velocity of A and *ob* the velocity of B. The velocity of B relative to A, that is the velocity at which B seems to be travelling to an observer on A, is given by *ab*, and by measurement is 160 m/s in a direction E 22° N. The velocity of A relative to B, that is, the velocity at which A seems to be travelling to an observer on B, is given by *ba* and by measurement is 160 m/s in a direction W 22° S

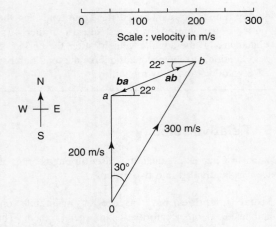

Fig. 29.3

Problem 8.   Two cars are travelling on horizontal roads in straight lines, car A at 70 km/h at N 10° E and car B at 50 km/h at W 60° N. Determine, by drawing a vector diagram to scale, the velocity of car A relative to car B.

With reference to Fig. 29.4(a), *oa* represents the velocity of car A relative to a fixed point 0, and *ob* represents the velocity of car B relative to a fixed point 0. The velocity of car A relative to car B is given by *ba* and by measurement is **45 km/h in a direction of E 35° N**

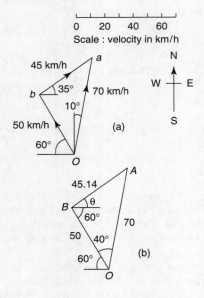

Fig. 29.4

Problem 9.   Verify the result obtained in Problem 8 by calculation.

The triangle shown in Fig. 29.4(b) is similar to the vector diagram shown in Fig. 29.4(a). Angle *BOA* is 40°. Using the cosine rule:

$$BA^2 = 50^2 + 70^2 - 2 \times 50 \times 70 \times \cos 40°$$

from which, $BA = \mathbf{45.14}$

Using the sine rule:

$$\frac{50}{\sin \angle BAO} = \frac{45.14}{\sin 40°}$$

from which,

$$\sin \angle BAO = \frac{50 \sin 40°}{45.14} = 0.7120$$

Hence, angle $BAO = 45.40°$; thus,
angle $ABO = 180° - (40° + 45.40°) = 94.60°$, and
angle $\theta = 94.60° - 60° = 34.60°$

Thus *ba* is **45.14 km/h in a direction E 34.60° N by calculation**

Problem 10.   A crane is moving in a straight line with a constant horizontal velocity of 2 m/s. At the same time it is lifting a load at a vertical velocity of 5 m/s. Calculate the velocity of the load relative to a fixed point on the earth's surface.

A vector diagram depicting the motion of the crane and load is shown in Fig. 29.5. *oa* represents the velocity of the crane relative to a fixed point on the earth's surface and *ab* represents the velocity of the load relative to the crane. The velocity of the load relative to the fixed point on the earth's surface is *ob*. By Pythagoras' theorem:

$$ob^2 = oa^2 + ab^2$$

$$= 4 + 25 = 29$$

Hence $ob = \sqrt{29} = 5.385$ m/s

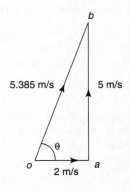

Fig. 29.5

$$\text{Tan } \theta = \frac{5}{2} = 2.5$$

hence

$$\theta = \tan^{-1} 2.5 = 68.20°$$

i.e. the velocity of the load relative to a fixed point on the earth's surface is **5.385 m/s in a direction 68.20° to the motion of the crane**.

Now try the following exercise

Exercise 135   Further problems on relative velocity (Answers on page 309)

1. A ship is sailing due east with a uniform speed of 7.5 m/s relative to the sea. If the tide has a velocity 2 m/s in a north-westerly direction, find the velocity of the ship relative to the sea bed.

2. A lorry is moving along a straight road at a constant speed of 54 km/h. The tip of its windscreen wiper blade has a linear velocity, when in a vertical position, of 4 m/s. Find the velocity of the tip of the wiper blade relative to the road when in this vertical position.

3. A fork-lift truck is moving in a straight line at a constant speed of 5 m/s and at the same time a pallet is being lowered at a constant speed of 2 m/s. Determine the velocity of the pallet relative to the earth.

Exercise 136   Short answer questions on linear and angular motion

1. State and define the unit of angular displacement.

2. Write down the formula connecting an angle, arc length and the radius of a circle.

3. Define linear velocity and state its unit.

4. Define angular velocity and state its unit.

5. Write down a formula connecting angular velocity and revolutions per second in coherent units.

6. State the formula connecting linear and angular velocity.

7. Define linear acceleration and state its unit.

8. Define angular acceleration and state its unit.

9. Write down the formula connecting linear and angular acceleration.

10. Define a scalar quantity and give two examples.

11. Define a vector quantity and give two examples.

Exercise 137   Multi-choice questions on linear and angular motion (Answers on page 309)

1. Angular displacement is measured in:

    (a) degrees   (b) radians   (c) rev/s   (d) metres

2. An angle of $\frac{3\pi}{4}$ radians is equivalent to:

    (a) 270°   (b) 67.5°   (c) 135°   (d) 2.356°

3. An angle of 120° is equivalent to:

    (a) $\frac{2\pi}{3}$ rad   (b) $\frac{\pi}{3}$ rad   (c) $\frac{3\pi}{4}$ rad   (d) $\frac{1}{3}$ rad

4. An angle of 2 rad at the centre of a circle subtends an arc length of 40 mm at the circumference of the circle. The radius of the circle is:

    (a) 40π mm   (b) 80 mm

    (c) 20 mm   (d) $(40/\pi)$ mm

5. A point on a wheel has a constant angular velocity of 3 rad/s. The angle turned through in 15 seconds is:

    (a) 45 rad   (b) 10π rad   (c) 5 rad   (d) 90π rad

6. An angular velocity of 60 revolutions per minute is the same as

    (a) $(1/2\pi)$ rad/s   (b) 120π rad/s

    (c) $(30/\pi)$ rad/s   (d) 2π rad/s

7. A wheel of radius 15 mm has an angular velocity of 10 rad/s. A point on the rim of the wheel has a linear velocity of:

    (a) 300π mm/s   (b) 2/3 mm/s

    (c) 150 mm/s   (d) 1.5 mm/s

8. The shaft of an electric motor is rotating at 20 rad/s and its speed is increased uniformly to 40 rad/s in 5 s. The angular acceleration of the shaft is:

    (a) 4000 rad/s^2 (b) 4 rad/s^2

    (c) 160 rad/s^2 (d) 12 rad/s^2

9. A point on a flywheel of radius 0.5 m has a uniform linear acceleration of 2 m/s^2. Its angular acceleration is:

    (a) 2.5 rad/s^2 (b) 0.25 rad/s^2

    (c) 1 rad/s^2 (d) 4 rad/s^2

Questions 10 to 13 refer to the following data.

A car accelerates uniformly from 10 m/s to 20 m/s over a distance of 150 m. The wheels of the car each have a radius of 250 mm.

10. The time the car is accelerating is:

    (a) 0.2 s (b) 15 s (c) 10 s (d) 5 s

11. The initial angular velocity of each of the wheels is:

    (a) 20 rad/s (b) 40 rad/s

    (c) 2.5 rad/s (d) 0.04 rad/s

12. The angular acceleration of each of the wheels is:

    (a) 1 rad/s^2 (b) 0.25 rad/s^2

    (c) 400 rad/s^2 (d) 4 rad/s^2

13. The linear acceleration of a point on each of the wheels is:

    (a) 1 m/s^2 (b) 4 m/s^2 (c) 3 m/s^2 (d) 100 m/s^2

30

Friction

At the end of this chapter you should be able to:

- understand dynamic or sliding friction
- appreciate factors which affect the size and direction of frictional forces
- define coefficient of friction, μ
- perform calculations involving $F = \mu N$
- state practical applications of friction
- state advantages and disadvantages of frictional forces

30.1 Introduction to friction

When an object, such as a block of wood, is placed on a floor and sufficient force is applied to the block, the force being parallel to the floor, the block slides across the floor. When the force is removed, motion of the block stops; thus there is a force which resists sliding. This force is called **dynamic** or **sliding friction**. A force may be applied to the block which is insufficient to move it. In this case, the force resisting motion is called the **static friction** or **striction**. Thus there are two categories into which a frictional force may be split:

(i) dynamic or sliding friction force which occurs when motion is taking place, and

(ii) static friction force which occurs before motion takes place.

There are three factors which affect the size and direction of frictional forces.

(i) The size of the frictional force depends on the type of surface (a block of wood slides more easily on a polished metal surface than on a rough concrete surface).

(ii) The size of the frictional force depends on the size of the force acting at right angles to the surfaces in contact, called the **normal force**; thus, if the weight of a block of wood is doubled, the frictional force is doubled when it is sliding on the same surface.

(iii) The direction of the frictional force is always opposite to the direction of motion. Thus the frictional force opposes motion, as shown in Fig. 30.1.

30.2 Coefficient of friction

The **coefficient of friction**, μ, is a measure of the amount of friction existing between two surfaces. A low value of coefficient of friction indicates that the force required for sliding to occur is less than the force required when the coefficient of friction is high. The value of the coefficient

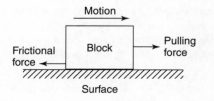

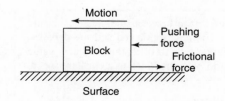

Fig. 30.1

of friction is given by:

$$\mu = \frac{\text{frictional force } (F)}{\text{normal force } (N)}$$

Transposing gives: frictional force = $\mu \times$ normal force, i.e.

$$\boxed{F = \mu N}$$

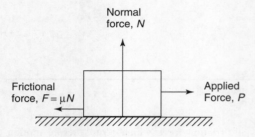

Fig. 30.2

The direction of the forces given in this equation is as shown in Fig. 30.2 The coefficient of friction is the ratio of a force to a force, and hence has no units. Typical values for the coefficient of friction when sliding is occurring, i.e. the dynamic coefficient of friction, are:

For polished oiled metal surfaces	less than 0.1
For glass on glass	0.4
For rubber on tarmac	close to 1.0

Problem 1. A block of steel requires a force of 10.4 N applied parallel to a steel plate to keep it moving with constant velocity across the plate. If the normal force between the block and the plate is 40 N, determine the dynamic coefficient of friction.

As the block is moving at constant velocity, the force applied must be that required to overcome frictional forces, i.e. frictional force, $F = 10.4$ N; the normal force is 40 N, and since $F = \mu N$,

$$\mu = \frac{F}{N} = \frac{10.4}{40} = 0.26$$

i.e. **the dynamic coefficient of friction is 0.26**

Problem 2. The surface between the steel block and plate of problem 1 is now lubricated and the dynamic coefficient of friction falls to 0.12. Find the new value of force required to push the block at a constant speed.

The normal force depends on the weight of the block and remains unaltered at 40 N. The new value of the dynamic

coefficient of friction is 0.12 and since the frictional force $F = \mu N$, $F = 0.12 \times 40 = 4.8$ N

The block is sliding at constant speed, thus the force required to overcome the frictional force is also 4.8 N, i.e. **the required applied force is 4.8 N**

Problem 3. The material of a brake is being tested and it is found that the dynamic coefficient of friction between the material and steel is 0.91. Calculate the normal force when the frictional force is 0.728 kN

The dynamic coefficient of friction, $\mu = 0.91$ and the frictional force,

$$F = 0.728 \text{ kN} = 728 \text{ N}$$

Since $F = \mu N$, then normal force,

$$N = \frac{F}{\mu} = \frac{728}{0.91} = 800 \text{ N}$$

i.e. **the normal force is 800 N**

Now try the following exercise

Exercise 138   Further problems on the coefficient of friction (Answers on page 309)

1. The coefficient of friction of a brake pad and a steel disc is 0.82. Determine the normal force between the pad and the disc if the frictional force required is 1025 N
2. A force of 0.12 kN is needed to push a bale of cloth along a chute at a constant speed. If the normal force between the bale and the chute is 500 N, determine the dynamic coefficient of friction.
3. The normal force between a belt and its driver wheel is 750 N. If the static coefficient of friction is 0.9 and the dynamic coefficient of friction is 0.87, calculate (a) the maximum force which can be transmitted and (b) maximum force which can be transmitted when the belt is running at constant speed.

30.3   Applications of friction

In some applications, a low coefficient of friction is desirable, for example, in bearings, pistons moving within cylinders, on ski runs, and so on. However, for such applications as force being transmitted by belt drives and braking systems, a high value of coefficient is necessary.

Problem 4. State three advantages and three disadvantages of frictional forces.

Instances where frictional forces are an advantage include:

(i) Almost all fastening devices rely on frictional forces to keep them in place once secured, examples being screws, nails, nuts, clips and clamps.

(ii) Satisfactory operation of brakes and clutches rely on frictional forces being present.

(iii) In the absence of frictional forces, most accelerations along a horizontal surface are impossible; for example, a person's shoes just slip when walking is attempted and the tyres of a car just rotate with no forward motion of the car being experienced.

Disadvantages of frictional forces include:

(i) Energy is wasted in the bearings associated with shafts, axles and gears due to heat being generated.

(ii) Wear is caused by friction, for example, in shoes, brake lining materials and bearings.

(iii) Energy is wasted when motion through air occurs (it is much easier to cycle with the wind rather than against it).

Problem 5. Discuss briefly two design implications which arise due to frictional forces and how lubrication may or may not help.

(i) Bearings are made of an alloy called white metal, which has a relatively low melting point. When the rotating shaft rubs on the white metal bearing, heat is generated by friction, often in one spot and the white metal may melt in this area, rendering the bearing useless. Adequate lubrication (oil or grease) separates the shaft from the white metal, keeps the coefficient of friction small and prevents damage to the bearing. For very large bearings, oil is pumped under pressure into the bearing and the oil is used to remove the heat generated, often passing through oil coolers before being re-circulated. Designers should ensure that the heat generated by friction can be dissipated.

(ii) Wheels driving belts, to transmit force from one place to another, are used in many workshops. The coefficient of friction between the wheel and the belt must be high, and it may be increased by dressing the belt with a tar-like substance. Since frictional force is proportional to the normal force, a slipping belt is made more efficient by tightening it, thus increasing the normal and hence the frictional force. Designers should incorporate some belt tension mechanism into the design of such a system.

Problem 6. Explain what is meant by the terms (a) the limiting or static coefficient of friction and (b) the sliding or dynamic coefficient of friction.

(a) When an object is placed on a surface and a force is applied to it in a direction parallel to the surface, if no movement takes place, then the applied force is balanced exactly by the frictional force. As the size of the applied force is increased, a value is reached such that the object is just on the point of moving. The limiting or static coefficient of friction is given by the ratio of this applied force to the normal force, where the normal force is the force acting at right angles to the surfaces in contact.

(b) Once the applied force is sufficient to overcome the striction its value can be reduced slightly and the object moves across the surface. A particular value of the applied force is then sufficient to keep the object moving at a constant velocity. The sliding or dynamic coefficient of friction is the ratio of the applied force, to maintain constant velocity, to the normal force.

Now try the following exercises

Exercise 139   Short answer questions on friction

1. The ...... of frictional force depends on the ...... of surfaces in contact.

2. The ...... of frictional force depends on the size of the ...... to the surfaces in contact.

3. The .... of frictional force is always ...... to the direction of motion.

4. The coefficient of friction between surfaces should be a ...... value for materials concerned with bearings.

5. The coefficient of friction should have a ...... value for materials concerned with braking systems.

6. The coefficient of dynamic or sliding friction is given by $\dfrac{\,.\,.\,.\,.\,.\,.\,}{.\,.\,.\,.\,.\,.}$

7. The coefficient of static or limiting friction is given by $\dfrac{\,.\,.\,.\,.\,.\,.\,}{.\,.\,.\,.\,.\,.}$ when ...... is just about to take place.

8. Lubricating surfaces in contact result in a ...... of the coefficient of friction.

9. Briefly discuss the factors affecting the size and direction of frictional forces.

10. Name three practical applications where a low value of coefficient of friction is desirable and state briefly how this is achieved in each case.

11. Name three practical applications where a high value of coefficient of friction is required when transmitting forces and discuss how this is achieved.

12. For an object on a surface, two different values of coefficient of friction are possible. Give the names of these two coefficients of friction and state how their values may be obtained.

Exercise 140   Multi-choice   questions   on   friction
                    (Answers on page 309)

1. A block of metal requires a frictional force F to keep it moving with constant velocity across a surface. If the coefficient of friction is μ, then the normal force N is given by:

   (a) $\dfrac{\mu}{F}$ (b) μF (c) $\dfrac{F}{\mu}$ (d) F

2. The unit of the linear coefficient of friction is:

   (a) newtons   (b) radians

   (c) dimensionless   (d) newtons/metre

   Questions 3 to 7 refer to the statements given below. Select the statement required from each group given.

   (a) The coefficient of friction depends on the type of surfaces in contact.
   (b) The coefficient of friction depends on the force acting at right angles to the surfaces in contact.
   (c) The coefficient of friction depends on the area of the surfaces in contact.
   (d) Frictional force acts in the opposite direction to the direction of motion.
   (e) Frictional force acts in the direction of motion.
   (f) A low value of coefficient of friction is required between the belt and the wheel in a belt drive system.
   (g) A low value of coefficient of friction is required for the materials of a bearing.

   (h) The dynamic coefficient of friction is given by (normal force)/(frictional force) at constant speed.
   (i) The coefficient of static friction is given by (applied force) ÷ (frictional force) as sliding is just about to start.
   (j) Lubrication results in a reduction in the coefficient of friction.

3. Which statement is false from (a), (b), (f) and (i)?

4. Which statement is false from (b), (e), (g) and (j)?

5. Which statement is true from (c), (f), (h) and (i)?

6. Which statement is false from (b), (c), (e) and (j)?

7. Which statement is false from (a), (d), (g) and (h)?

8. The normal force between two surfaces is 100 N and the dynamic coefficient of friction is 0.4. The force required to maintain a constant speed of sliding is:

   (a) 100.4 N   (b) 40 N   (c) 99.6 N   (d) 250 N

9. The normal force between two surfaces is 50 N and the force required to maintain a constant speed of sliding is 25 N. The dynamic coefficient of friction is:

   (a) 25   (b) 2   (c) 75   (d) 0.5

10. The maximum force which can be applied to an object without sliding occurring is 60 N, and the static coefficient of friction is 0.3. The normal force between the two surfaces is:

    (a) 200 N (b) 18 N   (c) 60.3 N   (d) 59.7 N

31

Simple machines

At the end of this chapter you should be able to:

- define a simple machine
- define force ratio, movement ratio, efficiency and limiting efficiency
- understand and perform calculations with pulley systems
- understand and perform calculations with a simple screw-jack
- understand and perform calculations with gear trains
- understand and perform calculations with levers

31.1 Machines

A machine is a device which can change the magnitude or line of action, or both magnitude and line of action of a force. A simple machine usually amplifies an input force, called the **effort**, to give a larger output force, called the **load**. Some typical examples of simple machines include pulley systems, screw-jacks, gear systems and lever systems.

31.2 Force ratio, movement ratio and efficiency

The **force ratio** or **mechanical advantage** is defined as the ratio of load to effort, i.e.

$$\boxed{\text{Force ratio} = \frac{\text{load}}{\text{effort}}} \tag{1}$$

Since both load and effort are measured in newtons, force ratio is a ratio of the same units and thus is a dimension-less quantity.

The **movement ratio** or **velocity ratio** is defined as the ratio of the distance moved by the effort to the distance moved by the load, i.e.

$$\boxed{\textbf{Movement ratio} = \frac{\textbf{distance moved by the effort}}{\textbf{distance moved by the load}}} \tag{2}$$

Since the numerator and denominator are both measured in metres, movement ratio is a ratio of the same units and thus is a dimension-less quantity.

The **efficiency of a simple machine** is defined as the ratio of the force ratio to the movement ratio, i.e.

$$\text{Efficiency} = \frac{\text{force ratio}}{\text{movement ratio}}$$

Since the numerator and denominator are both dimension-less quantities, efficiency is a dimension-less quantity. It is usually expressed as a percentage, thus:

$$\boxed{\textbf{Efficiency} = \frac{\textbf{force ratio}}{\textbf{movement ratio}} \times \textbf{100\%}} \tag{3}$$

Due to the effects of friction and inertia associated with the movement of any object, some of the input energy to a machine is converted into heat and losses occur. Since losses occur, the energy output of a machine is less than the energy input, thus the mechanical efficiency of any machine cannot reach 100%.

For simple machines, the relationship between effort and load is of the form: $F_e = aF_l + b$, where F_e is the effort, F_l is the load and a and b are constants.

From equation (1), force ratio $= \dfrac{\text{load}}{\text{effort}} = \dfrac{F_l}{F_e} = \dfrac{F_l}{aF_l + b}$

Dividing both numerator and denominator by F_l gives:

$$\frac{F_l}{aF_l + b} = \frac{1}{a + \dfrac{b}{F_l}}$$

When the load is large, F_l is large and $\dfrac{b}{F_l}$ is small compared with a. The force ratio then becomes approximately equal to $\dfrac{1}{a}$ and is called the **limiting force ratio**, i.e.

$$\boxed{\text{limiting ratio} = \frac{1}{a}}$$

The limiting efficiency of a simple machine is defined as the ratio of the limiting force ratio to the movement ratio, i.e.

$$\boxed{\text{Limiting efficiency} = \frac{1}{a \times \text{movement ratio}} \times 100\%}$$

where a is the constant for the law of the machine:

$$F_e = aF_l + b$$

Due to friction and inertia, the limiting efficiency of simple machines is usually well below 100%.

Problem 1. A simple machine raises a load of 160 kg through a distance of 1.6 m. The effort applied to the machine is 200 N and moves through a distance of 16 m. Taking g as 9.8 m/s^2, determine the force ratio, movement ratio and efficiency of the machine.

From equation (1),

$$\textbf{force ratio} = \frac{\text{load}}{\text{effort}} = \frac{160 \text{ kg}}{200 \text{ N}} = \frac{160 \times 9.8 \text{ N}}{200 \text{ N}} = \textbf{7.84}$$

From equation (2),

$$\textbf{movement ratio} = \frac{\text{distance moved by the effort}}{\text{distance moved by the load}}$$

$$= \frac{16 \text{ m}}{1.6 \text{ m}} = \textbf{10}$$

From equation (3),

$$\textbf{efficiency} = \frac{\text{force ratio}}{\text{movement ratio}} \times 100\% = \frac{7.84}{10} \times 100$$

$$= \textbf{78.4\%}$$

Problem 2. For the simple machine of Problem 1, determine: (a) the distance moved by the effort to move

---

the load through a distance of 0.9 m, (b) the effort which would be required to raise a load of 200 kg, assuming the same efficiency, (c) the efficiency if, due to lubrication, the effort to raise the 160 kg load is reduced to 180 N.

(a) Since the movement ratio is 10, then from equation (2), **distance moved by the effort**

$$= 10 \times \text{distance moved by the load}$$

$$= 10 \times 0.9 = \textbf{9 m}$$

(b) Since the force ratio is 7.84, then from equation (1),

$$\textbf{effort} = \frac{\text{load}}{7.84} = \frac{200 \times 9.8}{7.84} = \textbf{250 N}$$

(c) The new force ratio is given by

$$\frac{\text{load}}{\text{effort}} = \frac{160 \times 9.8}{180} = 8.711$$

Hence **the new efficiency after lubrication**

$$= \frac{8.711}{10} \times 100 = \textbf{87.11\%}$$

Problem 3. In a test on a simple machine, the effort/load graph was a straight line of the form $F_e = aF_l + b$. Two values lying on the graph were at $F_e = 10$ N, $F_l = 30$ N, and at $F_e = 74$ N, $F_l = 350$ N. The movement ratio of the machine was 17. Determine: (a) the limiting force ratio, (b) the limiting efficiency of the machine.

(a) The equation $F_e = aF_l + b$ is of the form $y = mx + c$, where m is the gradient of the graph. The slope of the line passing through points (x_1, y_1) and (x_2, y_2) of the graph $y = mx + c$ is given by:

$$m = \frac{y_2 - y_1}{x_2 - x_1} \quad \text{(see Chapter 9, page 48)}$$

Thus for $F_e = aF_l + b$, the slope a is given by:

$$a = \frac{74 - 10}{350 - 30} = \frac{64}{320} = 0.2$$

The **limiting force ratio** is $\dfrac{1}{a}$, that is $\dfrac{1}{0.2} = \textbf{5}$

(b) The **limiting efficiency** $= \dfrac{1}{a \times \text{movement ratio}} \times 100$

$$= \frac{1}{0.2 \times 17} \times 100 = \textbf{29.4\%}$$

Now try the following exercise

Exercise 141   Further problems on force ratio, move-ment ratio and efficiency (Answers on page 309)

1. A simple machine raises a load of 825 N through a distance of 0.3 m. The effort is 250 N and moves through a distance of 3.3 m. Determine: (a) the force ratio, (b) the movement ratio, (c) the efficiency of the machine at this load.

2. The efficiency of a simple machine is 50%. If a load of 1.2 kN is raised by an effort of 300 N, determine the movement ratio.

3. An effort of 10 N applied to a simple machine moves a load of 40 N through a distance of 100 mm, the efficiency at this load being 80%. Calculate: (a) the movement ratio, (b) the distance moved by the effort.

4. The effort required to raise a load using a simple machine, for various values of load is as shown:

Load (N)	2050	4120	7410	8240	10 300
Effort (N)	252	340	465	505	580

If the movement ratio for the machine is 30, determine (a) the law of the machine, (b) the limiting force ratio, (c) the limiting efficiency.

5. For the data given in question 4, determine the values of force ratio and efficiency for each value of the load. Hence plot graphs of effort, force ratio and efficiency to a base of load. From the graphs, determine the effort required to raise a load of 6 kN and the efficiency at this load.

31.3 Pulleys

A **pulley system** is a simple machine. A single-pulley system, shown in Fig. 31.1(a), changes the line of action of the effort, but does not change the magnitude of the force. A two-pulley system, shown in Fig. 31.1(b), changes both the line of action and the magnitude of the force.

Theoretically, each of the ropes marked (i) and (ii) share the load equally, thus the theoretical effort is only half of the load, i.e. the theoretical force ratio is 2. In practice the actual force ratio is less than 2 due to losses. A three-pulley system is shown in Fig. 31.1(c). Each of the ropes marked (i), (ii) and (iii) carry one-third of the load, thus the theoretical force ratio is 3. In general, for a multiple pulley system having a total of n pulleys, the theoretical force ratio is n. Since the theoretical efficiency of a pulley system (neglecting losses) is 100 and since from equation (3),

$$\text{efficiency} = \frac{\text{force ratio}}{\text{movement ratio}} \times 100\%$$

it follows that when the force ratio is n,

$$100 = \frac{n}{\text{movement ratio}} \times 100$$

that is, the movement ratio is also n

> *Problem 4.*   A load of 80 kg is lifted by a three-pulley system similar to that shown in Fig. 31.1(c) and the applied effort is 392 N. Calculate (a) the force ratio, (b) the movement ratio, (c) the efficiency of the system. Take g to be 9.8 m/s^2

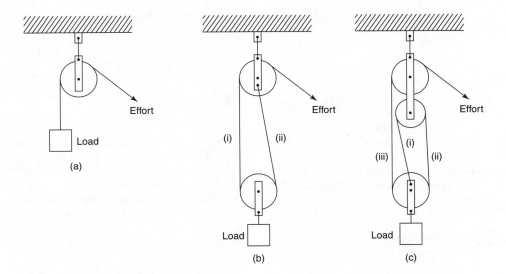

Fig. 31.1

(a) From equation (1), the force ratio is given by $\dfrac{load}{effort}$
The load is 80 kg, i.e. (80×9.8) N, hence

$$\textbf{force ratio} = \frac{80 \times 9.8}{392} = \textbf{2}$$

(b) From above, for a system having n pulleys, the movement ratio is n. Thus for a three-pulley system, the **movement ratio is 3**

(c) From equation (3),

$$\textbf{efficiency} = \frac{force\ ratio}{movement\ ratio} \times 100$$

$$= \frac{2}{3} \times 100 = \textbf{66.67\%}$$

Problem 5. A pulley system consists of two blocks, each containing three pulleys and connected as shown in Fig. 31.2. An effort of 400 N is required to raise a load of 1500 N. Determine (a) the force ratio, (b) the movement ratio, (c) the efficiency of the pulley system.

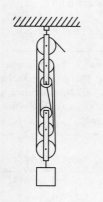

Fig. 31.2

(a) From equation (1),

$$\textbf{force ratio} = \frac{load}{effort} = \frac{1500}{400} = \textbf{3.75}$$

(b) An n-pulley system has a movement ratio of n, hence this 6-pulley system has a **movement ratio of 6**

(c) From equation (3),

$$\textbf{efficiency} = \frac{force\ ratio}{movement\ ratio} \times 100$$

$$= \frac{3.75}{6} \times 100 = \textbf{62.5\%}$$

Now try the following exercise

Exercise 142   Further problems on pulleys (Answers on page 309)

1. A pulley system consists of four pulleys in an upper block and three pulleys in a lower block. Make a sketch of this arrangement showing how a movement ratio of 7 may be obtained. If the force ratio is 4.2, what is the efficiency of the pulley.

2. A three-pulley lifting system is used to raise a load of 4.5 kN. Determine the effort required to raise this load when losses are neglected. If the actual effort required is 1.6 kN, determine the efficiency of the pulley system at this load.

31.4   The screw-jack

A **simple screw-jack** is shown in Fig. 31.3 and is a simple machine since it changes both the magnitude and the line of action of a force.

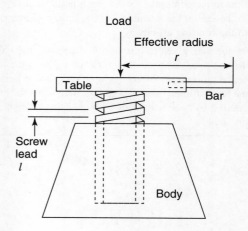

Fig. 31.3

The screw of the table of the jack is located in a fixed nut in the body of the jack. As the table is rotated by means of a bar, it raises or lowers a load placed on the table. For a single-start thread, as shown, for one complete revolution of the table, the effort moves through a distance $2\pi r$, and the load moves through a distance equal to the lead of the screw, say, l

$$\textbf{Movement ratio} = \frac{2\pi r}{l}$$

the same shaft as B and attached to it is a gear C with 60 teeth, meshing with a gear D on the output shaft having 120 teeth. Calculate the movement and force ratios if the overall efficiency of the gears is 72%.

3. A compound gear train is as shown in Fig. 30.6. The movement ratio is 6 and the numbers of teeth on gears A, C and D are 25, 100 and 60, respectively. Determine the number of teeth on gear B and the force ratio when the efficiency is 60%.

31.6 Levers

A **lever** can alter both the magnitude and the line of action of a force and is thus classed as a simple machine. There are three types or orders of levers, as shown in Fig. 31.7.

A lever of the first order has the fulcrum placed between the effort and the load, as shown in Fig. 31.7(a).

A lever of the second order has the load placed between the effort and the fulcrum, as shown in Fig. 31.7(b).

A lever of the third order has the effort applied between the load and the fulcrum, as shown in Fig. 31.7(c).

Problems on levers can largely be solved by applying the principle of moments (see Chapter 28). Thus for the lever shown in Fig. 31.7(a), when the lever is in equilibrium,

anticlockwise moment = clockwise moment

i.e. $$a \times F_l = b \times F_e$$

Thus, **force ratio** $= \dfrac{F_l}{F_e} = \dfrac{b}{a}$

$$= \frac{\textbf{distance of effort from fulcrum}}{\textbf{distance of load from fulcrum}}$$

Problem 9. The load on a first-order lever, similar to that shown in Fig. 31.7(a), is 1.2 kN. Determine the effort, the force ratio and the movement ratio when the distance between the fulcrum and the load is 0.5 m and the distance between the fulcrum and effort is 1.5 m. Assume the lever is 100% efficient.

Applying the principle of moments, for equilibrium:

anticlockwise moment = clockwise moment

i.e. $1200\,\text{N} \times 0.5\,\text{m} = \text{effort} \times 1.5\,\text{m}$

Hence, **effort** $= \dfrac{1200 \times 0.5}{1.5} = \textbf{400 N}$

force ratio $= \dfrac{F_l}{F_e} = \dfrac{1200}{400} = \textbf{3}$

Alternatively, **force ratio** $= \dfrac{b}{a} = \dfrac{1.5}{0.5} = \textbf{3}$

This result shows that to lift a load of, say, 300 N, an effort of 100 N is required.

Since, from equation (3),

$$\text{efficiency} = \frac{\text{force ratio}}{\text{movement ratio}} \times 100\%$$

then, **movement ratio** $= \dfrac{\text{force ratio}}{\text{efficiency}} \times 100$

$$= \frac{3}{100} \times 100 = \textbf{3}$$

This result shows that to raise the load by, say, 100 mm, the effort has to move 300 mm.

Problem 10. A second-order lever, AB, is in a horizontal position. The fulcrum is at point C. An effort of 60 N applied at B just moves a load at point D, when BD is 0.5 m and BC is 1.25 m. Calculate the load and the force ratio of the lever.

A second-order lever system is shown in Fig. 31.7(b). Taking moments about the fulcrum as the load is just moving, gives:

anticlockwise moment = clockwise moment

i.e. $60\,\text{N} \times 1.25\,\text{m} = \text{load} \times 0.75\,\text{m}$

Thus, **load** $= \dfrac{60 \times 1.25}{0.75} = \textbf{100 N}$

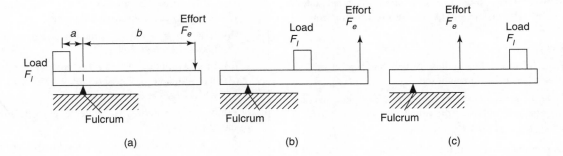

Fig. 31.7

From equation (1),

$$\textbf{force ratio} = \frac{\text{load}}{\text{effort}} = \frac{100}{60} = 1\frac{2}{3}$$

Alternatively,

$$\text{force ratio} = \frac{\text{distance of effort from fulcrum}}{\text{distance of load from fulcrum}}$$

$$= \frac{1.25}{0.75} = 1\frac{2}{3}$$

Now try the following exercise

Exercise 145   Further problems on levers (Answers on page 309)

1. In a second-order lever system, the force ratio is 2.5. If the load is at a distance of 0.5 m from the fulcrum, find the distance that the effort acts from the fulcrum if losses are negligible.

2. A lever *AB* is 2 m long and the fulcrum is at a point 0.5 m from *B*. Find the effort to be applied at *A* to raise a load of 0.75 kN at *B* when losses are negligible.

3. The load on a third-order lever system is at a distance of 750 mm from the fulcrum and the effort required to just move the load is 1 kN when applied at a distance of 250 mm from the fulcrum. Determine the value of the load and the force ratio if losses are negligible.

Exercise 146   Short answer questions on simple machines

1. State what is meant by a simple machine.
2. Define force ratio.
3. Define movement ratio.
4. Define the efficiency of a simple machine in terms of the force and movement ratios.
5. State briefly why the efficiency of a simple machine cannot reach 100%.
6. With reference to the law of a simple machine, state briefly what is meant by the term 'limiting force ratio'.
7. Define limiting efficiency.
8. Explain why a four-pulley system has a force ratio of 4 when losses are ignored.
9. Give the movement ratio for a screw-jack in terms of the effective radius of the effort and the screw lead.
10. Explain the action of an idler gear.
11. Define the movement ratio for a two-gear system in terms of the teeth on the wheels.

12. Show that the action of an idler wheel does not affect the movement ratio of a gear system.
13. State the relationship between the speed of the first gear and the speed of the last gear in a compound train of four gears, in terms of the teeth on the wheels.
14. Define the force ratio of a first-order lever system in terms of the distances of the load and effort from the fulcrum.
15. Use sketches to show what is meant by: (a) a first-order, (b) a second-order, (c) a third-order lever system. Give one practical use for each type of lever.

Exercise 147   Multi-choice questions on simple machines (Answers on page 309)

A simple machine requires an effort of 250 N moving through 10 m to raise a load of 1000 N through 2 m. Use this data to find the correct answers to questions 1 to 3, selecting these answers from:

(a) 0.25   (b) 4   (c) 80%   (d) 20%

(e) 100   (f) 5   (g) 100%   (h) 0.2   (i) 25%

1. Find the force ratio.
2. Find the movement ratio.
3. Find the efficiency.

The law of a machine is of the form $F_e = aF_l + b$. An effort of 12 N is required to raise a load of 40 N and an effort of 6 N is required to raise a load of 16 N. The movement ratio of the machine is 5. Use this data to find the correct answers to questions 4 to 6, selecting these answers from:

(a) 80%   (b) 4   (c) 2.8   (d) 0.25

(e) $\frac{1}{2.8}$ (f) 25%   (g) 100%   (h) 2   (i) 25%

4. Determine the constant '*a*'
5. Find the limiting force ratio.
6. Find the limiting efficiency.
7. Which of the following statements is false?
   (a) A single-pulley system changes the line of action of the force but does not change the magnitude of the force, when losses are neglected
   (b) In a two-pulley system, the force ratio is $\frac{1}{2}$ when losses are neglected
   (c) In a two-pulley system, the movement ratio is 2
   (d) The efficiency of a two-pulley system is 100% when losses are neglected
8. Which of the following statements concerning a screw-jack is false?
   (a) A screw-jack changes both the line of action and the magnitude of the force

(b) For a single-start thread, the distance moved in 5 revolutions of the table is $5l$, where l is the lead of the screw

(c) The distance moved by the effort is $2\pi r$, where r is the effective radius of the effort

(d) The movement ratio is given by $\dfrac{2\pi r}{5l}$

9. In a simple gear train, a follower has 50 teeth and the driver has 30 teeth. The movement ratio is:

    (a) 0.6   (b) 20   (c) 1.67   (d) 80

10. Which of the following statements is true?

    (a) An idler wheel between a driver and a follower is used to make the direction of the follower opposite to that of the driver

    (b) An idler wheel is used to change the movement ratio

    (c) An idler wheel is used to change the force ratio

    (d) An idler wheel is used to make the direction of the follower the same as that of the driver

11. Which of the following statements is false?

    (a) In a first-order lever, the fulcrum is between the load and the effort

    (b) In a second-order lever, the load is between the effort and the fulcrum

    (c) In a third-order lever, the effort is applied between the load and the fulcrum

    (d) The force ratio for a first-order lever system is given by: $\dfrac{\text{distance of load from fulcrum}}{\text{distance of effort from fulcrum}}$

12. In a second-order lever system, the load is 200 mm from the fulcrum and the effort is 500 mm from the fulcrum. If losses are neglected, an effort of 100 N will raise a load of:

    (a) 100 N   (b) 250 N   (c) 400 N   (d) 40 N

Assignment 11

This assignment covers the material contained in chapters 29 to 31. The marks for each question are shown in brackets at the end of each question.

1. A train is travelling at 90 km/h and has wheels of diameter 1600 mm.

    (a) Find the angular velocity of the wheels in both rad/s and rev/min

    (b) If the speed remains constant for 2 km, determine the number of revolutions made by a wheel, assuming no slipping occurs   (9)

2. The speed of a shaft increases uniformly from 200 revolutions per minute to 700 revolutions per minute in 12 s. Find the angular acceleration, correct to 3 significant figures.   (5)

3. The shaft of an electric motor, initially at rest, accelerates uniformly for 0.3 s at 20 rad/s². Determine the angle (in radians) turned through by the shaft in this time.   (4)

4. The material of a brake is being tested and it is found that the dynamic coefficient of friction between the material and steel is 0.90. Calculate the normal force when the frictional force is 0.630 kN   (5)

5. A simple machine raises a load of 120 kg through a distance of 1.2 m. The effort applied to the machine is 150 N and moves through a distance of 12 m. Taking g as 10 m/s², determine the force ratio, movement ratio and efficiency of the machine.   (6)

6. A load of 30 kg is lifted by a three-pulley system and the applied effort is 140 N. Calculate, taking g to be 9.8 m/s², (a) the force ratio, (b) the movement ratio, (c) the efficiency of the system.   (5)

7. A screw-jack is being used to support the axle of a lorry, the load on it being 5.6 kN. The screw-jack has an eifort of effective radius 318 mm and a single-start square thread, having a lead of 5 mm. Determine the efficiency of the jack if an effort of 70 N is required to raise the car axle.   (6)

8. A driver gear on a shaft of a motor has 32 teeth and meshes with a follower having 96 teeth. If the speed of the motor is 1410 revolutions per minute, find the speed of rotation of the follower.   (4)

9. The load on a first-order lever is 1.5 kN. Determine the effort, the force ratio and the movement ratio when the distance between the fulcrum and the load is 0.4 m and the distance between the fulcrum and effort is 1.6 m. Assume the lever is 100% efficient.   (6)

32

The effects of forces on materials

At the end of this chapter you should be able to:

- define force and state its unit
- recognise a tensile force and state relevant practical examples
- recognise a compressive force and state relevant practical examples
- recognise a shear force and state relevant practical examples
- define stress and state its unit
- calculate stress σ from $\sigma = \dfrac{F}{A}$
- define strain
- calculate strain ε from $\varepsilon = \dfrac{x}{l}$
- define elasticity, plasticity and elastic limit
- state Hooke's law
- define Young's modulus of elasticity E and stiffness
- appreciate typical values for E
- calculate E from $E = \dfrac{\sigma}{\varepsilon}$
- perform calculations using Hooke's law
- plot a load/extension graph from given data
- define ductility, brittleness and malleability, with examples of each

32.1 Introduction

A **force** exerted on a body can cause a change in either the shape or the motion of the body. The unit of force is the **newton, N**.

No solid body is perfectly rigid and when forces are applied to it, changes in dimensions occur. Such changes are not always perceptible to the human eye since they are so small. For example, the span of a bridge will sag under the weight of a vehicle and a spanner will bend slightly when tightening a nut. It is important for engineers and designers to appreciate the effects of forces on materials, together with their mechanical properties.

The three main types of mechanical force that can act on a body are:

(i) tensile, (ii) compressive, and (iii) shear

32.2 Tensile force

Tension is a force which tends to stretch a material, as shown in Fig. 32.1. Examples include:

(i) the rope or cable of a crane carrying a load is in tension

(ii) rubber bands, when stretched, are in tension

(iii) a bolt; when a nut is tightened, a bolt is under tension

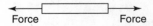
Force   Force

Fig. 32.1

A tensile force, i.e. one producing tension, increases the length of the material on which it acts.

32.3 Compressive force

Compression is a force which tends to squeeze or crush a material, as shown in Fig. 32.2. Examples include:

Fig. 32.2

(i) a pillar supporting a bridge is in compression

(ii) the sole of a shoe is in compression

(iii) the jib of a crane is in compression

A compressive force, i.e. one producing compression, will decrease the length of the material on which it acts.

32.4 Shear force

Shear is a force which tends to slide one face of the material over an adjacent face. Examples include:

(i) a rivet holding two plates together is in shear if a tensile force is applied between the plates – as shown in Fig. 32.3

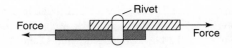

Fig. 32.3

(ii) a guillotine cutting sheet metal, or garden shears, each provide a shear force

(iii) a horizontal beam is subject to shear force

(iv) transmission joints on cars are subject to shear forces

A shear force can cause a material to bend, slide or twist.

Problem 1. Figure. 32.4(a) represents a crane and Fig. 32.4(b) a transmission joint. State the types of forces acting labelled A to F.

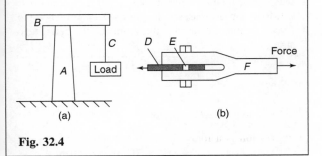

Fig. 32.4

(a) For the crane, A, a supporting member, is in **compression**, B, a horizontal beam, is in **shear**, and C, a rope, is in **tension**.

(b) For the transmission joint, parts D and F are in **tension**, and E, the rivet or bolt, is in **shear**.

32.5 Stress

Forces acting on a material cause a change in dimensions and the material is said to be in a state of **stress**. Stress is the ratio of the applied force F to cross-sectional area A of the material. The symbol used for tensile and compressive stress is σ (Greek letter sigma). The unit of stress is the **Pascal, Pa**, where $1\,\text{Pa} = 1\,\text{N/m}^2$. Hence

$$\sigma = \frac{F}{A}\ \textbf{Pa}$$

where F is the force in newtons and A is the cross-sectional area in square metres. For tensile and compressive forces, the cross-sectional area is that which is at right angles to the direction of the force. For a shear force the shear stress is equal to $\frac{F}{A}$, where the cross-sectional area A is that which is parallel to the direction of the force. The symbol used for shear stress is the Greek letter tau, τ

Problem 2. A rectangular bar having a cross-sectional area of $75\,\text{mm}^2$ has a tensile force of $15\,\text{kN}$ applied to it. Determine the stress in the bar.

Cross-sectional area $A = 75\,\text{mm}^2 = 75 \times 10^{-6}\,\text{m}^2$ and force $F = 15\,\text{kN} = 15 \times 10^3\,\text{N}$.

Stress in bar, $\sigma = \dfrac{F}{A} = \dfrac{15 \times 10^3\,\text{N}}{75 \times 10^{-6}\,\text{m}^2}$

$$= 0.2 \times 10^9\,\text{Pa} = \textbf{200\,MPa}$$

Problem 3. A circular wire has a tensile force of $60.0\,\text{N}$ applied to it and this force produces a stress of $3.06\,\text{MPa}$ in the wire. Determine the diameter of the wire.

Force $F = 60.0\,\text{N}$ and stress $\sigma = 3.06\,\text{MPa} = 3.06 \times 10^6\,\text{Pa}$.

Since $\sigma = \dfrac{F}{A}$ then area,

$$A = \frac{F}{\sigma} = \frac{60.0\,\text{N}}{3.06 \times 10^6\,\text{Pa}} = 19.61 \times 10^{-6}\,\text{m}^2$$

$$= 19.61\,\text{mm}^2$$

Cross-sectional area $A = \dfrac{\pi d^2}{4}$; hence $19.61 = \dfrac{\pi d^2}{4}$

from which, $\qquad d^2 = \dfrac{4 \times 19.61}{\pi}$

from which, $\qquad d = \sqrt{\left(\dfrac{4 \times 19.61}{\pi}\right)}$

i.e. **diameter of wire $= 5.0$ mm**

Now try the following exercise

Exercise 148　Further problems on stress
(Answers on page 310)

1. A rectangular bar having a cross-sectional area of 80 mm^2 has a tensile force of 20 kN applied to it. Determine the stress in the bar.
2. A circular cable has a tensile force of 1 kN applied to it and the force produces a stress of 7.8 MPa in the cable. Calculate the diameter of the cable.
3. A square-sectioned support of side 12 mm is loaded with a compressive force of 10 kN. Determine the compressive stress in the support.
4. A bolt having a diameter of 5 mm is loaded so that the shear stress in it is 120 MPa. Determine the value of the shear force on the bolt.
5. A tube of outside diameter 60 mm and inside diameter 40 mm is subjected to a load of 60 kN. Determine the stress in the tube.

32.6　Strain

The fractional change in a dimension of a material produced by a force is called the **strain**. For a tensile or compressive force, strain is the ratio of the change of length to the original length. The symbol used for strain is ε (Greek epsilon). For a material of length l metres which changes in length by an amount x metres when subjected to stress,

$$\boxed{\varepsilon = \frac{x}{l}}$$

Strain is dimension-less and is often expressed as a percentage, i.e.

$$\text{percentage strain} = \frac{x}{l} \times 100$$

For a shear force, strain is denoted by the symbol γ (Greek letter gamma) and, with reference to Fig. 32.5, is given by:

$$\gamma = \frac{x}{l}$$

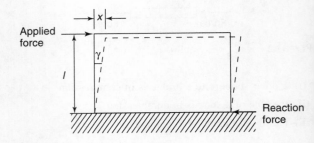

Fig. 32.5

Problem 4. A bar 1.60 m long contracts by 0.1 mm when a compressive load is applied to it. Determine the strain and the percentage strain.

$$\textbf{Strain } \varepsilon = \frac{\text{contraction}}{\text{original length}} = \frac{0.1 \text{ mm}}{1.60 \times 10^3 \text{ mm}}$$

$$= \frac{0.1}{1600} = \textbf{0.0000625}$$

Percentage strain $= 0.0000625 \times 100 = \textbf{0.00625\%}$

Problem 5. A wire of length 2.50 m has a percentage strain of 0.012% when loaded with a tensile force. Determine the extension of the wire.

Original length of wire $= 2.50$ m $= 2500$ mm and

$$\text{strain} = \frac{0.012}{100} = 0.00012$$

$$\text{Strain } \varepsilon = \frac{\text{extension } x}{\text{original length } l}$$

hence,

extension $x = \varepsilon l = (0.00012)(2500) = \textbf{0.3 mm}$

Problem 6. (a) A rectangular metal bar has a width of 10 mm and can support a maximum compressive stress of 20 MPa; determine the minimum breadth of the bar when loaded with a force of 3 kN. (b) If the bar in (a) is 2 m long and decreases in length by 0.25 mm when the force is applied, determine the strain and the percentage strain.

(a) Since stress $\sigma = \dfrac{\text{force } F}{\text{area } A}$ then,

$$\text{area } A = \frac{F}{\sigma} = \frac{3000 \text{ N}}{20 \times 10^6 \text{ Pa}}$$

$$= 150 \times 10^{-6} \text{ m}^2 = 150 \text{ mm}^2$$

Cross-sectional area = width × breadth, hence

$$\textbf{breadth} = \frac{\text{area}}{\text{width}} = \frac{150}{10} = \textbf{15 mm}$$

(b) **Strain** $\varepsilon = \dfrac{\text{contraction}}{\text{original length}}$

$$= \frac{0.25}{2000} = \textbf{0.000125}$$

Percentage strain $= 0.000125 \times 100 = \textbf{0.0125\%}$

Problem 7. A rectangular block of plastic material 500 mm long by 20 mm wide by 300 mm high has its lower face glued to a bench and a force of 200 N is applied to the upper face and in line with it. The upper face moves 15 mm relative to the lower face. Determine (a) the shear stress, and (b) the shear strain in the upper face, assuming the deformation is uniform.

(a) Shear stress

$$\tau = \frac{\text{force}}{\text{area parallel to the force}}$$

Area of any face parallel to the force

$$= 500\,\text{mm} \times 20\,\text{mm}$$

$$= (0.5 \times 0.02)\,\text{m}^2$$

$$= 0.01\,\text{m}^2$$

Hence, **shear stress,**

$$\tau = \frac{200\,\text{N}}{0.01\,\text{m}^2} = \textbf{20 000 Pa} \quad \text{or} \quad \textbf{20 kPa}$$

(b) Shear strain

$$\gamma = \frac{x}{l} \quad \text{(see side view in Fig. 32.6)}$$

$$= \frac{15}{300} = \textbf{0.05} \quad \text{or 5\%}$$

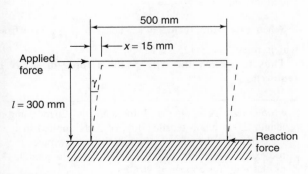

Fig. 32.6

Now try the following exercise

Exercise 149   Further problems on strain (Answers on page 310)

1. A wire of length 4.5 m has a percentage strain of 0.050% when loaded with a tensile force. Determine the extension in the wire.

2. A metal bar 2.5 m long extends by 0.05 mm when a tensile load is applied to it. Determine (a) the strain, (b) the percentage strain.

3. A pipe has an outside diameter of 25 mm, an inside diameter of 15 mm and length 0.40 m and it supports a compressive load of 40 kN. The pipe shortens by 0.5 mm when the load is applied. Determine (a) the compressive stress, (b) the compressive strain in the pipe when supporting this load.

32.7 Elasticity and elastic limit

Elasticity is the ability of a material to return to its original shape and size on the removal of external forces.

Plasticity is the property of a material of being permanently deformed by a force without breaking. Thus if a material does not return to the original shape, it is said to be plastic.

Within certain load limits, mild steel, copper, polythene and rubber are examples of elastic materials; lead and plasticine are examples of plastic materials.

If a tensile force applied to a uniform bar of mild steel is gradually increased and the corresponding extension of the bar is measured, then provided the applied force is not too large, a graph depicting these results is likely to be as shown in Fig. 32.7. Since the graph is a straight line, **extension is directly proportional to the applied force.**

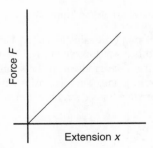

Fig. 32.7

If the applied force is large, it is found that the material no longer returns to its original length when the force is removed. The material is then said to have passed its elastic limit and the resulting graph of force/extension is no longer a straight line. Stress, $\sigma = \dfrac{F}{A}$, from section 32.5, and since, for a particular bar, area A can be considered as a constant, then $F \propto \sigma$. Strain $\varepsilon = \dfrac{x}{l}$, from section 32.6, and since for a particular bar l is constant, then $x \propto \varepsilon$. Hence for stress applied to a material below the elastic limit a graph of stress/strain will be as shown in Fig. 32.8, and is a similar shape to the force/extension graph of Fig. 32.7.

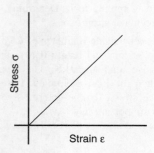

Fig. 32.8

32.8   Hooke's law

Hooke's law states:

Within the elastic limit, the extension of a material is proportional to the applied force.

It follows, from section 32.7, that:

Within the elastic limit of a material, the strain produced is directly proportional to the stress producing it.

Young's modulus of elasticity

Within the elastic limit, stress α strain, hence

stress = (a constant) × strain

This constant of proportionality is called **Young's modulus of elasticity** and is given the symbol E. The value of E may be determined from the gradient of the straight line portion of the stress/strain graph. The dimensions of E are pascals (the same as for stress, since strain is dimension-less).

$$E = \frac{\sigma}{\varepsilon}\ \text{Pa}$$

Some **typical values** for Young's modulus of elasticity, E, include:

Aluminium 70 GPa (i.e. 70×10^9 Pa), brass 90 GPa, copper 96 GPa, diamond 1200 GPa, mild steel 210 GPa, lead 18 GPa, tungsten 410 GPa, cast iron 110 GPa, zinc 85 GPa.

Stiffness

A material having a large value of Young's modulus is said to have a high value of stiffness, where stiffness is defined as:

$$\text{Stiffness} = \frac{\text{force } F}{\text{extension } x}$$

For example, mild steel is much stiffer than lead.

Since $E = \dfrac{\sigma}{\varepsilon}, \quad \sigma = \dfrac{F}{A}$

and $\varepsilon = \dfrac{x}{l}, \quad$ then

$$E = \frac{\frac{F}{A}}{\frac{x}{l}} = \frac{Fl}{Ax} = \left(\frac{F}{x}\right)\left(\frac{l}{A}\right)$$

i.e. $\boxed{E = (\text{stiffness}) \times \left(\dfrac{l}{A}\right)}$

Stiffness $\left(= \dfrac{F}{x}\right)$ is also the gradient of the force/extension graph, hence

$$E = (\text{gradient of force/extension graph}) \left(\frac{l}{A}\right)$$

Since l and A for a particular specimen are constant, the greater Young's modulus the greater the stiffness.

Problem 8. A wire is stretched 2 mm by a force of 250 N. Determine the force that would stretch the wire 5 mm, assuming that the elastic limit is not exceeded.

Hooke's law states that extension x is proportional to force F, provided that the elastic limit is not exceeded, i.e. $x \propto F$ or $x = kF$ where k is a constant.
When $x = 2$ mm, $F = 250$ N, thus $2 = k(250)$, from which, constant $k = \dfrac{2}{250} = \dfrac{1}{125}$

When $x = 5$ mm, then $5 = kF$ i.e. $5 = \left(\dfrac{1}{125}\right)(F)$ from which, force $F = 5(125) = 625$ N

Thus to stretch the wire 5 mm a force of 625 N is required.

Problem 9. A copper rod of diameter 20 mm and length 2.0 m has a tensile force of 5 kN applied to it. Determine (a) the stress in the rod, (b) by how much the rod extends when the load is applied. Take the modulus of elasticity for copper as 96 GPa.

(a) Force $F = 5\,\text{kN} = 5000\,\text{N}$ and cross-sectional area

$$A = \frac{\pi d^2}{4} = \frac{\pi (0.020)^2}{4} = 0.000314\,\text{m}^2$$

Stress, $\sigma = \dfrac{F}{A} = \dfrac{5000\,\text{N}}{0.000314\,\text{m}^2} = 15.92 \times 10^6\,\text{Pa}$

$$= \textbf{15.92 MPa}$$

(b) Since $E = \dfrac{\sigma}{\varepsilon}$ then

$$\text{strain } \varepsilon = \frac{\sigma}{E} = \frac{15.92 \times 10^6\,\text{Pa}}{96 \times 10^9\,\text{Pa}} = 0.000166$$

Strain $\varepsilon = \dfrac{x}{l}$, hence extension, $x = \varepsilon l$

$= (0.000166)(2.0) = 0.000332\,\text{m}$ i.e. **extension of rod is 0.332 mm**

Problem 10. A bar of thickness 15 mm and having a rectangular cross-section carries a load of 120 kN. Determine the minimum width of the bar to limit the maximum stress to 200 MPa. The bar, which is 1.0 m long, extends by 2.5 mm when carrying a load of 120 kN. Determine the modulus of elasticity of the material of the bar.

Force, $F = 120\,\text{kN} = 120\,000\,\text{N}$ and cross-sectional area $A = (15x)10^{-6}\,\text{m}^2$, where x is the width of the rectangular bar in millimetres.

Stress $\sigma = \dfrac{F}{A}$, from which,

$$A = \frac{F}{\sigma} = \frac{120\,000\,\text{N}}{200 \times 10^6\,\text{Pa}} = 6 \times 10^{-4}\,\text{m}^2$$

$$= 6 \times 10^2\,\text{mm}^2 = 600\,\text{mm}^2$$

Hence $600 = 15x$, from which,

$$\textbf{width of bar}, x = \frac{600}{15} = \textbf{40 mm}$$

Extension of bar $= 2.5\,\text{mm} = 0.0025\,\text{m}$, hence

$$\text{strain } \varepsilon = \frac{x}{l} = \frac{0.0025}{1.0} = 0.0025$$

Modulus of elasticity,

$$E = \frac{\text{stress}}{\text{strain}} = \frac{200 \times 10^6}{0.0025} = 80 \times 10^9 = \textbf{80 GPa}.$$

Problem 11. In an experiment to determine the modulus of elasticity of a sample of mild steel, a wire is loaded and the corresponding extension noted. The results of the experiment are as shown.

Load (N)	0	40	110	160	200	250	290	340
Extension (mm)	0	1.2	3.3	4.8	6.0	7.5	10.0	16.2

Draw the load/extension graph.

The mean diameter of the wire is 1.3 mm and its length is 8.0 m. Determine the modulus of elasticity E of the sample, and the stress at the elastic limit.

A graph of load/extension is shown in Fig. 32.9.

$$E = \frac{\sigma}{\varepsilon} = \frac{\frac{F}{A}}{\frac{x}{l}} = \left(\frac{F}{x}\right)\left(\frac{l}{A}\right)$$

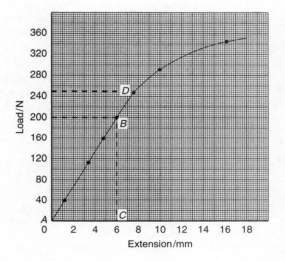

Fig. 32.9

$\dfrac{F}{x}$ is the gradient of the straight line part of the load/extension graph. Gradient,

$$\frac{F}{x} = \frac{BC}{AC} = \frac{200\,\text{N}}{6 \times 10^{-3}\,\text{m}} = 33.33 \times 10^3\,\text{N/m}$$

Modulus of elasticity $= (\text{gradient of graph})\left(\dfrac{l}{A}\right)$

Length of specimen, $l = 8.0\,\text{m}$ and

cross-sectional area $A = \dfrac{\pi d^2}{4} = \dfrac{\pi (0.0013)^2}{4}$

$$= 1.327 \times 10^{-6}\,\text{m}^2$$

Hence **modulus of elasticity,**

$$E = (33.33 \times 10^3)\left(\frac{8.0}{1.327 \times 10^{-6}}\right) = \textbf{201 GPa}$$

The elastic limit is at point D in Fig. 31.9 where the graph no longer follows a straight line. This point corresponds to a load of 250 N as shown.

$$\textbf{Stress at elastic limit} = \frac{\text{force}}{\text{area}} = \frac{250}{1.327 \times 10^{-6}}$$

$$= 188.4 \times 10^6\,\text{Pa}$$

$$= \textbf{188.4 MPa}$$

Now try the following exercise

**Exercise 150   Further problems on Hooke's law
(Answers on page 310)**

1. A wire is stretched 1.5 mm by a force of 300 N. Determine the force that would stretch the wire 4 mm, assuming the elastic limit of the wire is not exceeded.

2. A rubber band extends 50 mm when a force of 300 N is applied to it. Assuming the band is within the elastic limit, determine the extension produced by a force of 60 N.

3. A force of 25 kN applied to a piece of steel produces an extension of 2 mm. Assuming the elastic limit is not exceeded, determine (a) the force required to produce an extension of 3.5 mm, (b) the extension when the applied force is 15 kN.

4. A test to determine the load/extension graph for a specimen of copper gave the following results:

Load (kN)	8.5	15.0	23.5	30.0
Extension (mm)	0.04	0.07	0.11	0.14

   Plot the load/extension graph, and from the graph determine (a) the load at an extension of 0.09 mm, and (b) the extension corresponding to a load of 12.0 N

5. A circular bar is 2.5 m long and has a diameter of 60 mm. When subjected to a compressive load of 30 kN it shortens by 0.20 mm. Determine Young's modulus of elasticity for the material of the bar.

6. A bar of thickness 20 mm and having a rectangular cross-section carries a load of 82.5 kN. Determine (a) the minimum width of the bar to limit the maximum stress to 150 MPa, (b) the modulus of elasticity of the material of the bar if the 150 mm long bar extends by 0.8 mm when carrying a load of 200 kN.

7. A metal rod of cross-sectional area 100 mm^2 carries a maximum tensile load of 20 kN. The modulus of elasticity for the material of the rod is 200 GPa. Determine the percentage strain when the rod is carrying its maximum load.

32.9   Ductility, brittleness and malleability

Ductility is the ability of a material to be plastically deformed by elongation, without fracture. This is a property which enables a material to be drawn out into wires. For ductile materials such as mild steel, copper and gold, large extensions can result before fracture occurs with increasing tensile force. Ductile materials usually have a percentage elongation value of about 15% or more.

Brittleness is the property of a material manifested by fracture without appreciable prior plastic deformation. Brittleness is a lack of ductility, and brittle materials such as cast iron, glass, concrete, brick and ceramics, have virtually no plastic stage, the elastic stage being followed by immediate fracture. Little or no 'waist' occurs before fracture in a brittle material undergoing a tensile test.

Malleability is the property of a material whereby it can be shaped when cold by hammering or rolling. A malleable material is capable of undergoing plastic deformation without fracture. Examples of malleable materials include lead, gold, putty and mild steel.

Problem 12.   Sketch typical load/extension curves for (a) an elastic non-metallic material, (b) a brittle material and (c) a ductile material. Give a typical example of each type of material.

(a) A typical load/extension curve for an elastic non-metallic material is shown in Fig. 32.10(a), and an example of such a material is polythene.

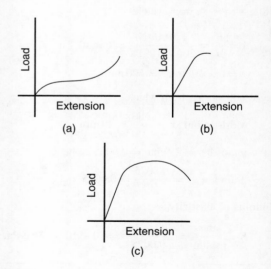

Fig. 32.10

(b) A typical load/extension curve for a brittle material is shown in Fig. 32.10(b), and an example of such a material is cast iron.

(c) A typical load/extension curve for a ductile material is shown in Fig. 32.10(c), and an example of such a material is mild steel.

Now try the following exercises

Exercise 151 Short answer questions on the effects of forces on materials

1. Name three types of mechanical force that can act on a body.

2. What is a tensile force? Name two practical examples of such a force.

3. What is a compressive force? Name two practical examples of such a force.

4. Define a shear force and name two practical examples of such a force.

5. Define elasticity and state two examples of elastic materials.

6. Define plasticity and state two examples of plastic materials.

7. State Hooke's law.

8. What is the difference between a ductile and a brittle material?

9. Define stress. What is the symbol used for
   (a) a tensile stress?   (b) a shear stress?

10. Strain is the ratio $\dfrac{\cdots\cdots}{\cdots\cdots}$

11. The ratio $\dfrac{\text{stress}}{\text{strain}}$ is called ....

12. State the units of (a) stress   (b) strain
    (c) Young's modulus of elasticity.

13. Stiffness is the ratio $\dfrac{\cdots\cdots}{\cdots\cdots}$

14. Sketch on the same axes a typical load/extension graph for a ductile and a brittle material.

15. Define (a) ductility   (b) brittleness   (c) malleability.

Exercise 152 Multi-choice questions on the effects of forces on materials (Answers on page 310)

1. The unit of strain is:
   (a) pascals   (b) metres
   (c) dimension-less   (d) newtons

2. The unit of stiffness is:
   (a) newtons   (b) pascals
   (c) newtons per metre   (d) dimension-less

3. The unit of Young's modulus of elasticity is:
   (a) pascals   (b) metres
   (c) dimension-less   (d) newtons

4. A wire is stretched 3 mm by a force of 150 N. Assuming the elastic limit is not exceeded, the force that will stretch the wire 5 mm is:
   (a) 150 N   (b) 250 N   (c) 90 N   (d) 450 N

5. For the wire in question 4, the extension when the applied force is 450 N is:
   (a) 1 mm   (b) 3 mm   (c) 9 mm   (d) 12 mm

6. Due to the forces acting, a horizontal beam is in:
   (a) tension   (b) compression   (c) shear

7. Due to forces acting, a pillar supporting a bridge is in:
   (a) tension   (b) compression   (c) shear

8. Which of the following statements is false?
   (a) Elasticity is the ability of a material to return to its original dimensions after deformation by a load.
   (b) Plasticity is the ability of a material to retain any deformation produced in it by a load.
   (c) Ductility is the ability to be permanently stretched without fracturing.
   (d) Brittleness is the lack of ductility and a brittle material has a long plastic stage.

9. A circular rod of cross-sectional area 100 mm² has a tensile force of 100 kN applied to it. The stress in the rod is:
   (a) 1 MPa   (b) 1 GPa   (c) 1 kPa   (d) 100 MPa

10. A metal bar 5.0 m long extends by 0.05 mm when a tensile load is applied to it. The percentage strain is:
    (a) 0.1   (b) 0.01   (c) 0.001   (d) 0.0001

An aluminium rod of length 1.0 m and cross-sectional area 500 mm² is used to support a load of 5 kN which causes the rod to contract by 100 μm. For questions 11 to 13, select the correct answer from the following list:
   (a) 100 MPa   (b) 0.001   (c) 10 kPa
   (d) 100 GPa   (e) 0.01   (f) 10 MPa
   (g) 10 GPa   (h) 0.0001   (i) 10 Pa

11. The stress in the rod.
12. The strain in the rod.
13. Young's modulus of elasticity.

33

Linear momentum and impulse

At the end of this chapter you should be able to:

- define momentum and state its unit
- state Newton's first law of conservation of energy
- calculate momentum given mass and velocity
- state Newton's second law of motion
- define impulse and appreciate when impulsive forces occur
- state Newton's third law of motion
- calculate impulse and impulsive force
- use the equation of motion $v^2 = u^2 + 2as$ in calculations

33.1 Linear momentum

The **momentum** of a body is defined as the product of its mass and its velocity, i.e. **momentum = mu**, where m = mass (in kg) and u = velocity (in m/s). The unit of momentum is kg m/s.

Since velocity is a vector quantity, **momentum is a vector quantity**, i.e. it has both magnitude and direction.

Newton's first law of motion states:

a body continues in a state of rest or in a state of uniform motion in a straight line unless acted on by some external force.

Hence the momentum of a body remains the same provided no external forces act on it.

The principle of conservation of momentum for a closed system (i.e. one on which no external forces act) may be stated as:

the total linear momentum of a system is a constant.

The total momentum of a system before collision in a given direction is equal to the total momentum of the system after collision in the same direction. In Fig. 33.1, masses m_1 and m_2 are travelling in the same direction with velocity $u_1 > u_2$. A collision will occur, and applying the principle of conservation of momentum:

total momentum before impact = total momentum after impact

i.e.
$$m_1 u_1 + m_2 u_2 = m_1 v_1 + m_2 v_2$$

where v_1 and v_2 are the velocities of m_1 and m_2 after impact.

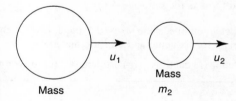

Fig. 33.1

Problem 1. Determine the momentum of a pile driver of mass 400 kg when it is moving downwards with a speed of 12 m/s.

Momentum = mass × velocity = 400 kg × 12 m/s

$$= \textbf{4800 kg m/s downwards}$$

Problem 2. A cricket ball of mass 150 g has a momentum of 4.5 kg m/s. Determine the velocity of the ball in km/h.

Momentum = mass × velocity, hence

$$\text{velocity} = \frac{\text{momentum}}{\text{mass}} = \frac{4.5 \text{ kg m/s}}{150 \times 10^{-3} \text{ kg}} = 30 \text{ m/s}$$

$$30 \text{ m/s} = 30 \times 3.6 \text{ km/h}$$

$$= \textbf{108 km/h}$$

$$= \textbf{velocity of cricket ball}$$

Problem 3. Determine the momentum of a railway wagon of mass 50 tonnes moving at a velocity of 72 km/h.

Momentum = mass × velocity

Mass = 50 t = 50 000 kg (since 1 t = 1000 kg) and

velocity = 72 km/h = $\frac{72}{3.6}$ m/s = 20 m/s.

Hence, **momentum** = 50 000 kg × 20 m/s

$$= 1\,000\,000 \text{ kg m/s} = \textbf{10}^6 \textbf{ kg m/s}$$

Problem 4. A wagon of mass 10 t is moving at a speed of 6 m/s and collides with another wagon of mass 15 t, which is stationary. After impact, the wagons are coupled together. Determine the common velocity of the wagons after impact.

Mass $m_1 = 10 \text{ t} = 10\,000$ kg, $m_2 = 15\,000$ kg and velocity $u_1 = 6$ m/s, $u_2 = 0$

Total momentum before impact

$$= m_1 u_1 + m_2 u_2 = (10\,000 \times 6) + (15\,000 \times 0)$$

$$= 60\,000 \text{ kg m/s}$$

Let the common velocity of the wagons after impact be v m/s since total momentum before impact = total momentum after impact:

$$60\,000 = m_1 v + m_2 v$$

$$= v(m_1 + m_2) = v(25\,000)$$

Hence $v = \frac{60\,000}{25\,000} = 2.4$ m/s

i.e. **the common velocity after impact is 2.4 m/s in the direction in which the 10 t wagon is initially travelling.**

Problem 5. A body has a mass of 30 g and is moving with a velocity of 20 m/s. It collides with a second body which has a mass of 20 g and which is moving with a velocity of 15 m/s. Assuming that the bodies both have the same velocity after impact, determine this common velocity, (a) when the initial velocities have the same

line of action and the same sense, and (b) when the initial velocities have the same line of action but are opposite in sense.

Mass $m_1 = 30 \text{ g} = 0.030$ kg, $m_2 = 20 \text{ g} = 0.020$ kg, velocity $u_1 = 20$ m/s and $u_2 = 15$ m/s.

(a) When the velocities have the same line of action and the same sense, both u_1 and u_2 are considered as positive values.

Total momentum before impact

$$= m_1 u_1 + m_2 u_2 = (0.030 \times 20) + (0.020 \times 15)$$

$$= 0.60 + 0.30 = 0.90 \text{ kg m/s}$$

Let the common velocity after impact be v m/s

Total momentum before impact

$$= \text{total momentum after impact}$$

i.e. $0.90 = m_1 v + m_2 v = v(m_1 + m_2)$

$$0.90 = v(0.030 + 0.020)$$

from which, **common velocity**, $v = \frac{0.90}{0.050} = \textbf{18 m/s}$ **in the direction in which the bodies are initially travelling.**

(b) When the velocities have the same line of action but are opposite in sense, one is considered as positive and the other negative. Taking the direction of mass m_1 as positive gives: velocity $u_1 = +20$ m/s and $u_2 = -15$ m/s.

Total momentum before impact

$$= m_1 u_1 + m_2 u_2 = (0.030 \times 20) + (0.020 \times -15)$$

$$= 0.60 - 0.30 = +0.30 \text{ kg m/s}$$

and since it is positive this indicates a momentum in the same direction as that of mass m_1. If the common velocity after impact is v m/s then

$$0.30 = v(m_1 + m_2) = v(0.050)$$

from which, **common velocity**, $v = \frac{0.30}{0.050} = \textbf{6 m/s}$ in **the direction that the 30 g mass is initially travelling.**

Now try the following exercise

Exercise 153 Further problems on linear momentum
(Answers on page 310)

(Where necessary, take g as 9.81 m/s²)

1. Determine the momentum in a mass of 50 kg having a velocity of 5 m/s.

2. A milling machine and its component have a combined mass of 400 kg. Determine the momentum of the table and component when the feed rate is 360 mm/min.

3. The momentum of a body is 160 kg m/s when the velocity is 2.5 m/s. Determine the mass of the body.

4. Calculate the momentum of a car of mass 750 kg moving at a constant velocity of 108 km/h.

5. A football of mass 200 g has a momentum of 5 kg m/s. What is the velocity of the ball in km/h.

6. A wagon of mass 8 t is moving at a speed of 5 m/s and collides with another wagon of mass 12 t, which is stationary. After impact, the wagons are coupled together. Determine the common velocity of the wagons after impact.

7. A car of mass 800 kg was stationary when hit head-on by a lorry of mass 2000 kg travelling at 15 m/s. Assuming no brakes are applied and the car and lorry move as one, determine the speed of the wreckage immediately after collision.

8. A body has a mass of 25 g and is moving with a velocity of 30 m/s. It collides with a second body which has a mass of 15 g and which is moving with a velocity of 20 m/s. Assuming that the bodies both have the same speed after impact, determine their common velocity (a) when the speeds have the same line of action and the same sense, and (b) when the speeds have the same line of action but are opposite in sense.

33.2 Impulse and impulsive forces

Newton's second law of motion states:

the rate of change of momentum is directly proportional to the applied force producing the change, and takes place in the direction of this force.

In the SI system, the units are such that:

the applied force = rate of change of momentum

$$= \frac{\text{change of momentum}}{\text{time taken}} \qquad (1)$$

When a force is suddenly applied to a body due to either a collision with another body or being hit by an object such as a hammer, the time taken in equation (1) is very small and difficult to measure. In such cases, the total effect of the force is measured by the change of momentum it produces.

Forces which act for very short periods of time are called **impulsive forces**. The product of the impulsive force and the time during which it acts is called the **impulse** of the force and is equal to the change of momentum produced by the impulsive force, i.e.

> **impulse = applied force × time**
> **= change in linear momentum**

Examples where impulsive forces occur include when a gun recoils and when a free-falling mass hits the ground. Solving problems associated with such occurrences often requires the use of the equation of motion: $v^2 = u^2 + 2as$, from Chapter 29.

When a pile is being hammered into the ground, the ground resists the movement of the pile and this resistance is called a **resistive force**.

Newton's third law of motion may be stated as:

for every force there is an equal and opposite force.

The force applied to the pile is the resistive force; the pile exerts an equal and opposite force on the ground.

In practice, when impulsive forces occur, energy is not entirely conserved and some energy is changed into heat, noise, and so on.

Problem 6. The average force exerted on the work-piece of a press-tool operation is 150 kN, and the tool is in contact with the work-piece for 50 ms. Determine the change in momentum.

From above, change of linear momentum

$$= \text{applied force} \times \text{time} \ (= \text{impulse})$$

Hence, **change in momentum of work-piece**

$$= 150 \times 10^3 \, \text{N} \times 50 \times 10^{-3} \, \text{s}$$

$$= \textbf{7500 kg m/s} \qquad (\text{since } 1 \, \text{N} = 1 \, \text{kg m/s}^2)$$

Problem 7. A force of 15 N acts on a body of mass 4 kg for 0.2 s. Determine the change in velocity.

Impulse = applied force × time

$$= \text{change in linear momentum}$$

i.e. 15 N × 0.2 s = mass × change in velocity

$$= 4 \, \text{kg} \times \text{change in velocity}$$

from which, **change in velocity** $= \dfrac{15 \, \text{N} \times 0.2 \, \text{s}}{4 \, \text{kg}} = \textbf{0.75 m/s}$

$$(\text{since } 1 \, \text{N} = 1 \, \text{kg m/s}^2)$$

Problem 8. A mass of 8 kg is dropped vertically on to a fixed horizontal plane and has an impact velocity of 10 m/s. The mass rebounds with a velocity of 6 m/s. If the mass-plane contact time is 40 ms, calculate (a) the impulse, and (b) the average value of the impulsive force on the plane.

a) Impulse = change in momentum = $m(u_1 - v_1)$

where u_1 = impact velocity = 10 m/s and
v_1 = rebound velocity = -6 m/s (v_1 is negative since it acts in the opposite direction to u_1).

Thus, **impulse** = $m(u_1 - v_1) = 8$ kg$(10 - -6)$ m/s

$$= 8 \times 16 = \mathbf{128\,kg\,m/s}$$

b) **Impulsive force** $= \dfrac{\text{impulse}}{\text{time}} = \dfrac{128\,\text{kg m/s}}{40 \times 10^{-3}\,\text{s}}$

$$= \mathbf{3200\,N} \quad \text{or} \quad \mathbf{3.2\,kN}$$

Problem 9. The hammer of a pile-driver of mass 1 t falls a distance of 1.5 m on to a pile. The blow takes place in 25 ms and the hammer does not rebound. Determine the average applied force exerted on the pile by the hammer.

Initial velocity, $u = 0$, acceleration due to gravity, $= 9.81$ m/s^2 and distance, $s = 1.5$ m

Using the equation of motion: $v^2 = u^2 + 2gs$

then $$v^2 = 0^2 + 2(9.81)(1.5)$$

from which, impact velocity, $v = \sqrt{(2)(9.81)(1.5)}$

$$= 5.425\,\text{m/s}$$

Neglecting the small distance moved by the pile and hammer after impact,

momentum lost by hammer = the change of momentum

$$= mv = 1000\,\text{kg} \times 5.425\,\text{m/s}$$

rate of change of momentum $= \dfrac{\text{change of momentum}}{\text{change of time}}$

$$= \dfrac{1000 \times 5.425}{25 \times 10^{-3}} = 217\,000\,\text{N}$$

Since the impulsive force is the rate of change of momentum, the average force exerted on the pile is **217 kN**

Problem 10. A mass of 40 g having a velocity of 15 m/s collides with a rigid surface and rebounds with a velocity of 5 m/s. The duration of the impact is 0.20 ms. Determine (a) the impulse, and (b) the impulsive force at the surface.

Mass $m = 40$ g $= 0.040$ kg, initial velocity, $u = 15$ m/s and final velocity, $v = -5$ m/s (negative since the rebound is in the opposite direction to velocity u)

a) Momentum before impact $= mu = 0.040 \times 15$

$$= 0.6\,\text{kg m/s}$$

Momentum after impact $= mv = 0.040 \times -5$

$$= -0.2\,\text{kg m/s}$$

Impulse = change of momentum

$$= 0.6 - (-0.2) = \mathbf{0.8\,kg\,m/s}$$

(b) **Impulsive force** $= \dfrac{\text{change of momentum}}{\text{change of time}}$

$$= \dfrac{0.8\,\text{kg m/s}}{0.20 \times 10^{-3}\,\text{s}} = \mathbf{4000\,N} \quad \text{or} \quad \mathbf{4\,kN}$$

Now try the following exercises

Exercise 154 **Further problems on impulse and impulsive forces (Answers on page 310)**

(Where necessary, take g as 9.81 m/s^2)

1. The sliding member of a machine tool has a mass of 200 kg. Determine the change in momentum when the sliding speed is increased from 10 mm/s to 50 mm/s.

2. A force of 48 N acts on a body of mass 8 kg for 0.25 s. Determine the change in velocity.

3. The speed of a car of mass 800 kg is increased from 54 km/h to 63 km/h in 2 s. Determine the average force in the direction of motion necessary to produce the change in speed.

4. A 10 kg mass is dropped vertically on to a fixed horizontal plane and has an impact velocity of 15 m/s. The mass rebounds with a velocity of 5 m/s. If the contact time of mass and plane is 0.025 s, calculate (a) the impulse, and (b) the average value of the impulsive force on the plane.

5. The hammer of a pile driver of mass 1.2 t falls 1.4 m on to a pile. The blow takes place in 20 ms and the hammer does not rebound. Determine the average applied force exerted on the pile by the hammer.

6. A tennis ball of mass 60 g is struck from rest with a racket. The contact time of ball on racket is 10 ms and the ball leaves the racket with a velocity of 25 m/s. Calculate (a) the impulse, and (b) the average force exerted by a racket on the ball.

Exercise 155 **Short answer questions on linear momentum and impulse**

1. Define momentum
2. State Newton's first law of motion
3. State the principle of the conservation of momentum
4. State Newton's second law of motion
5. Define impulse

6. What is meant by an impulsive force?

7. State Newton's third law of motion

Exercise 156 Multi-choice questions on linear momentum and impulse (Answers on page 310)

1. A mass of 100 g has a momentum of 100 kg m/s. The velocity of the mass is:

   (a) 10 m/s (b) 10^2 m/s (c) 10^{-3} m/s (d) 10^3 m/s.

2. A rifle bullet has a mass of 50 g. The momentum when the muzzle velocity is 108 km/h is:

   (a) 54 kg m/s (b) 1.5 kg m/s
   (c) 15 000 kg m/s (d) 21.6 kg m/s

   A body P of mass 10 kg has a velocity of 5 m/s and the same line of action as a body Q of mass 2 kg and having a velocity of 25 m/s. The bodies collide, and their velocities are the same after impact. In questions 3 to 6, select the correct answer from the following:

   (a) 25/3 m/s (b) 360 kg m/s (c) 0
   (d) 30 m/s (e) 160 kg m/s (f) 100 kg m/s
   (g) 20 m/s

3. Determine the total momentum of the system before impact when P and Q have the same sense.

4. Determine the total momentum of the system before impact when P and Q have the opposite sense.

5. Determine the velocity of P and Q after impact if their sense is the same before impact.

6. Determine the velocity of P and Q after impact if their sense is opposite before impact.

7. A force of 100 N acts on a body of mass 10 kg for 0.1 s. The change in velocity of the body is:

   (a) 1 m/s (b) 100 m/s (c) 0.1 m/s (d) 0.01 m/s

   A vertical pile of mass 200 kg is driven 100 mm into the ground by the blow of a 1 t hammer which falls through 1.25 m. In questions 8 to 11, take g as 10 m/s^2 and select the correct answer from the following:

   (a) 25 m/s (b) 25/6 m/s (c) 5 kg m/s
   (d) 0 (e) 1 kg m/s (f) 5000 kg m/s
   (g) 5 m/s (h) 125 m/s

8. Calculate the velocity of the hammer immediately before impact.

9. Calculate the momentum of the hammer just before impact.

10. Calculate the momentum of the hammer and pile immediately after impact assuming they have the same velocity.

11. Calculate the velocity of the hammer and pile immediately after impact assuming they have the same velocity.

Assignment 12

This assignment covers the material contained in Chapters 32 and 33. The marks for each question are shown in brackets at the end of each question.

1. A metal bar having a cross-sectional area of 80 mm^2 has a tensile force of 20 kN applied to it. Determine the stress in the bar. (4)

2. (a) A rectangular metal bar has a width of 16 mm and can support a maximum compressive stress of 15 MPa; determine the minimum breadth of the bar when loaded with a force of 6 kN.

   (b) If the bar in (a) is 1.5 m long and decreases in length by 0.18 mm when the force is applied, determine the strain and the percentage strain. (7)

3. A wire is stretched 2.50 mm by a force of 400 N. Determine the force that would stretch the wire 3.50 mm, assuming that the elastic limit is not exceeded. (5)

4. A copper tube has an internal diameter of 140 mm and an outside diameter of 180 mm and is used to support a load of 4 kN. The tube is 600 mm long before the load is applied. Determine, in micrometres, by how much the tube contracts when loaded, taking the modulus of elasticity for copper as 96 GPa. (8)

5. Determine the momentum of a lorry of mass 10 tonnes moving at a velocity of 81 km/h. (4)

6. A ball of mass 50 g is moving with a velocity of 4 m/s when it strikes a stationary ball of mass 25 g. The velocity of the 50 g ball after impact is 2.5 m/s in the same direction as before impact. Determine the velocity of the 25 g ball after impact. (8)

7. A force of 24 N acts on a body of mass 6 kg for 150 ms. Determine the change in velocity. (4)

8. The hammer of a pile-driver of mass 800 kg falls a distance of 1.0 m on to a pile. The blow takes place in 20 ms and the hammer does not rebound. Determine (a) the velocity of impact (b) the momentum lost by the hammer (c) the average applied force exerted on the pile by the hammer. (10)

Torque

At the end of this chapter you should be able to:

- define a couple
- define a torque and state its unit
- calculate torque given force and radius
- calculate work done given torque and angle turned through
- calculate power given torque and angle turned through
- appreciate kinetic energy $= I\omega^2/2$ where I is the moment of inertia
- appreciate that torque $T = I\alpha$ where α is the angular acceleration
- calculate torque given I and α
- calculate kinetic energy given I and ω
- understand power transmission by means of belt and pulley
- perform calculations involving torque, power and efficiency of belt drives

4.1   Couple and torque

When two equal forces act on a body as shown in Fig. 34.1, they cause the body to rotate, and the system of forces is called a **couple**.

The turning moment of a couple is called a **torque**, T. In Fig. 34.1, torque = magnitude of either force × perpendicular distance between the forces, i.e.

$$\boxed{T = Fd}$$

The unit of torque is the **newton metre, Nm**.

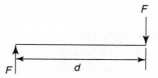

Fig. 34.1

When a force F newtons is applied at a radius r metres from the axis of, say, a nut to be turned by a spanner, as shown in Fig. 34.2, the torque T applied to the nut is given by:

$$T = Fr \ \textbf{Nm}$$

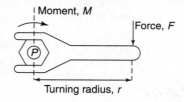

Fig. 34.2

Problem 1.   Determine the torque when a pulley wheel of diameter 300 mm has a force of 80 N applied at the rim.

Torque $T = Fr$, where force $F = 80$ N and radius

$$r = \frac{300}{2} = 150\,\text{mm} = 0.15\,\text{m}$$

Hence, **torque, $T = (80)(0.15) = 12\,\text{Nm}$**.

Problem 2. Determine the force applied tangentially to a bar of a screwjack at a radius of 800 mm, if the torque required is 600 Nm.

Torque, $T =$ force $\times$ radius, from which

$$\text{force} = \frac{\text{torque}}{\text{radius}} = \frac{600 \text{ Nm}}{800 \times 10^{-3} \text{ m}} = \textbf{750 N}$$

Problem 3. The circular hand-wheel of a valve of diameter 500 mm has a couple applied to it composed of two forces, each of 250 N. Calculate the torque produced by the couple.

Torque produced by couple, $T = Fd$, where force $F = 250$ N and distance between the forces, $d = 500$ mm $= 0.5$ m.

Hence, **torque,** $T = (250)(0.5) = \textbf{125 Nm}$.

Now try the following exercise

Exercise 157   Further problems on torque (Answers on page 310)

1. Determine the torque developed when a force of 200 N is applied tangentially to a spanner at a distance of 350 mm from the centre of the nut.

2. During a machining test on a lathe, the tangential force on the tool is 150 N. If the torque on the lathe spindle is 12 Nm, determine the diameter of the work-piece.

34.2   Work done and power transmitted by a constant torque

Figure 34.3(a) shows a pulley wheel of radius r metres attached to a shaft and a force F newtons applied to the rim at point P.

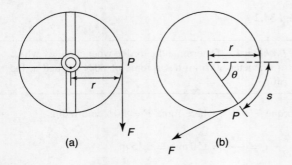

(a)   (b)

Fig. 34.3

Figure 34.3(b) shows the pulley wheel having turned through an angle θ radians as a result of the force F being applied. The force moves through a distance s, where arc length $s = r\theta$.

Work done $=$ force $\times$ distance moved by force

$$= F \times r\theta = Fr\theta \text{ Nm} = Fr\theta \text{ J}$$

However, Fr is the torque T, hence,

$$\boxed{\textbf{work done} = T\theta \textbf{ joules}}$$

$$\text{Average power} = \frac{\text{work done}}{\text{time taken}} = \frac{T\theta}{\text{time taken}}$$

for a constant torque T

However, (angle θ)/(time taken) $=$ angular velocity, ω rad/s Hence,

$$\boxed{\textbf{power, } P = T\omega \textbf{ watts}}$$

Angular velocity, $\omega = 2\pi n$ rad/s where n is the speed in rev/s. Hence,

$$\boxed{\textbf{power, } P = 2\pi nT \textbf{ watts}}$$

Problem 4. A constant force of 150 N is applied tangentially to a wheel of diameter 140 mm. Determine the work done, in joules, in 12 revolutions of the wheel.

Torque $T = Fr$, where $F = 150$ N and radius

$$r = \frac{140}{2} = 70 \text{ mm} = 0.070 \text{ m}$$

Hence, torque $T = (150)(0.070) = 10.5$ Nm

Work done $= T\theta$ joules, where torque, $T = 10.5$ Nm and angular displacement, $\theta = 12$ revolutions $= 12 \times 2\pi$ rad $= 24\pi$ rad.

Hence, **work done** $= (10.5)(24\pi) = \textbf{792 J}$

Problem 5. Calculate the torque developed by a motor whose spindle is rotating at 1000 rev/min and developing a power of 2.50 kW

Power $P = 2\pi nT$ (from above), from which, torque,

$$T = \frac{P}{2\pi n} \text{ Nm}$$

where power, $P = 2.50$ kW $= 2500$ W and speed, $n = 1000/60$ rev/s. Thus,

$$\textbf{torque, } T = \frac{P}{2\pi n} = \frac{2500}{2\pi \left(\dfrac{1000}{60} \right)} = \frac{2500 \times 60}{2\pi \times 1000}$$

$$= \textbf{23.87 Nm}$$

Problem 6. An electric motor develops a power of 3.75 kW and a torque of 12.5 Nm. Determine the speed of rotation of the motor in rev/min.

Power, $P = 2\pi n T$, from which,

$$\text{speed } n = \frac{P}{2\pi T} \text{ rev/s}$$

where   power, $P = 3.75\,\text{kW} = 3750\,\text{W}$

and   torque $T = 12.5\,\text{Nm}$

Hence,   speed $n = \dfrac{3750}{2\pi(12.5)} = 47.75\,\text{rev/s}$

The speed of rotation of the motor $= 47.75 \times 60$

$$= \textbf{2865 rev/min}$$

Problem 7. In a turning-tool test, the tangential cutting force is 50 N. If the mean diameter of the work-piece is 40 mm, calculate (a) the work done per revolution of the spindle, (b) the power required when the spindle speed is 300 rev/min.

(a) Work done $= T\theta$, where $T = Fr$. Force $F = 50\,\text{N}$, radius $r = (40/2) = 20\,\text{mm} = 0.02\,\text{m}$, and angular displacement, $\theta = 1\,\text{rev} = 2\pi\,\text{rad}$.

Hence, **work done per revolution of spindle**
$= Fr\theta = (50)(0.02)(2\pi) = \textbf{6.28 J}$

(b) Power, $P = 2\pi n T$, where torque, $T = Fr = (50)(0.02)$ $= 1\,\text{Nm}$, and speed, $n = (300/60) = 5\,\text{rev/s}$.

Hence,   **power required**, $P = 2\pi(5)(1) = \textbf{31.42 W}$

Problem 8. A motor connected to a shaft develops a torque of 5 kNm. Determine the number of revolutions made by the shaft if the work done is 9 MJ.

Work done $= T\theta$, from which, angular displacement,

$$\theta = \frac{\text{work done}}{\text{torque}}$$

Work done $= 9\,\text{MJ} = 9 \times 10^6\,\text{J}$

and   torque $= 5\,\text{kNm} = 5000\,\text{Nm}$

Hence, angular displacement,

$$\theta = \frac{9 \times 10^6}{5000} = 1800\,\text{rad}$$

$2\pi\,\text{rad} = 1\,\text{rev}$,

hence, **the number of revolutions made by the shaft**

$$= \frac{1800}{2\pi} = \textbf{286.5 revs}$$

Now try the following exercise

(Answers on page 310)

Exercise 158   Further problems on work done and power transmitted by a constant torque

1. A constant force of 4 kN is applied tangentially to the rim of a pulley wheel of diameter 1.8 m attached to a shaft. Determine the work done, in joules, in 15 revolutions of the pulley wheel.

2. A motor connected to a shaft develops a torque of 3.5 kNm. Determine the number of revolutions made by the shaft if the work done is 11.52 MJ.

3. A wheel is turning with an angular velocity of 18 rad/s and develops a power of 810 W at this speed. Determine the torque developed by the wheel.

4. Determine the angular velocity of a shaft when the power available is 2.75 kW and the torque is 200 Nm.

5. The drive shaft of a ship supplies a torque of 400 kNm to its propeller at 400 rev/min. Determine the power delivered by the shaft.

6. A wheel is rotating at 1720 rev/min and develops a power of 600 W at this speed. Calculate (a) the torque,   (b) the work done, in joules, in a quarter of an hour.

34.3   Kinetic energy and moment of inertia

The tangential velocity v of a particle of mass m moving at an angular velocity ω rad/s at a radius r metres (see Fig. 34.4) is given by: $v = \omega r$ m/s.

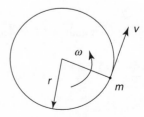

Fig. 34.4

The kinetic energy of a particle of mass m is given by:

Kinetic energy $= \frac{1}{2}\,\text{mv}^2$ (from Chapter 12)

$$= \tfrac{1}{2}\,\text{m}(\omega r)^2$$

$$= \tfrac{1}{2}\,\textbf{m}\boldsymbol{\omega}^2\textbf{r}^2 \textbf{ joules}$$

The total kinetic energy of a system of masses rotating at different radii about a fixed axis but with the same angular velocity, as shown in Fig. 34.5, is given by:

$$\text{Total kinetic energy} = \frac{1}{2}m_1\omega^2 r_1^2 + \frac{1}{2}m_2\omega^2 r_2^2 + \frac{1}{2}m_3\omega^2 r_3^2$$

$$= (m_1 r_1^2 + m_2 r_2^2 + m_3 r_3^2)\frac{\omega^2}{2}$$

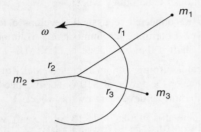

Fig. 34.5

In general, this may be written as:

$$\textbf{Total kinetic energy} = (\Sigma mr^2)\frac{\omega^2}{2} = \mathbf{I}\frac{\omega^2}{2}$$

where $I(= \Sigma mr^2)$ is called the **moment of inertia** of the system about the axis of rotation and has units of $kg\,m^2$.

The moment of inertia of a system is a measure of the amount of work done to give the system an angular velocity of ω rad/s, or the amount of work which can be done by a system turning at ω rad/s.

From Section 34.2, work done $= T\theta$, and if this work is available to increase the kinetic energy of a rotating body of moment of inertia I, then:

$$T\theta = I\left(\frac{\omega_2^2 - \omega_1^2}{2}\right)$$

where ω_1 and ω_2 are the initial and final angular velocities, i.e.

$$T\theta = I\left(\frac{\omega_2 + \omega_1}{2}\right)(\omega_2 - \omega_1)$$

However,

$$\left(\frac{\omega_2 + \omega_1}{2}\right)$$

is the mean angular velocity, i.e. $\dfrac{\theta}{t}$, where t is the time, and $(\omega_2 - \omega_1)$ is the change in angular velocity, i.e. αt, where a is the angular acceleration. Hence,

$$T\theta = I\left(\frac{\theta}{t}\right)(\alpha t)$$

from which,

$$\boxed{\textbf{torque } T = \mathbf{I}\alpha}$$

where I is the moment of inertia in $kg\,m^2$, α is the angular acceleration in rad/s^2 and T is the torque in Nm.

Problem 9. A shaft system has a moment of inertia of $37.5\,kg\,m^2$. Determine the torque required to give it an angular acceleration of $5.0\,rad/s^2$

Torque, $T = I\alpha$, where moment of inertia $I = 37.5\,kg\,m^2$ and angular acceleration, $\alpha = 5.0\,rad/s^2$. Hence,

$$\textbf{torque, } T = (37.5)(5.0) = \mathbf{187.5\,Nm}$$

Problem 10. A shaft has a moment of inertia of $31.4\,kg\,m^2$. What angular acceleration of the shaft would be produced by an accelerating torque of 495 Nm?

Torque, $T = I\alpha$, from which, angular acceleration, $\alpha = \dfrac{T}{I}$ where torque, $T = 495\,Nm$ and moment of inertia $I = 31.4\,kg\,m^2$

Hence, **angular acceleration,** $\alpha = \dfrac{495}{31.4} = \mathbf{15.76\,rad/s^2}$

Problem 11. A body of mass 100 g is fastened to a wheel and rotates in a circular path of 500 mm in diameter. Determine the increase in kinetic energy of the body when the speed of the wheel increases from 450 rev/min to 750 rev/min.

From above, kinetic energy $= I\dfrac{\omega^2}{2}$

Thus, increase in kinetic energy $= I\left(\dfrac{\omega_2^2 - \omega_1^2}{2}\right)$

where moment of inertia, $I = mr^2$, mass, $m = 100\,g = 0.1\,k$ and radius,

$$r = \frac{500}{2} = 250\,mm = 0.25\,m$$

Initial angular velocity,

$$\omega_1 = 450\,\text{rev/min} = \frac{450 \times 2\pi}{60}\,rad/s$$

$$= 47.12\,rad/s,$$

and final angular velocity,

$$\omega_2 = 750\,\text{rev/min} = \frac{750 \times 2\pi}{60}\,rad/s$$

$$= 78.54\,rad/s$$

Thus, **increase in kinetic energy**

$$= I\left(\frac{\omega_2^2 - \omega_1^2}{2}\right) = (mr^2)\left(\frac{\omega_2^2 - \omega_1^2}{2}\right)$$

$$= (0.1)(0.25^2)\left(\frac{78.54^2 - 47.12^2}{2}\right) = \mathbf{12.34\,J}$$

Problem 12. A system consists of three small masses rotating at the same speed about the same fixed axis. The masses and their radii of rotation are: 15 g at 250 mm, 20 g at 180 mm and 30 g at 200 mm. Determine (a) the moment of inertia of the system about the given axis (b) the kinetic energy in the system if the speed of rotation is 1200 rev/min.

(a) Moment of inertia of the system, $I = \Sigma mr^2$

i.e. $I = [(15 \times 10^{-3}\,\text{kg})(0.25\,\text{m})^2]$
$+ [(20 \times 10^{-3}\,\text{kg})(0.18\,\text{m})^2]$
$+ [(30 \times 10^{-3}\,\text{kg})(0.20\,\text{m})^2]$

$= (9.375 \times 10^{-4}) + (6.48 \times 10^{-4}) + (12 \times 10^{-4})$

$= 27.855 \times 10^{-4}\,\text{kg m}^2$

$= \mathbf{2.7855 \times 10^{-3}\,kg\,m^2}$

(b) Kinetic energy $= I\dfrac{\omega^2}{2}$, where moment of inertia,
$I = 2.7855 \times 10^{-3}\,\text{kg m}^2$ and angular velocity,

$$\omega = 2\pi n = 2\pi\left(\frac{1200}{60}\right)\text{rad/s} = 40\pi\,\text{rad/s}$$

Hence, **kinetic energy in the system**

$$= (2.7855 \times 10^{-3})\frac{(40\pi)^2}{2} = \mathbf{21.99\,J}$$

Now try the following exercise

Exercise 159   Further problems on kinetic energy and moment of inertia (Answer on page 310)

1. A shaft system has a moment of inertia of 51.4 kg m². Determine the torque required to give it an angular acceleration of 5.3 rad/s²

2. A shaft has an angular acceleration of 20 rad/s² and produces an accelerating torque of 600 Nm. Determine the moment of inertia of the shaft.

3. A uniform torque of 3.2 kNm is applied to a shaft while it turns through 25 revolutions. Assuming no frictional or other resistance's, calculate the increase in kinetic energy of the shaft (i.e. the work done. If

the shaft is initially at rest and its moment of inertia is 24.5 kg m², determine its rotational speed, in rev/min, at the end of the 25 revolutions.

4. A small body, of mass 82 g, is fastened to a wheel and rotates in a circular path of 456 mm diameter. Calculate the increase in kinetic energy of the body when the speed of the wheel increases from 450 rev/min to 950 rev/min.

5. A system consists of three small masses rotating at the same speed about the same fixed axis. The masses and their radii of rotation are: 16 g at 256 mm, 23 g at 192 mm and 31 g at 176 mm. Determine (a) the moment of inertia of the system about the given axis, and (b) the kinetic energy in the system if the speed of rotation is 1250 rev/min.

34.4   Power transmission and efficiency

A common and simple method of transmitting power from one shaft to another is by means of a **belt** passing over pulley wheels which are keyed to the shafts, as shown in Fig. 34.6. Typical applications include an electric motor driving a lathe or a drill, and an engine driving a pump or generator.

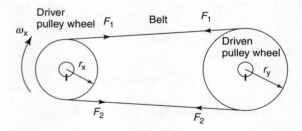

Fig. 34.6

For a belt to transmit power between two pulleys there must be a difference in tensions in the belt on either side of the driving and driven pulleys. For the direction of rotation shown in Fig. 34.6, $F_2 > F_1$

The torque T available at the driving wheel to do work is given by:

$$T = (F_2 - F_1)r_x\,\text{Nm}$$

and the available power P is given by:

$$P = T\omega = (F_2 - F_1)r_x\omega_x\,\text{watts}$$

From Section 34.3, the linear velocity of a point on the driver wheel, $v_x = r_x\omega_x$

Similarly, the linear velocity of a point on the driven wheel, $v_y = r_y\omega_y$

Assuming no slipping, $v_x = v_y$ i.e. $r_x \omega_x = r_y \omega_y$. Hence $r_x(2\pi n_x) = r_y(2\pi n_y)$ from which,

$$\boxed{\frac{r_x}{r_y} = \frac{n_y}{n_x}}$$

Percentage efficiency $= \dfrac{\text{useful work output}}{\text{energy output}} \times 100$

or $\boxed{\textbf{efficiency} = \dfrac{\textbf{power output}}{\textbf{power input}} \times \textbf{100\%}}$

Problem 13.   An electric motor has an efficiency of 75% when running at 1450 rev/min. Determine the output torque when the power input is 3.0 kW

Efficiency $= \dfrac{\text{power output}}{\text{power input}} \times 100\%$, hence

$75 = \dfrac{\text{power output}}{3000} \times 100$ from which,

power output $= \dfrac{75}{100} \times 3000 = 2250\,\text{W}$

From Section 34.2, power output, $P = 2\pi n T$, from which torque,

$$T = \frac{P}{2\pi n}$$

where $n = (1450/60)$ rev/s. Hence,

$$\textbf{output torque} = \frac{2250}{2\pi\left(\dfrac{1450}{60}\right)} = \textbf{14.82\,Nm}$$

Problem 14.   A 15 kW motor is driving a shaft at 1150 rev/min by means of pulley wheels and a belt. The tensions in the belt on each side of the driver pulley wheel are 400 N and 50 N. The diameters of the driver and driven pulley wheels are 500 mm and 750 mm respectively. Determine (a) the efficiency of the motor, (b) the speed of the driven pulley wheel.

(a) From above, power output from motor $= (F_2 - F_1)r_x\omega_x$
Force $F_2 = 400\,\text{N}$ and $F_1 = 50\,\text{N}$, hence $(F_2 - F_1) = 350\,\text{N}$, radius

$$r_x = \frac{500}{2} = 250\,\text{mm} = 0.25\,\text{m}$$

and angular velocity,

$$\omega_x = \frac{1150 \times 2\pi}{60}\ \text{rad/s}$$

Hence power output from motor $= (F_2 - F_1)r_x\omega_x$

$$= (350)(0.25)\left(\frac{1150 \times 2\pi}{60}\right) = 10.54\,\text{kW}$$

Power input $= 15\,\text{kW}$

Hence, **efficiency of the motor** $= \dfrac{\text{power output}}{\text{power input}}$

$$= \frac{10.54}{15} \times 100$$

$$= \textbf{70.27\%}$$

(b) From above, $\dfrac{r_x}{r_y} = \dfrac{n_y}{n_x}$

from which, **speed of driven pulley wheel,**

$$\mathbf{n_y} = \frac{n_x r_x}{r_y} = \frac{1150 \times 0.25}{\dfrac{0.750}{2}} = \textbf{767 rev/min}$$

Problem 15.   A crane lifts a load of mass 5 tonne to a height of 25 m. If the overall efficiency of the crane is 65% and the input power to the hauling motor is 100 kW, determine how long the lifting operation takes.

The increase in potential energy is the work done and is given by mgh (see Chapter 12), where mass, $m = 5\,\text{t} = 5000\,\text{kg}$, $g = 9.81\,\text{m/s}^2$ and height $h = 25\,\text{m}$.

Hence,   work done $= mgh = (5000)(9.81)(25) = 1.226\,\text{MJ}$

Input power $= 100\,\text{kW} = 100\,000\,\text{W}$

Efficiency $= \dfrac{\text{output power}}{\text{input power}} \times 100$

hence $65 = \dfrac{\text{output power}}{100\,000} \times 100$

from which,   output power $= \dfrac{65}{100} \times 100\,000 = 65\,000\,\text{W}$

$$= \frac{\text{work done}}{\text{time taken}}$$

Thus, **time taken for lifting operation** $= \dfrac{\text{work done}}{\text{output power}}$

$$= \frac{1.226 \times 10^6\,\text{J}}{65\,000\,\text{W}}$$

$$= \textbf{18.86\,s}$$

Problem 16.   The tool of a shaping machine has a mean cutting speed of 250 mm/s and the average cutting force on the tool in a certain shaping operation is 1.2 kN. If the power input to the motor driving the machine is 0.75 kW, determine the overall efficiency of the machine.

Velocity, $v = 250\,\text{mm/s} = 0.25\,\text{m/s}$ and force $F = 1.2\,\text{kN} = 1200\,\text{N}$

From Chapter 12, power output required at the cutting tool (i.e. power output), $P = \text{force} \times \text{velocity}$

$$= 1200\,\text{N} \times 0.25\,\text{m/s} = 300\,\text{W}$$

Power input $= 0.75\,\text{kW} = 750\,\text{W}$

Hence, **efficiency of the machine** $= \dfrac{\text{output power}}{\text{input power}} \times 100$

$$= \dfrac{300}{750} \times 100 = \textbf{40\%}$$

Problem 17. Calculate the input power of the motor driving a train at a constant speed of 72 km/h on a level track, if the efficiency of the motor is 80% and the resistance due to friction is 20 kN.

Force resisting motion $= 20\,\text{kN} = 20\,000\,\text{N}$ and

$$\text{velocity} = 72\,\text{km/h} = \frac{72}{3.6} = 20\,\text{m/s}$$

Output power from motor = resistive force × velocity of train (from Chapter 12) $= 20\,000 \times 20 = 400\,\text{kW}$

$$\text{Efficiency} = \frac{\text{power output}}{\text{power input}} \times 100$$

hence

$$80 = \frac{400}{\text{power input}} \times 100$$

from which, **power input** $= 400 \times \dfrac{100}{80} = \textbf{500\,kW}$

Now try the following exercises

Exercise 160 Further problems on power transmission and efficiency (Answers on page 310)

1. A motor has an efficiency of 72% when running at 2600 rev/min. If the output torque is 16 Nm at this speed, determine the power supplied to the motor.

2. The difference in tensions between the two sides of a belt round a driver pulley of radius 240 mm is 200 N. If the driver pulley wheel is on the shaft of an electric motor running at 700 rev/min and the power input to the motor is 5 kW, determine the efficiency of the motor. Determine also the diameter of the driven pulley wheel if its speed is to be 1200 rev/min.

3. A winch is driven by a 4 kW electric motor and is lifting a load of 400 kg to a height of 5.0 m. If

the lifting operation takes 8.6 s, calculate the overall efficiency of the winch and motor.

4. The average force on the cutting tool of a lathe is 750 N and the cutting speed is 400 mm/s. Determine the power input to the motor driving the lathe if the overall efficiency is 55%.

5. A ship's anchor has a mass of 5 tonne. Determine the work done in raising the anchor from a depth of 100 m. If the hauling gear is driven by a motor whose output is 80 kW and the efficiency of the haulage is 75%, determine how long the lifting operation takes.

Exercise 161 Short answer questions on torque

1. In engineering, what is meant by a couple?

2. Define torque.

3. State the unit of torque.

4. State the relationship between work, torque T and angular displacement θ.

5. State the relationship between power P, torque T and angular velocity ω.

6. Define moment of inertia and state the symbol used.

7. State the unit of moment of inertia.

8. State the relationship between torque, moment of inertia and angular acceleration.

9. State one method of power transmission commonly used.

10. Define efficiency.

Exercise 162 Multi-choice questions on torque (Answers on page 310)

1. The unit of torque is:

    (a) N (b) Pa (c) N/m (d) Nm

2. The unit of work is:

    (a) N (b) J (c) W (d) N/m

3. The unit of power is:

    (a) N (b) J (c) W (d) N/m

4. The unit of the moment of inertia is:

    (a) $\text{kg}\,\text{m}^2$ (b) kg (c) kg/m^2 (d) Nm

5. A force of 100 N is applied to the rim of a pulley wheel of diameter 200 mm. The torque is:

    (a) 2 Nm (b) 20 kNm (c) 10 Nm (d) 20 Nm

6. The work done on a shaft to turn it through 5π radians is 25π J. The torque applied to the shaft is:

(a) 0.2 Nm   (b) $125\pi^2$ Nm

(c) 30π Nm   (d) 5 Nm

7. A 5 kW electric motor is turning at 50 rad/s. The torque developed at this speed is:

   (a) 100 Nm   (b) 250 Nm   (c) 0.01 Nm   (d) 0.1 Nm

8. The force applied tangentially to a bar of a screw-jack at a radius of 500 mm if the torque required is 1 kNm is:

   (a) 2 N   (b) 2 kN   (c) 500 N   (d) 0.5 N

9. A 10 kW motor developing a torque of $(200/\pi)$ Nm is running at a speed of:

   (a) $(\pi/20)$ rev/s   (b) 50π rev/s

   (c) 25 rev/s   (d) $(20/\pi)$ rev/s

10. A shaft and its associated rotating parts has a moment of inertia of 50 kg m^2. The angular acceleration of the shaft to produce an accelerating torque of 5 kNm is:

(a) 10 rad/s^2 (b) 250 rad/s^2

(c) 0.01 rad/s^2 (d) 100 rad/s^2

11. A motor has an efficiency of 25% when running at 3000 rev/min. If the output torque is 10 Nm, the power input is:

   (a) 4π kW   (b) 0.25π kW

   (c) 15π kW   (d) 75π kW

12. In a belt-pulley wheel system, the effective tension in the belt is 500 N and the diameter of the driver wheel is 200 mm. If the power output from the driving motor is 5 kW, the driver pulley wheel turns at:

   (a) 50 rad/s   (b) 2500 rad/s

   (c) 100 rad/s   (d) 0.1 rad/s

35

Thermal expansion

At the end of this chapter you should be able to:

- appreciate that expansion and contraction occurs with change of temperature
- describe practical applications where expansion and contraction must be allowed for
- understand the expansion and contraction of water
- define the coefficient of linear expansion α
- recognise typical values for the coefficient of linear expansion
- calculate the new length l_2, after expansion or contraction, using $l_2 = l_1[1 + \alpha(t_2 - t_1)]$
- define the coefficient of superficial expansion β
- calculate the new surface area A_2, after expansion or contraction, using $A_2 = A_1[1 + \beta(t_2 - t_1)]$
- appreciate that $\beta \approx 2\alpha$
- define the coefficient of cubic expansion γ
- recognise typical values for the coefficient of cubic expansion
- appreciate that $\gamma \approx 3\alpha$
- calculate the new volume V_2, after expansion or contraction, using $V_2 = V_1[1 + \gamma(t_2 - t_1)]$

35.1 Introduction

When heat is applied to most materials, **expansion** occurs in all directions. Conversely, if heat energy is removed from a material (i.e. the material is cooled) **contraction** occurs in all directions. The effects of expansion and contraction each depend on the **change of temperature** of the material.

35.2 Practical applications of thermal expansion

Some practical applications where expansion and contraction of solid materials must be allowed for include:

(i) Overhead electrical transmission lines are hung so that they are slack in summer, otherwise their contraction in winter may snap the conductors or bring down pylons.

(ii) Gaps need to be left in lengths of railway lines to prevent buckling in hot weather (except where these are continuously welded).

(iii) Ends of large bridges are often supported on rollers to allow them to expand and contract freely.

(iv) Fitting a metal collar to a shaft or a steel tyre to a wheel is often achieved by first heating them so that they expand, fitting them in position, and then cooling them so that the contraction holds them firmly in place; this is known as a 'shrink-fit'. By a similar method hot rivets are used for joining metal sheets.

(v) The amount of expansion varies with different materials. Figure 35.1(a) shows a bimetallic strip at room temperature (i.e. two different strips of metal riveted together). When heated, brass expands more than steel, and since the two metals are riveted together the bimetallic strip is forced into an arc as shown in Fig. 35.1(b). Such a movement can be arranged to make or break an electric circuit and bimetallic strips are used, in particular, in thermostats (which are temperature-operated switches) used to control central heating systems, cookers, refrigerators, toasters, irons, hot-water and alarm systems.

(vi) Motor engines use the rapid expansion of heated gases to force a piston to move.

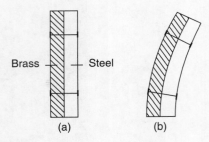

Brass — Steel

(a)   (b)

Fig. 35.1

(vii) Designers must predict, and allow for, the expansion of steel pipes in a steam-raising plant so as to avoid damage and consequent danger to health.

35.3 Expansion and contraction of water

Water is a liquid which at low temperature displays an unusual effect. If cooled, contraction occurs until, at about 4°C, the volume is at a minimum. As the temperature is further decreased from 4°C to 0°C expansion occurs, i.e. the volume increases. When ice is formed, considerable expansion occurs and it is this expansion which often causes frozen water pipes to burst.

A practical application of the expansion of a liquid is with thermometers, where the expansion of a liquid, such as mercury or alcohol, is used to measure temperature.

35.4 Coefficient of linear expansion

The amount by which unit length of a material expands when the temperature is raised one degree is called the **coefficient of linear expansion** of the material and is represented by α (Greek alpha).

The units of the coefficient of linear expansion are m/(mK), although it is usually quoted as just /K or K^{-1}. For example, copper has a coefficient of linear expansion value of $17 \times 10^{-6}\,K^{-1}$, which means that a 1 m long bar of copper expands by 0.000017 m if its temperature is increased by 1 K (or 1°C). If a 6 m long bar of copper is subjected to a temperature rise of 25 K then the bar will expand by $(6 \times 0.000017 \times 25)$ m, i.e. 0.00255 m or 2.55 mm. (Since the kelvin scale uses the same temperature interval as the Celsius scale, a **change** of temperature of, say, 50°C, is the same as a change of temperature of 50 K.)

If a material, initially of length l_1 and at a temperature of t_1 and having a coefficient of linear expansion α, has its temperature increased to t_2, then the new length l_2 of the material is given by:

New length = original length + expansion

i.e. $l_2 = l_1 + l_1\alpha(t_2 - t_1)$

i.e. $\boxed{l_2 = l_1[1 + \alpha(t_2 - t_1)]}$ (1

Some typical values for the coefficient of linear expansion include:

Aluminium	$23 \times 10^{-6}\,K^{-1}$
Concrete	$12 \times 10^{-6}\,K^{-1}$
Gold	$14 \times 10^{-6}\,K^{-1}$
Iron	$11-12 \times 10^{-6}\,K^{-1}$
Steel	$15-16 \times 10^{-6}\,K^{-1}$
Zinc	$31 \times 10^{-6}\,K^{-1}$
Brass	$18 \times 10^{-6}\,K^{-1}$
Copper	$17 \times 10^{-6}\,K^{-1}$
Invar (nickel-steel alloy)	$0.9 \times 10^{-6}\,K^{-1}$
Nylon	$100 \times 10^{-6}\,K^{-1}$
Tungsten	$4.5 \times 10^{-6}\,K^{-1}$

Problem 1. The length of an iron steam pipe is 20.0 m at a temperature of 18°C. Determine the length of the pipe under working conditions when the temperature is 300°C. Assume the coefficient of linear expansion of iron is $12 \times 10^{-6}\,K^{-1}$

Length $l_1 = 20.0$ m, temperature $t_1 = 18$°C, $t_2 = 300$°C and $\alpha = 12 \times 10^{-6}\,K^{-1}$

Length of pipe at 300°C is given by:

$l_2 = l_1[1 + \alpha(t_2 - t_1)]$

$= 20.0[1 + (12 \times 10^{-6})(300 - 18)]$

$= 20.0[1 + 0.003384] = 20.0[1.003384]$

$= \mathbf{20.06768\,m}$

i.e. an increase in length of 0.06768 m or **67.68 mm**.

In practice, allowances are made for such expansions. U shaped expansion joints are connected into pipelines carryin hot fluids to allow some 'give' to take up the expansion.

Problem 2. An electrical overhead transmission line has a length of 80.0 m between its supports at 15°C. Its length increases by 92 mm at 65°C. Determine the coefficient of linear expansion of the material of the line.

Length $l_1 = 80.0$ m, $l_2 = 80.0 + 92$ mm $= 80.092$ m temperature $t_1 = 15$°C and temperature $t_2 = 65$°C

Length $l_2 = l_1[1 + \alpha(t_2 - t_1)]$

i.e. $80.092 = 80.0[1 + \alpha(65 - 15)]$

$80.092 = 80.0 + (80.0)(\alpha)(50)$

i.e. $80.092 - 80.0 = (80.0)(\alpha)(50)$

Hence, the coefficient of linear expansion,

$$\alpha = \frac{0.092}{(80.0)(50)} = 0.000023$$

i.e. $\alpha = 23 \times 10^{-6}\,\text{K}^{-1}$

(which is aluminium – see above).

Problem 3. A measuring tape made of copper measures 5.0 m at a temperature of 288 K. Calculate the percentage error in measurement when the temperature has increased to 313 K. Take the coefficient of linear expansion of copper as $17 \times 10^{-6}\,\text{K}^{-1}$

Length $l_1 = 5.0$ m, temperature $t_1 = 288$ K, $t_2 = 313$ K and $\alpha = 17 \times 10^{-6}\,\text{K}^{-1}$

Length at 313 K is given by:

$$\begin{aligned}
\text{Length } l_2 &= l_1[1 + \alpha(t_2 - t_1)] \\
&= 5.0[1 + (17 \times 10^{-6})(313 - 288)] \\
&= 5.0[1 + (17 \times 10^{-6})(25)] \\
&= 5.0[1 + 0.000425] \\
&= 5.0[1.000425] = 5.002125\,\text{m}
\end{aligned}$$

i.e. the length of the tape has increased by 0.002125 m.

Percentage error in measurement at 313 K

$$= \frac{\text{increase in length}}{\text{original length}} \times 100\%$$

$$= \frac{0.002125}{5.0} \times 100 = \mathbf{0.0425\%}$$

Problem 4. The copper tubes in a boiler are 4.20 m long at a temperature of 20°C. Determine the length of the tubes (a) when surrounded only by feed water at 10°C, (b) when the boiler is operating and the mean temperature of the tubes is 320°C. Assume the coefficient of linear expansion of copper to be $17 \times 10^{-6}\,\text{K}^{-1}$

(a) Initial length, $l_1 = 4.20$ m, initial temperature, $t_1 = 20$°C, final temperature, $t_2 = 10$°C and $\alpha = 17 \times 10^{-6}\,\text{K}^{-1}$

Final length at 10°C is given by:

$$\begin{aligned}
l_2 &= l_1[1 + \alpha(t_2 - t_1)] \\
&= 4.20[1 + (17 \times 10^{-6})(10 - 20)] \\
&= 4.20[1 - 0.00017] = \mathbf{4.1993\,m}
\end{aligned}$$

i.e. the tube contracts by 0.7 mm when the temperature decreases from 20°C to 10°C.

(b) Length, $l_1 = 4.20$ m, $t_1 = 20$°C, $t_2 = 320$°C and $\alpha = 17 \times 10^{-6}\,\text{K}^{-1}$

Final length at 320°C is given by

$$\begin{aligned}
l_2 &= l_1[1 + \alpha(t_2 - t_1)] \\
&= 4.20[1 + (17 \times 10^{-6})(320 - 20)] \\
&= 4.20[1 + 0.0051] = \mathbf{4.2214\,m}
\end{aligned}$$

i.e. the tube extends by 21.4 mm when the temperature rises from 20°C to 320°C.

Now try the following exercise

Exercise 163 Further problems on the coefficient of linear expansion (Answers on page 310)

1. A length of lead piping is 50.0 m long at a temperature of 16°C. When hot water flows through it the temperature of the pipe rises to 80°C. Determine the length of the hot pipe if the coefficient of linear expansion of lead is $29 \times 10^{-6}\,\text{K}^{-1}$

2. A rod of metal is measured at 285 K and is 3.521 m long. At 373 K the rod is 3.523 m long. Determine the value of the coefficient of linear expansion for the metal

3. A copper overhead transmission line has a length of 40.0 m between its supports at 20°C. Determine the increase in length at 50°C if the coefficient of linear expansion of copper is $17 \times 10^{-6}\,\text{K}^{-1}$

4. A brass measuring tape measures 2.10 m at a temperature of 15°C. Determine (a) the increase in length when the temperature has increased to 40°C (b) the percentage error in measurement at 40°C. Assume the coefficient of linear expansion of brass to be $18 \times 10^{-6}\,\text{K}^{-1}$

5. A pendulum of a 'grandfather' clock is 2.0 m long and made of steel. Determine the change in length of the pendulum if the temperature rises by 15 K. Assume the coefficient of linear expansion of steel to be $15 \times 10^{-6}\,\text{K}^{-1}$

6. A temperature control system is operated by the expansion of a zinc rod which is 200 mm long at 15°C. If the system is set so that the source of heat supply is cut off when the rod has expanded by 0.20 mm, determine the temperature to which the system is limited. Assume the coefficient of linear expansion of zinc to be $31 \times 10^{-6}\,\text{K}^{-1}$

7. A length of steel railway line is 30.0 m long when the temperature is 288 K. Determine the increase in length of the line when the temperature is raised to 303 K. Assume the coefficient of linear expansion of steel to be $15 \times 10^{-6}\,\text{K}^{-1}$

8. A brass shaft is 15.02 mm in diameter and has to be inserted in a hole of diameter 15.0 mm. Determine by how much the shaft must be cooled to make this possible, without using force. Take the coefficient of linear expansion of brass as $18 \times 10^{-6} \, \text{K}^{-1}$

35.5 Coefficient of superficial expansion

The amount by which unit area of a material increases when the temperature is raised by one degree is called the **coefficient of superficial (i.e. area) expansion** and is represented by β (Greek beta).

If a material having an initial surface area A_1 at temperature t_1 and having a coefficient of superficial expansion β, has its temperature increased to t_2, then the new surface area A_2 of the material is given by:

New surface area = original surface area + increase in area

i.e. $A_2 = A_1 + A_1\beta(t_2 - t_1)$

i.e. $$\boxed{A_2 = A_1[1 + \beta(t_2 - t_1)]} \tag{2}$$

It is shown in Problem 5 below that the coefficient of superficial expansion is twice the coefficient of linear expansion, i.e. $\beta = 2\alpha$, to a very close approximation.

Problem 5. Show that for a rectangular area of material having dimensions l by b the coefficient of superficial expansion $\beta \approx 2\alpha$, where α is the coefficient of linear expansion.

Initial area, $A_1 = lb$. For a temperature rise of $1 \, \text{K}$, side l will expand to $(l + l\alpha)$ and side b will expand to $(b + b\alpha)$. Hence the new area of the rectangle, A_2, is given by:

$$A_2 = (l + l\alpha)(b + b\alpha) = l(1 + \alpha)b(1 + \alpha)$$
$$= lb(1 + \alpha)^2 = lb(1 + 2\alpha + \alpha^2)$$
$$\approx lb(1 + 2\alpha)$$

since α^2 is very small (see typical values in Section 35.4).

Hence $A_2 \approx A_1(1 + 2\alpha)$

For a temperature rise of $(t_2 - t_1) \, \text{K}$,

$$A_2 \approx A_1[1 + 2\alpha(t_2 - t_1)]$$

Thus from equation (2), $\beta \approx 2\alpha$

35.6 Coefficient of cubic expansion

The amount by which unit volume of a material increases for a one degree rise of temperature is called the **coefficient of cubic (or volumetric) expansion** and is represented by γ (Greek gamma).

If a material having an initial volume V_1 at temperature t_1 and having a coefficient of cubic expansion γ, has its temperature raised to t_2, then the new volume V_2 of the material is given by:

New volume = initial volume + increase in volume

i.e. $V_2 = V_1 + V_1\gamma(t_2 - t_1)$

i.e. $$\boxed{V_2 = V_1[1 + \gamma(t_2 - t_1)]} \tag{3}$$

It is shown in Problem 6 below that the coefficient of cubic expansion is three times the coefficient of linear expansion, i.e. $\gamma = 3\alpha$, to a very close approximation. A liquid has no definite shape and only its cubic or volumetric expansion need be considered. Thus with expansions in liquids, equation (3) is used.

Problem 6. Show that for a rectangular block of material having dimensions l, b and h, the coefficient of cubic expansion $\gamma \approx 3\alpha$, where α is the coefficient of linear expansion.

Initial volume, $V_1 = lbh$. For a temperature rise of $1 \, \text{K}$, side l expands to $(l + l\alpha)$, side b expands to $(b + b\alpha)$ and side h expands to $(h + h\alpha)$

Hence the new volume of the block V_2 is given by:

$$V_2 = (l + l\alpha)(b + b\alpha)(h + h\alpha)$$
$$= l(1 + \alpha)b(1 + \alpha)h(1 + \alpha)$$
$$= lbh(1 + \alpha)^3 = lbh(1 + 3\alpha + 3\alpha^2 + \alpha^3)$$
$$\approx lbh(1 + 3\alpha) \quad \text{since terms in } \alpha^2 \text{ and } \alpha^3 \text{ are very small}$$

Hence $V_2 \approx V_1(1 + 3\alpha)$

For a temperature rise of $(t_2 - t_1) \, \text{K}$,

$$V_2 \approx V_1[1 + 3\alpha(t_2 - t_1)]$$

Thus from equation (3), $\gamma \approx 3\alpha$

Some **typical values** for the coefficient of cubic expansion measured at 20°C (i.e. 293 K) include:

Ethyl alcohol	$1.1 \times 10^{-3} \, \text{K}^{-1}$
Paraffin oil	$9 \times 10^{-2} \, \text{K}^{-1}$
Mercury	$1.82 \times 10^{-4} \, \text{K}^{-1}$
Water	$2.1 \times 10^{-4} \, \text{K}^{-1}$

The coefficient of cubic expansion γ is only constant over a limited range of temperature.

Problem 7. A brass sphere has a diameter of 50 mm at a temperature of 289 K. If the temperature of the sphere is raised to 789 K, determine the increase in (a) the diameter (b) the surface area (c) the volume of the sphere. Assume the coefficient of linear expansion for brass is 18×10^{-6} K^{-1}

Initial diameter, $l_1 = 50$ mm, initial temperature, $t_1 = 289$ K, final temperature, $t_2 = 789$ K and $\alpha = 18 \times 10^{-6}$ K^{-1}

New diameter at 789 K is given by:

$$l_2 = l_1[1 + \alpha(t_2 - t_1)] \quad \text{from equation (1)}$$

i.e. $\quad l_2 = 50[1 + (18 \times 10^{-6})(789 - 289)]$

$$= 50[1 + 0.009]$$

$$= 50.45 \text{ mm}$$

Hence the increase in the diameter is 0.45 mm

Initial surface area of sphere,

$$A_1 = 4\pi r^2 = 4\pi \left(\frac{50}{2}\right)^2 = 2500\pi \text{ mm}^2$$

New surface area at 789 K is given by

$$A_2 = A_1[1 + \beta(t_2 - t_1)] \quad \text{from equation (2)}$$

i.e. $\quad A_2 = A_1[1 + 2\alpha(t_2 - t_1)]$ since $\beta = 2\alpha$, to a very close approximation

Thus $A_2 = 2500\pi[1 + 2(18 \times 10^{-6})(500)]$

$$= 2500\pi[1 + 0.018]$$

$$= 2500\pi + 2500\pi(0.018)$$

Hence increase in surface area

$$= 2500\pi(0.018) = \textbf{141.4 mm}^2$$

Initial volume of sphere,

$$V_1 = \frac{4}{3}\pi r^3 = \frac{4}{3}\pi \left(\frac{50}{2}\right)^3 \text{ mm}^3$$

New volume at 789 K is given by:

$$V_2 = V_1[1 + \gamma(t_2 - t_1)] \quad \text{from equation (3)}$$

i.e. $V_2 = V_1[1 + 3\alpha(t_2 - t_1)]$ since $\gamma = 3\alpha$, to a very close approximation

Thus $V_2 = \frac{4}{3}\pi(25)^3[1 + 3(18 \times 10^6)(500)]$

$$= \frac{4}{3}\pi(25)^3[1 + 0.027]$$

$$= \frac{4}{3}\pi(25)^3 + \frac{4}{3}\pi(25)^3(0.027)$$

Hence the increase in volume

$$= \frac{4}{3}\pi(25)^3(0.027) = \textbf{1767 mm}^3$$

Problem 8. Mercury contained in a thermometer has a volume of 476 mm^3 at 15°C. Determine the temperature at which the volume of mercury is 478 mm^3, assuming the coefficient of cubic expansion for mercury to be 1.8×10^{-4} K^{-1}

Initial volume, $V_1 = 476$ mm, final volume $V_2 = 478$ mm^3, initial temperature, $t_1 = 15$°C and $\gamma = 1.8 \times 10^{-4}$ K^{-1}

Final volume, $V_2 = V_1[1 + \gamma(t_2 - t_1)]$, from equation (3)

i.e. $\quad V_2 = V_1 + V_1\gamma(t_2 - t_1)$, from which

$$(t_2 - t_1) = \frac{V_2 - V_1}{V_1\gamma}$$

$$= \frac{478 - 476}{(476)(1.8 \times 10^{-4})}$$

$$= 23.34°C$$

Hence $t_2 = 23.34 + 15 = 38.34$°C

Hence the temperature at which the volume of mercury is 478 mm^3 is 38.34°C

Problem 9. A rectangular glass block has a length of 100 mm, width 50 mm and depth 20 mm at 293 K. When heated to 353 K its length increases by 0.054 mm. What is the coefficient of linear expansion of the glass? Find also (a) the increase in surface area (b) the change in volume resulting from the change of length.

Final length, $l_2 = l_1[1 + \alpha(t_2 - t_1)]$, from equation (1), hence increase in length is given by:

$$l_2 - l_1 = l_1\alpha(t_2 - t_1)$$

Hence $\quad 0.054 = (100)(\alpha)(353 - 293)$

from which, the coefficient of linear expansion is given by:

$$\alpha = \frac{0.054}{(100)(60)} = \textbf{9} \times \textbf{10}^{-6} \textbf{ K}^{-1}$$

(a) Initial surface area of glass,

$$A_1 = (2 \times 100 \times 50) + (2 \times 50 \times 20) + (2 \times 100 \times 20)$$

$$= 10\,000 + 2000 + 4000 = 16\,000 \text{ mm}^2$$

Final surface area of glass,

$$A_2 = A_1[1 + \beta(t_2 - t_1)] = A_1[1 + 2\alpha(t_2 - t_1)],$$

since $\beta = 2\alpha$ to a very close approximation

Hence, **increase in surface area**

$$= A_1(2\alpha)(t_2 - t_1)$$

$$= (16\,000)(2 \times 9 \times 10^{-6})(60)$$

$$= \textbf{17.28 mm}^2$$

(b) Initial volume of glass,

$$V_1 = 100 \times 50 \times 20 = 100\,000\,\text{mm}^3$$

Final volume of glass, $V_2 = V_1[1 + \gamma(t_2 - t_1)]$

$$= V_1[1 + 3\alpha(t_2 - t_1)],$$

since $\gamma = 3\alpha$ to a very close approximation.

Hence, **increase in volume of glass**

$$= V_1(3\alpha)(t_2 - t_1)$$
$$= (100\,000)(3 \times 9 \times 10^{-6})(60)$$
$$= \mathbf{162\,mm^3}$$

Now try the following exercises

Exercise 164   Further problems on the coefficients of superficial and cubic expansion (Answers on page 310)

1. A silver plate has an area of 800 mm² at 15°C. Determine the increase in the area of the plate when the temperature is raised to 100°C. Assume the coefficient of linear expansion of silver to be $19 \times 10^{-6}\,\text{K}^{-1}$

2. At 283 K a thermometer contains 440 mm³ of alcohol. Determine the temperature at which the volume is 480 mm³ assuming that the coefficient of cubic expansion of the alcohol is $12 \times 10^{-4}\,\text{K}^{-1}$

3. A zinc sphere has a radius of 30.0 mm at a temperature of 20°C. If the temperature of the sphere is raised to 420°C, determine the increase in: (a) the radius, (b) the surface area, (c) the volume of the sphere. Assume the coefficient of linear expansion for zinc to be $31 \times 10^{-6}\,\text{K}^{-1}$

4. A block of cast iron has dimensions of 50 mm by 30 mm by 10 mm at 15°C. Determine the increase in volume when the temperature of the block is raised to 75°C. Assume the coefficient of linear expansion of cast iron to be $11 \times 10^{-6}\,\text{K}^{-1}$

5. Two litres of water, initially at 20°C, is heated to 40°C. Determine the volume of water at 40°C if the coefficient of volumetric expansion of water within this range is $30 \times 10^{-5}\,\text{K}^{-1}$

6. Determine the increase in volume, in litres, of 3 m³ of water when heated from 293 K to boiling point if the coefficient of cubic expansion is $2.1 \times 10^{-4}\,\text{K}^{-1}$ (1 litre $\approx 10^{-3}\,\text{m}^3$)

7. Determine the reduction in volume when the temperature of 0.5 litre of ethyl alcohol is reduced from 40°C to -15°C. Take the coefficient of cubic expansion for ethyl alcohol as $1.1 \times 10^{-3}\,\text{K}^{-1}$

Exercise 165   Short answer questions on thermal expansion

1. When heat is applied to most solids and liquids ..... occurs.

2. When solids and liquids are cooled they usually ......

3. State three practical applications where the expansion of metals must be allowed for.

4. State a practical disadvantage where the expansion of metals occurs.

5. State one practical advantage of the expansion of liquids.

6. What is meant by the 'coefficient of expansion'?

7. State the symbol and the unit used for the coefficient of linear expansion.

8. Define the 'coefficient of superficial expansion' and state its symbol.

9. Describe how water displays an unexpected effect between 0°C and 4°C.

10. Define the 'coefficient of cubic expansion' and state its symbol.

Exercise 166   Multi-choice questions on thermal expansion (Answers on page 311)

1. When the temperature of a rod of copper is increased its length:

   (a) stays the same   (b) increases   (c) decreases

2. The amount by which unit length of a material increases when the temperature is raised one degree is called the coefficient of:

   (a) cubic expansion   (b) superficial expansion

   (c) linear expansion

3. The symbol used for volumetric expansion is:

   (a) γ (b) β (c) l (d) α

4. A material of length l_1, at temperature θ_1 K is subjected to a temperature rise of θ K. The coefficient of linear expansion of the material is α K⁻¹. The material expands by:

   (a) $l_2(1 + \alpha\theta)$ (b) $l_1\alpha(\theta - \theta_1)$

   (c) $l_1[1 + \alpha(\theta - \theta_1)]$ (d) $l_1\alpha\theta$

5. Some iron has a coefficient of linear expansion of $12 \times 10^{-6}\,\text{K}^{-1}$. A 100 mm length of iron piping is heated through 20 K. The pipe extends by:

   (a) 0.24 mm   (b) 0.024 mm

   (c) 2.4 mm   (d) 0.0024 mm

6. If the coefficient of linear expansion is A, the coefficient of superficial expansion is B and the coefficient of cubic expansion is C, which of the following is false?

(a) $C = 3A$ (b) $A = B/2$

(c) $B = \dfrac{3}{2}C$ (d) $A = C/3$

7. The length of a 100 mm bar of metal increases by 0.3 mm when subjected to a temperature rise of 100 K. The coefficient of linear expansion of the metal is:

(a) $3 \times 10^{-3}\,\text{K}^{-1}$ (b) $3 \times 10^{-4}\,\text{K}^{-1}$

(c) $3 \times 10^{-5}\,\text{K}^{-1}$ (d) $3 \times 10^{-6}\,\text{K}^{-1}$

8. A liquid has a volume V_1 at temperature θ_1. The temperature is increased to θ_2. If γ is the coefficient of cubic expansion, the increase in volume is given by:

(a) $V_1\gamma(\theta_2 - \theta_1)$ (b) $V_1\gamma\theta_2$

(c) $V_1 + V_1\gamma\theta_2$ (d) $V_1[1 + \gamma(\theta_2 - \theta_1)]$

9. Which of the following statements is false?

(a) Gaps need to be left in lengths of railway lines to prevent buckling in hot weather

(b) Bimetallic strips are used in thermostats, a thermostat being a temperature-operated switch

(c) As the temperature of water is decreased from 4°C to 0°C contraction occurs

(d) A change of temperature of 15°C is equivalent to a change of temperature of 15 K

10. The volume of a rectangular block of iron at a temperature t_1 is V_1. The temperature is raised to t_2 and the volume increases to V_2. If the coefficient of linear expansion of iron is α, then volume V_1 is given by:

(a) $V_2[1 + \alpha(t_2 - t_1)]$ (b) $\dfrac{V_2}{1 + 3\alpha(t_2 - t_1)}$

(c) $3V_2\alpha(t_2 - t_1)$ (d) $\dfrac{1 + \alpha(t_2 - t_1)}{V_2}$

36

The measurement of temperature

At the end of this chapter you should be able to:

- describe the construction, principle of operation and practical applications of the following temperature measuring devices:
  - (a) liquid-in-glass thermometer (including advantages of mercury, and sources of error)
  - (b) thermocouples (including advantages and sources of error)
  - (c) resistance thermometer (including limitations and advantages of platinum coil)
  - (d) thermistors
  - (e) pyrometers (total radiation and optical types, including advantages and disadvantages
- describe the principle of operation of
  - (a) temperature indicating paints and crayons
  - (b) bimetallic thermometers
  - (c) mercury-in-steel thermometer
  - (d) gas thermometer
- select the appropriate temperature measuring device for a particular application

36.1   Introduction

A change in temperature of a substance can often result in a change in one or more of its physical properties. Thus, although temperature cannot be measured directly, its effects can be measured. Some properties of substances used to determine changes in temperature include changes in dimensions, electrical resistance, state, type and volume of radiation and colour.

Temperature measuring devices available are many a varied. Those described in Sections 36.2 to 36.10 are the most often used in science and industry.

36.2   Liquid-in-glass thermometer

A **liquid-in-glass thermometer** uses the expansion of a l uid with increase in temperature as its principle of operatie

Construction

A typical liquid-in-glass thermometer is shown in Fig. 3e and consists of a sealed stem of uniform small-bore tubi called a capillary tube, made of glass, with a cylindri glass bulb formed at one end. The bulb and part of stem are filled with a liquid such as mercury or alcohol a the remaining part of the tube is evacuated. A temperatu scale is formed by etching graduations on the stem. A saf reservoir is usually provided, into which the liquid c expand without bursting the glass if the temperature is rai beyond the upper limit of the scale.

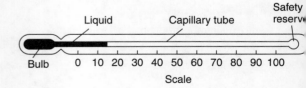

Fig. 36.1

Principle of operation

The operation of a liquid-in-glass thermometer depen on the liquid expanding with increase in temperature a contracting with decrease in temperature. The position of the end of the column of liquid in the tube is a measure of the temperature of the liquid in the bulb – shown as 15°C in Fig. 36.1, which is about room temperature. Two fixed points are needed to calibrate the thermometer, with the interval between these points being divided into 'degrees'. In the first thermometer, made by Celsius, the fixed points chosen were the temperature of melting ice (0°C) and that of boiling water at standard atmospheric pressure (100°C), in each case the blank stem being marked at the liquid level. The distance between these two points, called the fundamental interval, was divided into 100 equal parts, each equivalent to 1°C, thus forming the scale.

The **clinical thermometer**, with a limited scale around body temperature, the **maximum and/or minimum thermometer**, recording the maximum day temperature and minimum night temperature, and the **Beckman thermometer**, which is used only in accurate measurement of temperature change and has no fixed points, are particular types of liquid-in-glass thermometer which all operate on the same principle.

Advantages

The liquid-in-glass thermometer is simple in construction, relatively inexpensive, easy to use and portable, and is the most widely used method of temperature measurement having industrial, chemical, clinical and meteorological applications.

Disadvantages

Liquid-in-glass thermometers tend to be fragile and hence easily broken, can only be used where the liquid column is visible, cannot be used for surface temperature measurements, cannot be read from a distance and are unsuitable for high temperature measurements.

Advantages of mercury

The use of mercury in a thermometer has many advantages, for mercury:

(i) is clearly visible,

(ii) has a fairly uniform rate of expansion,

(iii) is readily obtainable in the pure state,

(iv) does not 'wet' the glass,

(v) is a good conductor of heat.

Mercury has a freezing point of −39°C and cannot be used as a thermometer below this temperature. Its boiling point is 357°C but before this temperature is reached some distillation of the mercury occurs if the space above the mercury is a vacuum. To prevent this, and to extend the upper temperature limits to over 500°C, an inert gas such as nitrogen under pressure is used to fill the remainder of the capillary tube.

Alcohol, often dyed red to be seen in the capillary tube, is considerably cheaper than mercury and has a freezing point of −113°C, which is considerably lower than for mercury. However it has a low boiling point at about 79°C.

Errors

Typical errors in liquid-in-glass thermometers may occur due to:

(i) the slow cooling rate of glass,

(ii) incorrect positioning of the thermometer,

(iii) a delay in the thermometer becoming steady (i.e. slow response time),

(iv) non-uniformity of the bore of the capillary tube, which means that equal intervals marked on the stem do not correspond to equal temperature intervals.

36.3 Thermocouples

Thermocouples use the e.m.f. set up when the junction of two dissimilar metals is heated.

Principle of operation

At the junction between two different metals, say, copper and constantan, there exists a difference in electrical potential, which varies with the temperature of the junction. This is known as the 'thermo-electric effect'. If the circuit is completed with a second junction at a different temperature, a current will flow round the circuit. This principle is used in the thermocouple. Two different metal conductors having their ends twisted together are shown in Fig. 36.2. If the two junctions are at different temperatures, a current I flows round the circuit.

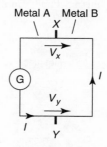

Fig. 36.2

The deflection on the galvanometer G depends on the difference in temperature between junctions X and Y and is caused by the difference between voltages V_x and V_y. The higher temperature junction is usually called the 'hot junction' and the lower temperature junction the 'cold junction'. If the cold junction is kept at a constant known temperature, the galvanometer can be calibrated to indicate the temperature of the hot junction directly. The cold junction is then known as the reference junction.

In many instrumentation situations, the measuring instrument needs to be located far from the point at which the measurements are to be made. Extension leads are then used, usually made of the same material as the thermocouple but of smaller gauge. The reference junction is then effectively moved to their ends. The thermocouple is used by positioning the hot junction where the temperature is required. The meter will indicate the temperature of the hot junction only if the reference junction is at 0°C for:

(temperature of hot junction)
= (temperature of the cold junction)
+ (temperature difference)

In a laboratory the reference junction is often placed in melting ice, but in industry it is often positioned in a thermostatically controlled oven or buried underground where the temperature is constant.

Construction

Thermocouple junctions are made by twisting together two wires of dissimilar metals before welding them. The construction of a typical copper-constantan thermocouple for industrial use is shown in Fig. 36.3. Apart from the actual junction the two conductors used must be insulated electrically from each other with appropriate insulation and is shown in Fig. 36.3 as twin-holed tubing. The wires and insulation are usually inserted into a sheath for protection from environments in which they might be damaged or corroded.

Applications

A copper-constantan thermocouple can measure temperature from −250°C up to about 400°C, and is used typically with

boiler flue gases, food processing and with sub-zero temperature measurement. An iron-constantan thermocouple can measure temperature from −200°C to about 850°C, and is used typically in paper and pulp mills, re-heat and annealing furnaces and in chemical reactors. A chromel-alumel thermocouple can measure temperatures from −200°C to about 1100°C and is used typically with blast furnace gases, brick kilns and in glass manufacture.

For the measurement of temperatures above 1100°C radiation pyrometers are normally used. However, thermocouples are available made of platinum-platinum/rhodium capable of measuring temperatures up to 1400°C, or tungsten molybdenum which can measure up to 2600°C.

Advantages

A thermocouple:

(i) has a very simple, relatively inexpensive construction,

(ii) can be made very small and compact,

(iii) is robust,

(iv) is easily replaced if damaged,

(v) has a small response time,

(vi) can be used at a distance from the actual measuring instrument and is thus ideal for use with automatic and remote-control systems.

Sources of error

Sources of error in the thermocouple which are difficult to overcome include:

(i) voltage drops in leads and junctions,

(ii) possible variations in the temperature of the cold junction,

(iii) stray thermoelectric effects, which are caused by the addition of further metals into the 'ideal' two-metal thermocouple circuit. Additional leads are frequently necessary for extension leads or voltmeter terminal connections.

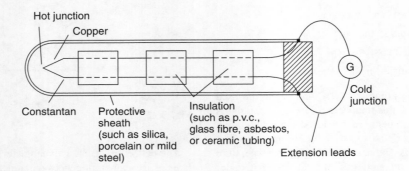

Hot junction
Copper
Constantan
Protective sheath (such as silica, porcelain or mild steel)
Insulation (such as p.v.c., glass fibre, asbestos, or ceramic tubing)
Extension leads
G
Cold junction

Fig. 36.3

A thermocouple may be used with a battery- or mains-operated electronic thermometer instead of a millivoltmeter. These devices amplify the small e.m.f.'s from the thermocouple before feeding them to a multi-range voltmeter calibrated directly with temperature scales. These devices have great accuracy and are almost unaffected by voltage drops in the leads and junctions.

Problem 1. A chromel-alumel thermocouple generates an e.m.f. of 5 mV. Determine the temperature of the hot junction if the cold junction is at a temperature of 15°C and the sensitivity of the thermocouple is 0.04 mV/°C.

Temperature difference for

$$5\,\text{mV} = \frac{5\,\text{mV}}{0.04\,\text{mV/°C}} = 125°C$$

Temperature at hot junction

= temperature of cold junction + temperature difference

= 15°C + 125°C = **140°C**

Now try the following exercise

Exercise 167   Further problem on the thermocouple (Answer on page 311)

1. A platinum-platinum/rhodium thermocouple generates an e.m.f. of 7.5 mV. If the cold junction is at a temperature of 20°C, determine the temperature of the hot junction. Assume the sensitivity of the thermocouple to be 6 μV/°C.

36.4 Resistance thermometers

Resistance thermometers use the change in electrical resistance caused by temperature change.

Construction

Resistance thermometers are made in a variety of sizes, shapes and forms depending on the application for which they are designed. A typical resistance thermometer is shown diagrammatically in Fig. 36.4. The most common metal used for the coil in such thermometers is platinum even though its sensitivity is not as high as other metals such as copper and nickel. However, platinum is a very stable metal and provides reproducible results in a resistance thermometer. A platinum resistance thermometer is often used as a calibrating device. Since platinum is expensive, connecting leads of another metal, usually copper, are used with the thermometer to connect it to a measuring circuit.

The platinum and the connecting leads are shown joined at *A* and *B* in Fig. 36.4, although sometimes this junction may be made outside of the sheath. However, these leads often come into close contact with the heat source which can introduce errors into the measurements. These may be eliminated by including a pair of identical leads, called dummy leads, which experience the same temperature change as the extension leads.

Principle of operation

With most metals a rise in temperature causes an increase in electrical resistance, and since resistance can be measured accurately this property can be used to measure temperature. If the resistance of a length of wire at 0°C is R_0, and its resistance at θ°C is R_θ, then $R_\theta = R_0(1 + \alpha\theta)$, where α is the temperature coefficient of resistance of the material (see Chapter 17).

Rearranging gives:

$$\text{temperature, } \theta = \frac{R_\theta - R_0}{\alpha R_0}$$

Values of R_0 and α may be determined experimentally or obtained from existing data. Thus, if R_θ can be measured, temperature θ can be calculated. This is the principle of

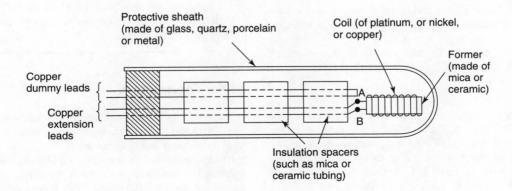

Copper dummy leads

Copper extension leads

Protective sheath (made of glass, quartz, porcelain or metal)

Coil (of platinum, or nickel, or copper)

Former (made of mica or ceramic)

Insulation spacers (such as mica or ceramic tubing)

Fig. 36.4

operation of a resistance thermometer. Although a sensitive ohmmeter can be used to measure R_θ, for more accurate determinations a Wheatstone bridge circuit is used as shown in Fig. 36.5 (see also Chapter 23). This circuit compares an unknown resistance R_θ with others of known values, R_1 and R_2 being fixed values and R_3 being variable. Galvanometer G is a sensitive centre-zero microammeter. R_3 is varied until zero deflection is obtained on the galvanometer, i.e. no current flows through G and the bridge is said to be 'balanced'.

At balance: $R_2R_\theta = R_1R_3$

from which, $$R_\theta = \frac{R_1R_3}{R_2}$$

and if R_1 and R_2 are of equal value, then $R_\theta = R_3$

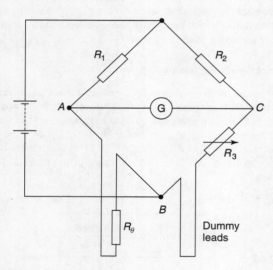

Fig. 36.5

A resistance thermometer may be connected between points A and B in Fig. 36.5 and its resistance R_θ at any temperature θ accurately measured. Dummy leads included in arm BC help to eliminate errors caused by the extension leads which are normally necessary in such a thermometer.

Limitations

Resistance thermometers using a nickel coil are used mainly in the range $-100°C$ to $300°C$, whereas platinum resistance thermometers are capable of measuring with greater accuracy temperatures in the range $-200°C$ to about $800°C$. This upper range may be extended to about $1500°C$ if high melting point materials are used for the sheath and coil construction.

Advantages and disadvantages of a platinum coil

Platinum is commonly used in resistance thermometers since it is chemically inert, i.e. un-reactive, resists corrosion and oxidation and has a high melting point of $1769°C$. A disadvantage of platinum is its slow response to temperature variation.

Applications

Platinum resistance thermometers may be used as calibrating devices or in applications such as heat treating and annealing processes and can be adapted easily for use with automatic recording or control systems. Resistance thermometers tend to be fragile and easily damaged especially when subjected to excessive vibration or shock.

Problem 2. A platinum resistance thermometer has a resistance of $25\,\Omega$ at $0°C$. When measuring the temperature of an annealing process a resistance value of $60\,\Omega$ is recorded. To what temperature does this correspond? Take the temperature coefficient of resistance of platinum as $0.0038/°C$.

$R_\theta = R_0(1 + \alpha\theta)$, where $R_0 = 25\,\Omega$, $R_\theta = 60\,\Omega$ and $\alpha = 0.0038/°C$. Rearranging gives:

$$\textbf{temperature}, \theta = \frac{R_\theta - R_0}{\alpha R_0} = \frac{60 - 25}{(0.0038)(25)}$$

$$= \mathbf{368.4°C}$$

Now try the following exercise

Exercise 168   Further problem on the resistance thermometer (Answer on page 311)

1. A platinum resistance thermometer has a resistance of $100\,\Omega$ at $0°C$. When measuring the temperature of a heat process a resistance value of $177\,\Omega$ is measured using a Wheatstone bridge. Given that the temperature coefficient of resistance of platinum is $0.0038/°C$, determine the temperature of the heat process, correct to the nearest degree.

36.5   Thermistors

A thermistor is a semi-conducting material – such as mixtures of oxides of copper, manganese, cobalt, etc. – in the form of a fused bead connected to two leads. As its temperature is increased its resistance rapidly decreases. Typical resistance/temperature curves for a thermistor and common metals are shown in Fig. 36.6. The resistance of a typical thermistor can vary from $400\,\Omega$ at $0°C$ to $100\,\Omega$ at $140°C$.

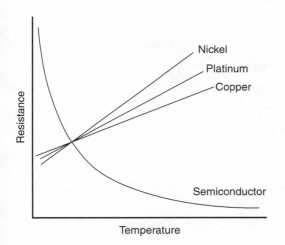

Fig. 36.6

Advantages

The main advantages of a thermistor are its high sensitivity and small size. It provides an inexpensive method of measuring and detecting small changes in temperature.

36.6 Pyrometers

A pyrometer is a device for measuring very high temperatures and uses the principle that all substances emit radiant energy when hot, the rate of emission depending on their temperature. The measurement of thermal radiation is therefore a convenient method of determining the temperature of hot sources and is particularly useful in industrial processes. There are two main types of pyrometer, namely the total radiation pyrometer and the optical pyrometer.

Pyrometers are very convenient instruments since they can be used at a safe and comfortable distance from the hot source. Thus applications of pyrometers are found in measuring the temperature of molten metals, the interiors of furnaces or the interiors of volcanoes. Total radiation pyrometers can also be used in conjunction with devices which record and control temperature continuously.

Total radiation pyrometer

A typical arrangement of a total radiation pyrometer is shown in Fig. 36.7. Radiant energy from a hot source, such as a furnace, is focused on to the hot junction of a thermocouple after reflection from a concave mirror. The temperature rise recorded by the thermocouple depends on the amount of radiant energy received, which in turn depends on the temperature of the hot source. The galvanometer G shown connected to the thermocouple records the current which results from the e.m.f. developed and may be calibrated to give a direct reading of the temperature of the hot source. The thermocouple is protected from direct radiation by a shield as shown and the hot source may be viewed through the sighting telescope. For greater sensitivity, a thermopile may be used, a **thermopile** being a number of thermocouples connected in series. Total radiation pyrometers are used to measure temperature in the range 700°C to 2000°C.

Optical pyrometers

When the temperature of an object is raised sufficiently two visual effects occur; the object appears brighter and there is a change in colour of the light emitted. These effects are used in the optical pyrometer where a comparison or matching is made between the brightness of the glowing hot source and the light from a filament of known temperature.

The most frequently used optical pyrometer is the disappearing filament pyrometer and a typical arrangement is shown in Fig. 36.8. A filament lamp is built into a telescope arrangement which receives radiation from a hot source, an image of which is seen through an eyepiece. A red filter is incorporated as a protection to the eye.

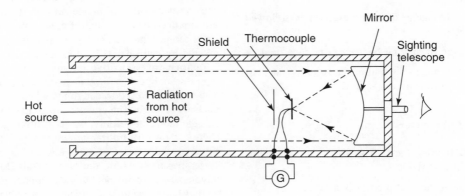

Fig. 36.7

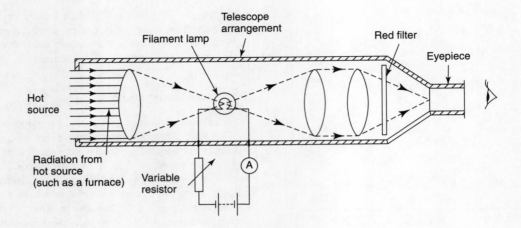

Fig. 36.8

The current flowing through the lamp is controlled by a variable resistor. As the current is increased the temperature of the filament increases and its colour changes. When viewed through the eyepiece the filament of the lamp appears superimposed on the image of the radiant energy from the hot source. The current is varied until the filament glows as brightly as the background. It will then merge into the background and seem to disappear. The current required to achieve this is a measure of the temperature of the hot source and the ammeter can be calibrated to read the temperature directly. Optical pyrometers may be used to measure temperatures up to, and even in excess of, 3000°C.

Advantages of pyrometers

(i) There is no practical limit to the temperature that a pyrometer can measure.

(ii) A pyrometer need not be brought directly into the hot zone and so is free from the effects of heat and chemical attack that can often cause other measuring devices to deteriorate in use.

(iii) Very fast rates of change of temperature can be followed by a pyrometer.

(iv) The temperature of moving bodies can be measured.

(v) The lens system makes the pyrometer virtually independent of its distance from the source.

Disadvantages of pyrometers

(i) A pyrometer is often more expensive than other temperature measuring devices.

(ii) A direct view of the heat process is necessary.

(iii) Manual adjustment is necessary.

(iv) A reasonable amount of skill and care is require in calibrating and using a pyrometer. For each new measuring situation the pyrometer must be re-calibrated

(v) The temperature of the surroundings may affect th reading of the pyrometer and such errors are difficul to eliminate.

36.7   Temperature indicating paints and crayons

Temperature indicating paints contain substances whic change their colour when heated to certain temperatures. Thi change is usually due to chemical decomposition, such a loss of water, in which the change in colour of the pair after having reached the particular temperature will be permanent one. However, in some types the original colou returns after cooling. Temperature indicating paints are use where the temperature of inaccessible parts of apparatus an machines is required. They are particularly useful in hea treatment processes where the temperature of the componer needs to be known before a quenching operation. There ar several such paints available and most have only a smal temperature range so that different paints have to be use for different temperatures. The usual range of temperature covered by these paints is from about 30°C to 700°C.

Temperature sensitive crayons consist of fusible solid compressed into the form of a stick. The melting point such crayons is used to determine when a given temperatu has been reached. The crayons are simple to use but indica a single temperature only, i.e. its melting point temperature There are over 100 different crayons available, each cove ing a particular range of temperature. Crayons are availabl for temperatures within the range of 50°C to 1400°C. Suc crayons are used in metallurgical applications such as pre heating before welding, hardening, annealing or tempering or in monitoring the temperature of critical parts of machine or for checking mould temperatures in the rubber and plastics industries.

36.8 Bimetallic thermometers

Bimetallic thermometers depend on the expansion of metal strips which operate an indicating pointer. Two thin metal strips of differing thermal expansion are welded or riveted together and the curvature of the bimetallic strip changes with temperature change. For greater sensitivity the strips may be coiled into a flat spiral or helix, one end being fixed and the other being made to rotate a pointer over a scale. Bimetallic thermometers are useful for alarm and over-temperature applications where extreme accuracy is not essential. If the whole is placed in a sheath, protection from corrosive environments is achieved but with a reduction in response characteristics. The normal upper limit of temperature measurement by this thermometer is about 200°C, although with special metals the range can be extended to about 400°C.

36.9 Mercury-in-steel thermometer

The **mercury-in-steel thermometer** is an extension of the principle of the mercury-in-glass thermometer. Mercury in a steel bulb expands via a small bore capillary tube into a pressure indicating device, say a Bourdon gauge, the position of the pointer indicating the amount of expansion and thus the temperature. The advantages of this instrument are that it is robust and, by increasing the length of the capillary tube, the gauge can be placed some distance from the bulb and can thus be used to monitor temperatures in positions which are inaccessible to the liquid-in-glass thermometer. Such thermometers may be used to measure temperatures up to 600°C.

36.10 Gas thermometers

The gas thermometer consists of a flexible U-tube of mercury connected by a capillary tube to a vessel containing gas. The change in the volume of a fixed mass of gas at constant pressure, or the change in pressure of a fixed mass of gas at constant volume, may be used to measure temperature. This thermometer is cumbersome and rarely used to measure temperature directly, but it is often used as a standard with which to calibrate other types of thermometer. With pure hydrogen the range of the instrument extends from −240°C to 1500°C and measurements can be made with extreme accuracy.

36.11 Choice of measuring device

Problem 3. State which device would be most suitable to measure the following:

(a) metal in a furnace, in the range 50°C to 1600°C

(b) the air in an office in the range 0°C to 40°C

(c) boiler flue gas in the range 15°C to 300°C

(d) a metal surface, where a visual indication is required when it reaches 425°C

(e) materials in a high-temperature furnace in the range 2000°C to 2800°C

(f) to calibrate a thermocouple in the range −100°C to 500°C

(g) brick in a kiln up to 900°C

(h) an inexpensive method for food processing applications in the range −25°C to −75°C

(a) Radiation pyrometer

(b) Mercury-in-glass thermometer

(c) Copper-constantan thermocouple

(d) Temperature sensitive crayon

(e) Optical pyrometer

(f) Platinum resistance thermometer or gas thermometer

(g) Chromel-alumel thermocouple

(h) Alcohol-in-glass thermometer

Now try the following exercises

Exercise 169 Short answer questions on the measurement of temperature

For each of the temperature measuring devices listed in 1 to 10, state very briefly its principle of operation and the range of temperatures that it is capable of measuring.

1. Mercury-in-glass thermometer
2. Alcohol-in-glass thermometer
3. Thermocouple
4. Platinum resistance thermometer
5. Total radiation pyrometer
6. Optical pyrometer
7. Temperature sensitive crayons
8. Bimetallic thermometer
9. Mercury-in-steel thermometer
10. Gas thermometer

Exercise 170   Multi-choice questions on the measurement of temperature (Answers on page 311)

1. The most suitable device for measuring very small temperature changes is a

   (a) thermopile   (b) thermocouple

   (c) thermistor

2. When two wires of different metals are twisted together and heat applied to the junction, an e.m.f. is produced. This effect is used in a thermocouple to measure:

   (a) e.m.f.   (b) temperature

   (c) expansion   (d) heat

3. A cold junction of a thermocouple is at room temperature of 15°C. A voltmeter connected to the thermocouple circuit indicates 10 mV. If the voltmeter is calibrated as 20°C/mV, the temperature of the hot source is:

   (a) 185°C   (b) 200°C   (c) 35°C   (d) 215°C

4. The e.m.f. generated by a copper-constantan thermometer is 15 mV. If the cold junction is at a temperature of 20°C, the temperature of the hot junction when the sensitivity of the thermocouple is 0.03 mV/°C is:

   (a) 480°C   (b) 520°C   (c) 20.45°C   (d) 500°C

   In questions 5 to 12, select the most appropriate temperature measuring device from this list.

   (a) copper-constantan thermocouple

   (b) thermistor

   (c) mercury-in-glass thermometer

   (d) total radiation pyrometer

   (e) platinum resistance thermometer

   (f) gas thermometer

   (g) temperature sensitive crayon

   (h) alcohol-in-glass thermometer

   (i) bimetallic thermometer

   (j) mercury-in-steel thermometer

   (k) optical pyrometer

5. Over-temperature alarm at about 180°C.

6. Food processing plant in the range −250°C to +250°C.

7. Automatic recording system for a heat treating process in the range 90°C to 250°C.

8. Surface of molten metals in the range 1000°C to 1800°C.

9. To calibrate accurately a mercury-in-glass thermometer.

10. Furnace up to 3000°C.

11. Inexpensive method of measuring very small changes in temperature.

12. Metal surface where a visual indication is required when the temperature reaches 520°C.

Assignment 13

This assignment covers the material contained in Chapters 34 to 36. The marks for each question are shown in brackets at the end of each question.

1. Determine the force applied tangentially to a bar of a screw-jack at a radius of 60 cm, if the torque required is 750 Nm. (3)

2. Calculate the torque developed by a motor whose spindle is rotating at 900 rev/min and developing a power of 4.20 kW (5)

3. A motor connected to a shaft develops a torque of 8 kNm. Determine the number of revolutions made by the shaft if the work done is 7.2 MJ (6)

4. Determine the angular acceleration of a shaft which has a moment of inertia of 32 kg m² produced by an accelerating torque of 600 Nm (5)

5. An electric motor has an efficiency of 72% when running at 1400 rev/min. Determine the output torque when the power input is 2.50 kW (5)

6. A copper overhead transmission line has a length of 60 m between its supports at 15°C. Calculate its length at 40°C if the coefficient of linear expansion of copper is $17 \times 10^{-6}\,\text{K}^{-1}$ (6)

7. A gold sphere has a diameter of 40 mm at a temperature of 285 K. If the temperature of the sphere is raised to 785 K, determine the increase in (a) the diameter (b) the surface area (c) the volume of the sphere. Assume the coefficient of linear expansion for gold is $14 \times 10^{-6}\,\text{K}^{-1}$ (12)

8. A platinum resistance thermometer has a resistance of 24 Ω at 0°C. When measuring the temperature of an annealing process a resistance value of 68 Ω is recorded. To what temperature does this correspond? Take the temperature coefficient of resistance of platinum as 0.0038/°C. (4)

9. State which device would be most suitable to measure the following:

   (a) materials in a high-temperature furnace in the range 1800°C to 3000°C

   (b) the air in a factory in the range 0°C to 35°C

   (c) an inexpensive method for food processing applications in the range −20°C to −80°C

   (d) boiler flue gas in the range 15°C to 250°C (4)

Formulae for mechanical applications

Formula	Formula Symbols	Units
Density $= \dfrac{\text{mass}}{\text{volume}}$	$\rho = \dfrac{m}{V}$	kg/m^3
Average velocity $= \dfrac{\text{distance travelled}}{\text{time taken}}$	$v = \dfrac{s}{t}$	m/s
Acceleration $= \dfrac{\text{change in velocity}}{\text{time taken}}$	$a = \dfrac{v - u}{t}$	m/s^2
Force $=$ mass $\times$ acceleration	$F = ma$	N
Weight $=$ mass $\times$ gravitational field	$W = mg$	N
Centripetal acceleration	$a = \dfrac{v^2}{r}$	m/s^2
Centripetal force	$F = \dfrac{mv^2}{r}$	N
Moment $=$ force $\times$ perpendicular distance	$M = Fd$	Nm
Angular velocity	$\omega = \dfrac{\theta}{t} = 2\pi n$	rad/s
Linear velocity	$v = \omega r$	m/s
Relationships between initial velocity u, final velocity v, displacement s, time t and constant acceleration a	$\begin{cases} s = ut + \dfrac{1}{2}at^2 \\ v^2 = u^2 + 2as \end{cases}$	m (m/s)^2
Relationships between initial angular velocity ω_1, final angular velocity ω_2, angle θ, time t and angular acceleration α	$\begin{cases} \theta = \omega_1 t + \dfrac{1}{2}\alpha t^2 \\ \omega_2^2 = \omega_1^2 + 2\alpha\theta \end{cases}$	rad (rad/s)^2
Frictional force $=$ coefficient of friction $\times$ normal force	$F = \mu N$	N
Force ratio $= \dfrac{\text{load}}{\text{effort}}$		

$$\text{Movement ratio} = \frac{\text{distance moved by effort}}{\text{distance moved by load}}$$

$$\text{Efficiency} = \frac{\text{force ratio}}{\text{movement ratio}}$$

$$\text{Stress} = \frac{\text{applied force}}{\text{cross-sectional area}} \qquad \sigma = \frac{F}{A} \qquad \text{Pa}$$

$$\text{Strain} = \frac{\text{change in length}}{\text{original length}} \qquad \varepsilon = \frac{x}{l}$$

$$\text{Young's modulus of elasticity} = \frac{\text{sress}}{\text{strain}} \qquad E = \frac{\sigma}{\varepsilon} \qquad \text{Pa}$$

$$\text{Stiffness} = \frac{\text{force}}{\text{extension}} \qquad \text{N/m}$$

$$\text{Momentum} = \text{mass} \times \text{velocity} \qquad \text{kg m/s}$$

$$\text{Impulse} = \text{applied force} \times \text{time} = \text{change in momentum} \qquad \text{kg m/s}$$

$$\text{Torque} = \text{force} \times \text{perpendicular distance} \qquad T = Fd \qquad \text{Nm}$$

$$\text{Power} = \text{torque} \times \text{angular velocity} \qquad P = T\omega = 2\pi n \qquad \text{W}$$

$$\text{Torque} = \text{moment of inertia} \times \text{angular acceleration} \qquad T = I\alpha \qquad \text{Nm}$$

$$\text{Kelvin temperature} = \text{degrees Celsius} + 273$$

$$\text{New length} = \text{original length} + \text{expansion} \qquad l_2 = l_1[1 + \alpha(t_2 - t_1)] \qquad \text{m}$$

$$\text{New surface area} = \text{original surface area} + \text{increase in area} \qquad A_2 = A_1[1 + \beta(t_2 - t_1)] \qquad \text{m}^2$$

$$\text{New volume} = \text{original volume} + \text{increase in volume} \qquad V_2 = V_1[1 + \gamma(t_2 - t_1)] \qquad \text{m}^3$$

Answer to Exercises

Exercise 1 (Page 4)

1. 19 **2.** −66 **3.** 565 **4.** −2136
5. −19 157 **6.** 1487 **7.** −225 **8.** 5914
9. (a) 1827 (b) 4158 **10.** (a) 8613 (b) 1752
11. (a) 1026 (b) 2233 **12.** (a) 48 (b) 89

Exercise 2 (Page 5)

1. (a) 2 (b) 210 **2.** (a) 3 (b) 180
3. (a) 5 (b) 210 **4.** (a) 15 (b) 6300
5. (a) 14 (b) 4 20 420 **6.** (a) 14 (b) 53 900

Exercise 3 (Page 6)

1. 59 **2.** 14 **3.** 88
4. 5 **5.** −107 **6.** 57

Exercise 4 (Page 9)

1. (a) $\dfrac{9}{10}$ (b) $\dfrac{3}{16}$ **2.** (a) $\dfrac{43}{77}$ (b) $\dfrac{47}{63}$

3. (a) $8\dfrac{51}{52}$ (b) $1\dfrac{9}{40}$ **4.** (a) $\dfrac{5}{12}$ (b) $\dfrac{3}{49}$

5. (a) $\dfrac{1}{13}$ (b) $\dfrac{5}{12}$ **6.** (a) $\dfrac{8}{15}$ (b) $\dfrac{12}{23}$

7. $-\dfrac{1}{9}$ **8.** $1\dfrac{1}{6}$

9. $5\dfrac{4}{5}$ **10.** $-\dfrac{13}{126}$

Exercise 5 (Page 9)

1. 91 mm to 221 mm
2. 81 cm to 189 cm to 351 cm
3. 17 g
4. 72 kg:27 kg
5. 5

Exercise 6 (Page 10)

1. 11.989 **2.** −31.265 **3.** 24.066
4. 10.906 **5.** 2.2446 **6.** (a) 24.81 (b) 24.812

7. (a) $\dfrac{13}{20}$ (b) $\dfrac{21}{25}$ (c) $\dfrac{1}{80}$ (d) $\dfrac{141}{500}$ (e) $\dfrac{3}{125}$

8. (a) $1\dfrac{41}{50}$ (b) $4\dfrac{11}{40}$ (c) $14\dfrac{1}{8}$ (d) $15\dfrac{7}{20}$ (e) $16\dfrac{17}{80}$

9. 0.44444 **10.** 0.62963 **11.** 1.563

Exercise 7 (Page 11)

1. (a) 5.7% (b) 37.4% (c) 128.5%
2. (a) 21.2% (b) 79.2% (c) 169%
3. (a) 496.4 t (b) 8.657 g (c) 20.73 s
4. 2.25%
5. (a) 14% (b) 15.67% (c) 5.36%
6. 37.8 g
7. 7.2%
8. A 0.6 kg, B 0.9 kg, C 0.5 kg

Exercise 8 (Page 13)

1. (a) 3^7 (b) 4^9 2. (a) 2^6 (b) 7^{10}
3. (a) 2 (b) 3^5 4. (a) 5^3 (b) 7^3
5. (a) 7^6 (b) 3^6 6. (a) 15^{15} (b) 17^8
7. (a) 2 (b) 3^6 8. (a) 3^4 (b) 1

Exercise 9 (Page 14)

1. (a) $\dfrac{1}{3 \times 5^2}$ (b) $\dfrac{1}{7^3 \times 3^7}$ 2. (a) $\dfrac{3^2}{2^5}$ (b) $\dfrac{1}{2^{10} \times 5^2}$

3. (a) 9 (b) ± 3 (c) $\pm\dfrac{1}{2}$ (d) $\pm\dfrac{2}{3}$

4. $\dfrac{25}{32}$ 5. 64 6. $4\dfrac{1}{2}$

Exercise 10 (Page 15)

1. (a) 7.39×10 (b) 1.9772×10^2
2. (a) 2.748×10^3 (b) 3.317×10^4
3. (a) 2.401×10^{-1} (b) 1.74×10^{-2}
4. (a) 1.7023×10^3 (b) 1.004×10
5. (a) 5×10^{-1} (b) 1.1875×10
   (c) 1.306×10^2 (d) 3.125×10^{-2}
6. (a) 1010 (b) 932.7 (c) $54\,100$ (d) 7
7. (a) 0.0389 (b) 0.6741 (c) 0.008

Exercise 11 (Page 16)

1. (a) 1.351×10^3 (b) 5.734×10^{-2}
2. (a) 1.7231×10^3 (b) 3.129×10^{-3}
3. (a) 1.35×10^2 (b) 1.1×10^5
4. (a) 2×10^2 (b) 1.5×10^{-3}
5. (a) $2.71 \times 10^3 \,\mathrm{kg\,m^{-3}}$ (b) 4.4×10^{-1}
   (c) $3.7673 \times 10^2 \,\Omega$ (d) $5.11 \times 10^{-1} \,\mathrm{MeV}$
   (e) $2.241 \times 10^{-2} \,\mathrm{m^3\,mol^{-1}}$

Exercise 12 (Page 19)

1. Order of magnitude error
2. Rounding-off error – should add 'correct to 4 significant figures' or 'correct to 1 decimal place'
3. Blunder
4. Order of magnitude error and rounding-off error – should be 0.0225, correct to 3 significant figures or 0.0225, correct to 4 decimal places

5. Measured values, hence $c = 55\,800\,\mathrm{Pa\,m^3}$
6. ≈ 30 (29.61, by calculator)
7. ≈ 2 (1.988, correct to 4 s.f., by calculator)
8. ≈ 10 (9.481, correct to 4 s.f., by calculator)

Exercise 13 (Page 20)

1. (a) 558.6   (b) 31.09
2. (a) 109.1   (b) 3.641
3. (a) 0.2489   (b) 500.5
4. (a) 0.05777 (b) 28.90
5. (a) 2515   (b) 146.0
6. (a) 18.63   (b) 0.1643
7. (a) 0.005559 (b) 1.900
8. (a) 6.248   (b) 0.9630
9. (a) 1.165   (b) 2.680

Exercise 14 (Page 21)

1. (a) 312 f   (b) \$123.42   (c) £68.29
   (d) £7   (e) 156.75 Dm
2. (a) 381 mm   (b) 56.35 km/h (c) 378.35 km
   (d) 52 lb 13 oz (e) 6.82 kg
   (f) 54.55 litres (g) 5.5 gallons

Exercise 15 (Page 22)

1. $C = 52.78\,\mathrm{mm}$ 2. $159\,\mathrm{m/s}$
3. $0.00502\,\mathrm{m}$ or $5.02\,\mathrm{mm}$ 4. $0.144\,\mathrm{J}$
5. 224.5 6. $14\,230\,\mathrm{kg/m^3}$ 7. $281.1\,\mathrm{m/s}$
8. $2.526\,\Omega$ 9. $508.1\,\mathrm{W}$ 10. $V = 2.61\,\mathrm{V}$
11. $F = 854.5$ 12. $t = 14.79\,\mathrm{s}$ 13. $E = 3.96\,\mathrm{J}$
14. $s = 17.25\,\mathrm{m}$

Exercise 16 (Page 24)

1. -16 2. -8 3. $4a$ 4. $a - 4b - c$ 5. $9d - 2e$
6. $3x - 5y + 5z$ 7. $3x^2 - xy - 2y^2$
8. $6a^2 - 13ab + 3ac - 5b^2 + bc$ 9. (i) $\dfrac{1}{3b}$ (ii) $2ab$

Exercise 17 (Page 25)

1. $x^5 y^4 z^3$, $13\dfrac{1}{2}$ 2. $a^2 b^{1/2} c^{-2}$, $\pm 4\dfrac{1}{2}$ 3. $a^3 b^{-2} c$, 9
4. $ab^6 c^{3/2}$ 5. $a^{-4} b^5 c^{11}$

Exercise 18 (Page 26)

1. $3x + y$ **2.** $3a + 5y$ **3.** $5(x - y)$

4. $x(3x - 3 + y)$ **5.** $-5p + 10q - 6r$

6. $a^2 + 3ab + 2b^2$ **7.** $3p^2 + pq - 2q^2$

8. (i) $x^2 - 4xy + 4y^2$ (ii) $9a^2 - 6ab + b^2$

9. $3ab + 7ac - 4bc$ **10.** $4 - a$

11. $11q - 2p$ **12.** (i) $p(b + 2c)$ (ii) $2q(q + 4n)$

13. (i) $7ab(3ab - 4)$ (ii) $2xy(y + 3x + 4x^2)$

Exercise 19 (Page 27)

1. $\frac{1}{2} + 6x$ **2.** $\frac{1}{5}$ **3.** $4a(1 - 2a)$

4. $a(4a + 1)$ **5.** $a(3 - 10a)$

6. $\frac{2}{3y} - 3y + 12$ **7.** $\frac{2}{3y} + 12\frac{1}{3} - 5y$

8. $\frac{2}{3y} + 12 - 13y$ **9.** $\frac{5}{y} + 1$ **10.** $\frac{1}{2}(x - 4)$

Exercise 20 (Page 29)

1. 1   **2.** 2   **3.** 6   **4.** -4 **5.** $1\frac{2}{3}$

6. 2   **7.** $\frac{1}{2}$ **8.** 0   **9.** 3   **10.** 2

11. -10 **12.** 6   **13.** -2 **14.** $2\frac{1}{2}$ **15.** 2

Exercise 21 (Page 31)

1. 5   **2.** -2 **3.** $-4\frac{1}{2}$ **4.** 15   **5.** -4

6. $5\frac{1}{3}$ **7.** -6 **8.** 9   **9.** $6\frac{1}{4}$ **10.** 4

11. 10   **12.** ± 12 **13.** ± 3 **14.** ± 4

Exercise 22 (Page 32)

1. $8\,\text{m/s}^2$ **2.** 3.472   **3.** (a) $1.8\,\Omega$ (b) $30\,\Omega$ **4.** $800\,\Omega$

Exercise 23 (Page 33)

1. 0.004   **2.** 30   **3.** $45°C$ **4.** 50   **5.** $12\,\text{m}, 8\,\text{m}$

Exercise 24 (Page 35)

1. $d = c - a - b - e$ **2.** $y = \frac{1}{3}(t - x)$

3. $r = \frac{c}{2\pi}$ **4.** $x = \frac{y - c}{m}$

5. $T = \frac{I}{PR}$ **6.** $R = \frac{E}{I}$

7. $r = \frac{s - a}{s}$ or $1 - \frac{a}{s}$ **8.** $C = \frac{5}{9}(F - 32)$

Exercise 25 (Page 36)

1. $x = \frac{d}{\lambda}(y + \lambda)$ or $d + \frac{yd}{\lambda}$

2. $f = \frac{3F - AL}{3}$ or $f = F - \frac{AL}{3}$

3. $E = \frac{Ml^2}{8yI}$

4. $t = \frac{R - R_0}{R_0\alpha}$

5. $R_2 = \frac{RR_1}{R_1 - R}$

6. $R = \frac{E - e - Ir}{I}$ or $R = \frac{E - e}{I} - r$

7. $l = \frac{t^2 g}{4\pi^2}$ **8.** $u = \sqrt{v^2 - 2as}$

Exercise 26 (Page 37)

1. $a = \sqrt{\left(\frac{xy}{m - n}\right)}$ **2.** $r = \frac{3(x + y)}{(1 - x - y)}$

3. $L = \frac{mrCR}{\mu - m}$ **4.** $b = \frac{c}{\sqrt{1 - a^2}}$

5. $v = \frac{uf}{u - f}, 30$ **6.** $t_2 = t_1 + \frac{Q}{mc}, 55$

7. $v = \sqrt{\left(\frac{2dgh}{0.03L}\right)}, 0.965$ **8.** $l = \frac{8S^2}{3d} + d, 2.725$

Exercise 27 (Page 41)

1. $a = 5, b = 2$ **2.** $x = 1, y = 1$

3. $s = 2, t = 3$ **4.** $x = 3, y = -2$

5. $m = 2\frac{1}{2}, n = \frac{1}{2}$ **6.** $a = 6, b = -1$

Exercise 28 (Page 42)

1. $p = -1, q = -2$ 2. $x = 4, y = 6$
3. $a = 2, b = 3$ 4. $s = 4, t = -1$

Exercise 29 (Page 43)

1. $a = \dfrac{1}{5}, b = 4$ 2. $u = 12, a = 4, v = 26$

3. $m = -\dfrac{1}{2}, c = 3$ 4. $a = 12, b = 0.40$

Exercise 30 (Page 44)

1. 14.5 2. $\dfrac{1}{2}$

3. (a) $4, -2$ (b) $-1, 0$ (c) $-3, -4$ (d) $0, 7$

4. (a) $\dfrac{3}{5}$ (b) -4 5. (a) and (c),   (b) and   (e)

6. (a) -1.1 (b) -1.4 7. $(2, 1)$

Exercise 31 (Page 51)

1. (a) $40°C$ (b) $128\,\Omega$
2. (a) 0.25 (b) 12 (c) $F = 0.25L + 12$
   (d) $89.5\,N$ (e) $592\,N$ (f) $212\,N$
3. (a) $22.5\,m/s$ (b) $6.43\,s$ (c) $v = 0.7t + 15.5$
4. (a) $\dfrac{1}{5}$ (b) 6 (c) $E = \dfrac{1}{5}L + 6$ (d) $12\,N$ (e) $65\,N$

Exercise 32 (Page 53)

1. $11.18\,cm$ 2. $24.11\,mm$ 3. $20.81\,km$ 4. $3.35\,m, 10\,cm$

Exercise 33 (Page 55)

1. $\sin Z = \dfrac{9}{41}$, $\cos Z = \dfrac{40}{41}$, $\tan X = \dfrac{40}{9}$, $\cos X = \dfrac{9}{41}$

2. $\sin A = \dfrac{3}{5}$, $\cos A = \dfrac{4}{5}$, $\tan A = \dfrac{3}{4}$, $\sin B = \dfrac{4}{5}$,
   $\cos B = \dfrac{3}{5}$, $\tan B = \dfrac{4}{3}$

3. $\sin X = \dfrac{15}{113}$, $\cos X = \dfrac{112}{113}$

4. (a) $\dfrac{15}{17}$ (b) $\dfrac{15}{17}$ (c) $\dfrac{8}{15}$

5. (a) 9.434 (b) -0.625 (c) $32°$

Exercise 34 (Page 56)

1. (a) 0.1321 (b) -0.8399
2. (a) 0.9307 (b) 0.2447
3. (a) -1.1671 (b) 1.1612
4. (a) 0.8660 (b) -0.1010 (c) 0.5865
5. $13.54°$, $13°32'$, $0.236\,rad$
6. $34.20°$, $34°12'$, $0.597\,rad$
7. $39.03°$, $39°2'$, $0.681\,rad$
8. (a) -0.8192 (b) -1.8040 (c) 0.6528

Exercise 35 (Page 57)

1. $BC = 3.50\,cm$, $AB = 6.10\,cm$, $\angle B = 55°$
2. $FE = 5\,cm$, $\angle E = 53°8'$, $\angle F = 36°52'$
3. $KL = 5.43\,cm$, $JL = 8.62\,cm$, $\angle J = 39°$, area $=$ $18.19\,cm^2$
4. $MN = 28.86\,mm$, $NO = 13.82\,mm$, $\angle O = 64°25'$, area $= 199.4\,mm^2$
5. $6.54\,m$

Exercise 36 (Page 60)

1. $C = 83°$, $a = 14.1\,mm$, $c = 28.9\,mm$, area $= 189\,mm^2$
2. $A = 52.04°$, $c = 7.568\,cm$, $a = 7.153\,cm$, area $= 25.66\,cm^2$
3. $D = 19.80°$, $E = 134.20°$, $e = 36.0\,cm$, area $= 134\,cm$
4. $E = 49.0°$, $F = 26.63°$, $f = 15.08\,mm$, area $= 185.6\,mm^2$

Exercise 37 (Page 61)

1. $p = 13.2\,cm$, $Q = 47.35°$, $R = 78.65°$, area $= 77.7\,cm^2$
2. $p = 6.127\,m$, $Q = 30.82°$, $R = 44.18°$, area $= 6.938\,m^2$
3. $X = 83.33°$, $Y = 52.62°$, $Z = 44.05°$, area $= 27.8\,cm$
4. $Z = 29.77°$, $Y = 53.52°$, $Z = 96.71°$, area $= 355\,mm$

Exercise 38 (Page 63)

1. $193\,km$ 2. (a) $11.4\,m$ (b) $17.55°$ 3. $163.4\,m$
4. $BF = 3.9\,m$, $EB = 4.0\,m$ 5. $x = 69.3\,mm$, $y = 142\,mm$

Exercise 39 (Page 65)

1. (b)	**2.** (d)	**3.** (b)	**4.** (b)	**5.** (a)
6. (c)	**7.** (c)	**8.** (d)	**9.** (b)	**10.** (c)
11. (a)	**12.** (d)	**13.** (a)	**14.** (b)	**15.** (a)
16. (d)	**17.** (a)	**18.** (b)	**19.** (d)	**20.** (d)
21. (a)	**22.** (a)	**23.** (a)	**24.** (a)	**25.** (a)
26. (d)	**27.** (b)	**28.** (c)	**29.** (d)	**30.** (b)
31. (a)	**32.** (c)	**33.** (b)	**34.** (b)	**35.** (c)
36. (c)	**37.** (b)	**38.** (d)	**39.** (c)	**40.** (c)
41. (a)	**42.** (a)	**43.** (c)	**44.** (b)	**45.** (d)
46. (b)	**47.** (d)	**48.** (b)	**49.** (d)	**50.** (a)
51. (a)	**52.** (b)	**53.** (a)	**54.** (a)	**55.** (d)
56. (d)	**57.** (c)	**58.** (b)	**59.** (c)	**60.** (c)

Exercise 40 (Page 74)

1. (a) 0.052 m   (b) 0.02 m^2 (c) 0.01 m^3
2. (a) 12.5 m^2 (b) 12.5 $\times$ 10^6 mm^2
3. (a) 37.5 m^3 (b) 37.5 $\times$ 10^9 mm^3
4. (a) 0.0063 m^3 (b) 6300 cm^3 (c) 6.3 $\times$ 10^6 mm^3

Exercise 41 (Page 76)

1. 11 400 kg/m^3 **2.** 1.5 kg **3.** 20 litres **4.** 8.9 **5.** 8000 kg/m^3
6. (a) 800 kg/m^3 (b) 0.0025 m^3 or 2500 cm^3

Exercise 42 (Page 76)

Answers found from within the text of the chapter, pages 73 to 75.

Exercise 43 (Page 76)

1. (c) | **2.** (d) | **3.** (b) | **4.** (a) | **5.** (b)
6. (c) | **7.** (b) | **8.** (b) | **9.** (c) | **10.** (a)

Exercise 44 (Page 80)

1. 75 kJ **2.** 20 kJ **3.** 50 kJ **4.** 1.6 J
5. (a) 2.5 J (b) 2.1 J **6.** 6.15 kJ

Exercise 45 (Page 82)

1. 75% **2.** 1.2 kJ **3.** 4 m **4.** 10 kJ

Exercise 46 (Page 84)

1. 600 kJ   **2.** 4 kW
3. (a) 500 N   (b) 625 W
4. (a) 10.8 MJ   (b) 12 kW   **5.** 160 m
6. (a) 72 MJ   (b) 100 MJ   **7.** 13.5 kW
8. 480 W   **9.** 100 kW   **10.** 90 W

Exercise 47 (Page 87)

1. 8.31 m **2.** 196.2 J **3.** 90 kJ **4.** 176.6 kJ, 176.6 kJ
5. 200 kJ, 2039 m **6.** 8 kJ, 14.0 m/s
7. 4.85 m/s (a) 2.83 m/s (b) 14.70 kN

Exercise 48 (Page 87)

Answers found from within the text of the chapter, pages 77 to 87.

Exercise 49 (Page 88)

1. (b)	**2.** (c)	**3.** (c)	**4.** (a)	**5.** (d)
6. (c)	**7.** (a)	**8.** (d)	**9.** (c)	**10.** (b)
11. (b)	**12.** (a)	**13.** (d)	**14.** (a)	**15.** (d)

Exercise 50 (Page 91)

1. 5 s   **2.** 3600 C   **3.** 13 min 20 s

Exercise 51 (Page 93)

1. 7 Ω **2.** 0.25 A   **3.** 2 mΩ, 5 mΩ **4.** 30 V

Exercise 52 (Page 95)

1. 0.4 A, 100 W **2.** 20 Ω, 2.88 kW, 57.6 kWh
3. 0.8 W **4.** 9.5 W **5.** 2.5 V
6. (a) 0.5 W (b) 1.33 W (c) 40 W **7.** 8 kW, 20 A
8. 5 kW **9.** £ 12.46 **10.** 3 kW, 90 kWh, £ 5.85

Exercise 53 (Page 96)

1. 3 A, 5 A

Exercise 54 (Page 96)

Answers found from within the text of the chapter, pages 90 to 96.

Exercise 55 (Page 97)

1. (d) **2.** (a) **3.** (c) **4.** (b) **5.** (d)
6. (d) **7.** (b) **8.** (c) **9.** (a) **10.** (a)
11. (c) **12.** (c) **13.** (b) **14.** (a) **15.** (c)
16. (b) **17.** (d) **18.** (d)

Exercise 56 (Page 102)

1. (a) 18 V, 2.88 Ω (b) 1.5 V, 0.02 Ω
2. (a) 2.17 V (b) 1.6 V (c) 0.7 V
3. 0.25 Ω **4.** 18 V, 1.8 Ω **5.** (a) 1 A (b) 21 V
6. (i) (a) 6 V (b) 2 V (ii) (a) 4 Ω (b) 0.25 Ω

Exercise 57 (Page 105)

Answers found from within the text of the chapter, pages 99 to 105.

Exercise 58 (Page 105)

1. (d) **2.** (a) **3.** (b) **4.** (c) **5.** (b)
6. (d) **7.** (d) **8.** (b) **9.** (c) **10.** (d)
11. (c) **12.** (a)

Exercise 59 (Page 109)

1. 2.5 mC **2.** 2 kV **3.** 2.5 μF **4.** 1.25 ms
5. 2 kV **6.** 750 kV/m **7.** 312.5 μC/m^2, 50 kV/m

Exercise 60 (Page 110)

1. 885 pF **2.** 0.885 mm **3.** 7 **4.** 2.97 mm **5.** 200 pF

Exercise 61 (Page 111)

1. (a) 8 μF (b) 1.5 μF **2.** 15 μF
3. (a) 0.06 μF (b) 0.25 μF **4.** 2.4 μF, 2.4 μF
5. (a) 1.2 μF (b) 100 V **6.** 4.2 μF each

Exercise 62 (Page 112)

1. (a) 0.02 μF (b) 0.4 mJ **2.** 20 J **3.** 550 V
4. (a) 0.04 mm (b) 361.6 cm^2 (c) 0.02 J (d) 1 kW

Exercise 63 (Page 115)

1. −150 V **2.** 144 ms **3.** 0.8 Wb/s **4.** 3.5 V **5.** −90 V

Exercise 64 (Page 116)

1. 62.5 J **2.** 0.18 mJ **3.** 40 H

Exercise 65 (Page 117)

1. (a) 7.2 H (b) 90 J (c) 180 V **2.** 4 H **3.** 40 ms
4. 12.5 H, 1.25 kV **5.** 12 500

Exercise 66 (Page 117)

Answers found from within the text of the chapter, pages 107 to 117.

Exercise 67 (Page 118)

1. (a) **2.** (b) **3.** (c) **4.** (a) **5.** (b)
6. (b) **7.** (c) **8.** (c) **9.** (c) **10.** (d)
11. (c) **12.** (b) **13.** (d) **14.** (a)

Exercise 68 (Page 120)

1. (a) 324 K (b) 195 K (c) 456 K
2. (a) 34°C (b) −36°C (c) 142°C

Exercise 69 (Page 121)

1. 2.1 MJ **2.** 390 kJ **3.** 5 kg **4.** 130 J/kg°C **5.** 65°C

Exercise 70 (Page 122)

1. Similar to Fig. 16.1, page 122 – see Section 16.4

Exercise 71 (Page 124)

1. 8.375 MJ **2.** 18.08 MJ **3.** 3.98 MJ **4.** 12.14 MJ

Exercise 72 (Page 124)

Answers found from within the text of the chapter, pages 119 to 124.

Exercise 73 (Page 124)

1. (d)	2. (b)	3. (a)	4. (c)	5. (b)
6. (b)	7. (b)	8. (a)	9. (c)	10. (b)

Exercise 74 (Page 133)

1. (a) $8.75\,\Omega$ (b) 5 m
2. (a) $5\,\Omega$ (b) $0.625\,\text{mm}^2$
3. $0.32\,\Omega$
4. $0.8\,\Omega$
5. $1.5\,\text{mm}^2$
6. $0.026\,\mu\Omega\text{m}$
7. $0.216\,\Omega$

Exercise 75 (Page 134)

1. $69\,\Omega$	2. $24.69\,\Omega$	3. $488\,\Omega$	4. $26.4\,\Omega$
5. $70°C$	6. $64.8\,\Omega$	7. $5.95\,\Omega$	

Exercise 76 (Page 135)

Answers found from within the text of the chapter, pages 131 to 134.

Exercise 77 (Page 135)

1. (c)	2. (d)	3. (b)	4. (d)
5. (d)	6. (c)	7. (b)	

Exercise 78 (Page 138)

1. (a) 22 V   (b) $11\,\Omega$ (c) $2.5\,\Omega$, $3.5\,\Omega$, $5\,\Omega$
2. 10 V, 0.5 A, $20\,\Omega$, $10\,\Omega$, $6\,\Omega$
3. 4 A, $2.5\,\Omega$
4. 45 V
5. (a) $1.2\,\Omega$ (b) 12 V

Exercise 79 (Page 144)

1. (a) $3\,\Omega$ (b) 3 A   (c) 2.25 A, 0.75 A   2. 2.5 A, $2.5\,\Omega$
3. (a) (i) $5\,\Omega$ (ii) $60\,\text{k}\Omega$ (iii) $28\,\Omega$ (iv) $6.3\,\text{k}\Omega$
   (b) (i) $1.2\,\Omega$ (ii) $13.33\,\text{k}\Omega$ (iii) $2.29\,\Omega$ (iv) $461.5\,\Omega$
4. $8\,\Omega$ 5. $27.5\,\Omega$ 6. $2.5\,\Omega$, 6 A   7. (a) 1.6 A   (b) $6\,\Omega$
8. (a) 30 V   (b) 42 V
9. $I_1 = 5\,\text{A}$, $I_2 = 2.5\,\text{A}$, $I_3 = 1.67\,\text{A}$,
   $I_4 = 0.83\,\text{A}$, $I_5 = 3\,\text{A}$, $I_6 = 2\,\text{A}$,
   $V_1 = 20\,\text{V}$, $V_2 = 5\,\text{V}$, $V_3 = 6\,\text{V}$ 10. 1.8 A

Exercise 80 (Page 146)

1. $400\,\Omega$ 2. (a) 70 V   (b) 210 V

Exercise 81 (Page 146)

Answers found from within the text of the chapter, pages 136 to 146.

Exercise 82 (Page 146)

1. (a)	2. (c)	3. (c)	4. (c)	5. (a)
6. (d)	7. (b)	8. (c)	9. (d)	

Exercise 83 (Page 152)

1. $I_3 = 2\,\text{A}$, $I_4 = -1\,\text{A}$, $I_6 = 3\,\text{A}$
2. (a) $I_1 = 4\,\text{A}$, $I_2 = -1\,\text{A}$, $I_3 = 13\,\text{A}$ (b) $I_1 = 40\,\text{A}$,
   $I_2 = 60\,\text{A}$, $I_3 = 120\,\text{A}$, $I_4 = 100\,\text{A}$, $I_5 = -80\,\text{A}$
3. 2.162 A,   42.07 W   4. 2.715 A,   7.410 V,   3.948 V
5. (a) 60.38 mA   (b) 15.10 mA
   (c) 45.28 mA   (d) 34.20 mW
6. $I_1 = 1.26\,\text{A}$, $I_2 = 0.74\,\text{A}$, $I_3 = 0.16\,\text{A}$,
   $I_4 = 1.42\,\text{A}$, $I_5 = 0.59\,\text{A}$

Exercise 84 (Page 153)

Answers found from within the text of the chapter, pages 149 to 152.

Exercise 85 (Page 153)

1. (a) **2.** (d) **3.** (c) **4.** (b) **5.** (c)

Exercise 86 (Page 159)

1. 1.5 T **2.** 2.7 mWb **3.** 32 cm **4.** 4 cm by 4 cm

Exercise 87 (Page 161)

1. 21.0 N, 14.8 N **2.** 4.0 A **3.** 0.80 T **4.** 0.582 N
5. (a) 14.2 mm   (b) towards the viewer

Exercise 88 (Page 163)

1. 8×10^{-19} N **2.** 10^6 m/s

Exercise 89 (Page 163)

Answers found from within the text of the chapter, pages 155 to 163   **14.** to the left.

Exercise 90 (Page 164)

1. (d) **2.** (d) **3.** (a) **4.** (a) **5.** (b)
6. (b) **7.** (d) **8.** (c) **9.** (d) **10.** (a)
11. (c) **12.** (c)

Exercise 91 (Page 169)

1. 0.135 V **2.** 25 m/s **3.** (a) 0 (b) 0.16 A
4. 1 T, 0.25 N **5.** 1.56 mV
6. (a) 48 V (b) 33.9 V (c) 24 V

Exercise 92 (Page 170)

1. 4.5 V **2.** 1.6 mH **3.** 250 V
4. (a) −180 V (b) 5.4 mWb

Exercise 93 (Page 172)

1. 96 **2.** 990 V **3.** 16 V **4.** 3000 turns
5. 15 V, 48 A **6.** 50 A **7.** 16 V, 45 A **8.** 225 V, 3 : 2

Exercise 94 (Page 172)

Answers found from within the text of the chapter, pages 166 to 172.

Exercise 95 (Page 172)

1. (c) **2.** (b) **3.** (c) **4.** (a) **5.** (d)
6. (a) **7.** (b) **8.** (c) **9.** (d) **10.** (a)
11. (b) **12.** (a) **13.** (d) **14.** (b) and (c)

Exercise 96 (Page 177)

1. (a) 0.4 s (b) 10 ms (c) 25 μs
2. (a) 200 Hz (b) 20 kHz (c) 5 Hz
3. 800 Hz

Exercise 97 (Page 180)

1. (a) 50 Hz (b) 5.5 A, 3.4 A (c) 2.8 A (d) 4.0 A
2. (a) (i) 100 Hz (ii) 2.50 A (iii) 2.88 A
       (iv) 1.15   (v) 1.74
   (b) (i) 250 Hz (ii) 20 V (iii) 20 V
       (iv) 1.0   (v) 1.0
   (c) (i) 125 Hz (ii) 18 A (iii) 19.56 A
       (iv) 1.09   (v) 1.23
   (d) (i) 250 Hz (ii) 25 V (iii) 50 V
       (iv) 2.0   (v) 2.0
3. (a) 150 V (b) 170 V
4. 212.1 V
5. 282.9 V, 180.2 V
6. 84.8 V, 76.4 V
7. 23.55 A, 16.65 A

Exercise 98 (Page 181)

Answers found from within the text of the chapter, pages 177 to 180.

Exercise 99 (Page 181)

1. (b) **2.** (c) **3.** (d) **4.** (d) **5.** (a)
6. (d) **7.** (c) **8.** (b) **9.** (c) **10.** (b)

Exercise 100 (Page 185)

1. (a) $2\,k\Omega$ (b) $10\,k\Omega$ (c) $25\,k\Omega$
2. (a) $18.18\,\Omega$ (b) $10.00\,m\Omega$ (c) $2.00\,m\Omega$
3. $39.98\,k\Omega$
4. (a) $50.10\,m\Omega$ in parallel   (b) $4.975\,k\Omega$ in series

Exercise 101 (Page 188)

1. (a) $0.250\,A$ (b) $0.238\,A$ (c) $2.83\,W$ (d) $56.64\,W$
2. (a) $900\,W$ (b) $904.5\,W$

Exercise 102 (Page 191)

1. (a) $41.7\,Hz$ (b) $176\,V$
2. (a) $0.56\,Hz$ (b) $8.4\,V$
3. (a) $7.14\,Hz$ (b) $220\,V$ (c) $77.78\,V$

Exercise 103 (Page 193)

1. $3\,k\Omega$ **2.** $1.525\,V$

Exercise 104 (Page 194)

Answers found from within the text of the chapter, pages 182 to 194.

Exercise 105 (Page 194)

1. (d)   **2.** (a) or (c)   **3.** (b)   **4.** (b)
5. (c)   **6.** (f)   **7.** (c)   **8.** (a)
9. (i)   **10.** (j)   **11.** (g)   **12.** (c)
13. (b)   **14.** (p)   **15.** (d)   **16.** (o)
17. (n)   **18.** (a)

Exercise 106 (Page 202)

1. (a) $72\,km/h$ (b) $20\,m/s$ **2.** $3.6\,km$ **3.** 3 days 19 hours

Exercise 107 (Page 203)

1. 0 to A, $30\,km/h$; A to B, $40\,km/h$; B to C, $10\,km/h$; 0 to C, $24\,km/h$; A to C, $20\,km/h$

Exercise 108 (Page 204)

1. $12.5\,km$ **2.** $4\,m/s$

Exercise 109 (Page 205)

Answers found from within the text of the chapter, pages 201 to 205

Exercise 110 (Page 206)

1. (c)   **2.** (g)   **3.** (d)   **4.** (c)   **5.** (e)
6. (b)   **7.** (a)   **8.** (b)   **9.** (d)   **10.** (a)
11. (a)   **12.** (b)   **13.** (c)   **14.** (d)

Exercise 111 (Page 208)

1. $50\,s$ **2.** $9.26 \times 10^{-4}\,m/s^2$
3. (a) $0.0231\,m/s^2$ (b) 0 (c) $-0.370\,m/s^2$
4. A to B, $-2\,m/s^2$, B to C, $2.5\,m/s^2$, C to D, $-1.5\,m/s^2$

Exercise 112 (Page 210)

1. $12.25\,m/s$ **2.** $10.2\,s$ **3.** $7.6\,m/s$
4. $39.6\,km/h$ **5.** 2 min 5 s   **6.** $1.35\,m/s$, $-0.45\,m/s^2$

Exercise 113 (Page 210)

Answers found from within the text of the chapter, pages 207 to 210.

Exercise 114 (Page 210)

1. (b)   **2.** (a)   **3.** (c)   **4.** (e)   **5.** (i)
6. (g)   **7.** (d)   **8.** (c)   **9.** (a)   **10.** (d)

Exercise 115 (Page 214)

1. $873\,N$ **2.** $1.87\,kN$, $6.55\,s$ **3.** $0.560\,m/s^2$
4. $16.5\,kN$ **5.** (a) $11.6\,kN$ (b) $19.6\,kN$ (c) $24.6\,kN$

The following answers belong to the top right column:

2. $4\,m/s$, $2.86\,m/s$, $3.33\,m/s$, $2.86\,m/s$, $3.2\,m/s$
3. $119\,km$
4. $333.33\,m/s$ or $1200\,km/h$
5. $2.6 \times 10^{13}\,km$

Exercise 116 (Page 216)

1. 988 N, 34.86 km/h   **2.** 1.49 m/s^2 **3.** 10 kg

Exercise 117 (Page 216)

Answers found from within the text of the chapter, pages 212 to 215.

Exercise 118 (Page 216)

1. (c)   **2.** (b)   **3.** (a)   **4.** (d)   **5.** (a)
6. (b)   **7.** (b)   **8.** (a)   **9.** (a)   **10.** (d)
11. (d)   **12.** (c)

Exercise 119 (Page 221)

1. 4.0 kN in the direction of the forces
2. 68 N in the direction of the 538 N force
3. 36.8 N at 20°
4. 12.6 N at 79°
5. 3.4 kN at −8°

Exercise 120 (Page 222)

1. 2.6 N at −20°   **2.** 15.7 N at −172°
3. 26.7 N at −82°   **4.** 0.86 kN at −100°
5. 15.5 N at −152°

Exercise 121 (Page 222)

1. 11.52 kN at 105°   **2.** 36.8 N at 20°
3. 3.4 kN at −8°   **4.** 15.7 N at −172°
5. 0.86 kN at −100°

Exercise 122 (Page 224)

1. 3.06 N at −45° to force *A*
2. 53.5 kN at 37° to force 1 (i.e. 117° to the horizontal)
3. 72 kN at −37° to force *P*
4. 43.18 N, 38.82° east of north

Exercise 123 (Page 225)

1. 5.86 N, 8.06 N
2. 9.96 kN, 7.77 kN
3. 100 N force at 263° to the 60 N force, 90 N force at 132° to the 60 N force

Exercise 124 (Page 228)

1. 10.08 N, 20.67 N
2. 9.09 N at 93.92°
3. 3.1 N at −45° to force *A*
4. 53.5 kN at 37° to force 1 (i.e. 117° to the horizontal)
5. 18.82 kN at 210.03° to the horizontal
6. 34.96 N at −10.23° to the horizontal

Exercise 125 (Page 228)

Answers found from within the text of the chapter, pages 218 to 228.
12. 15 N acting horizontally to the right
13. 10 N at 230°   **15.** 5 N at 45°

Exercise 126 (Page 229)

1. (b)   **2.** (a)   **3.** (b)   **4.** (d)   **5.** (b)
6. (c)   **7.** (b)   **8.** (b)   **9.** (c)   **10.** (d)
11. (c)   **12.** (d)   **13.** (d)   **14.** (a)

Exercise 127 (Page 232)

1. 4.5 Nm   **2.** 200 mm   **3.** 150 N

Exercise 128 (Page 234)

1. 50 mm, 3.8 kN   **2.** 560 N
3. $R_A = R_B = 25$ N   **4.** 5 N, 25 mm

Exercise 129 (Page 236)

1. 2 kN, 24 mm   **2.** 1.0 kN, 64 mm   **3.** 80 m
4. 36 kN   **5.** 50 N, 150 N
6. (a) $R_1 = 3$ kN, $R_2 = 12$ kN   (b) 15.5 kN

Exercise 130 (Page 237)

Answers found from within the text of the chapter, pages 231 to 236.

Exercise 131 (Page 237)

1. (a)   **2.** (c)   **3.** (a)   **4.** (d)   **5.** (a)
6. (d)   **7.** (c)   **8.** (a)   **9.** (d)   **10.** (c)

Exercise 132 (Page 240)

1. $\omega = 90$ rad/s, v $= 16.2$ m/s
2. $\omega = 40$ rad/s, v $= 10$ m/s

Exercise 133 (Page 241)

1. 275 rad/s, $8250/\pi$ rev/min
2. 0.4π rad/s^2, 0.1π m/s^2

Exercise 134 (Page 243)

1. 2.693 s   **2.** 222.8 rad/s^2 **3.** 187.5 revolutions
4. (a) 160 rad/s,   (b) 4 rad/s^2 (c) $12\,000/\pi$ rev

Exercise 135 (Page 245)

1. 6.248 m/s at E 13.08° N
2. 15.52 m/s at 14.93°
3. 5.385 m/s at $-21.80°$

Exercise 136 (Page 245)

Answers found from within the text of the chapter, pages 239 to 245.

Exercise 137 (Page 245)

1. (b)   **2.** (c)   **3.** (a)   **4.** (c)   **5.** (a)
6. (d)   **7.** (c)   **8.** (b)   **9.** (d)   **10.** (c)
11. (b)   **12.** (d)   **13.** (a)

Exercise 138 (Page 248)

1. 1250 N   **2.** 0.24   **3.** (a) 675 N   (b) 652.5 N

Exercise 139 (Page 249)

Answers found from within the text of the chapter, pages 247 to 249.

Exercise 140 (Page 250)

1. (c)   **2.** (c)   **3.** (f)   **4.** (e)   **5.** (i)
6. (c)   **7.** (h)   **8.** (b)   **9.** (d)   **10.** (a)

Exercise 141 (Page 253)

1. (a) 3.3   (b) 11   (c) 30%   **2.** 8
3. (a) 5   (b) 500 mm
4. (a) $F_e = 0.04\,F_l + 170$ (b) 25   (c) 83.3%
5. 410 N, 49%

Exercise 142 (Page 254)

1. 60%   **2.** 1.5 kN, 93.75%

Exercise 143 (Page 255)

1. 9.425 kN   **2.** 16.91 N

Exercise 144 (Page 256)

1. 6, 10 rev/s   **2.** 6, 4.32   **3.** 250, 3.6

Exercise 145 (Page 258)

1. 1.25 m   **2.** 250 N   **3.** 333.3 N, 1/3

Exercise 146 (Page 258)

Answers found from within the text of the chapter, pages 251 to 258.

Exercise 147 (Page 258)

1. (b)   **2.** (f)   **3.** (c)   **4.** (d)   **5.** (b)
6. (a)   **7.** (b)   **8.** (d)   **9.** (c)   **10.** (d)
11. (d)   **12.** (b)

Exercise 148 (Page 262)

1. 250 MPa　2. 12.78 mm　3. 69.44 MPa
4. 2.356 kN　5. 38.2 MPa

Exercise 149 (Page 263)

1. 2.25 mm
2. (a) 0.00002　(b) 0.002%
3. (a) 127.3 MPa　(b) 0.00125 or 0.125%

Exercise 150 (Page 266)

1. 800 N
2. 10 mm
3. (a) 43.75 kN　(b) 1.2 mm
4. (a) 19 kN　(b) 0.057 mm
5. 132.6 GPa
6. (a) 27.5 mm　(b) 68.2 GPa
7. 0.1%

Exercise 151 (Page 267)

Answers found from within the text of the chapter, pages 260 to 266

Exercise 152 (Page 267)

1. (c)　2. (c)　3. (a)　4. (b)　5. (c)
6. (c)　7. (b)　8. (d)　9. (b)　10. (c)
11. (f)　12. (h)　13. (d)

Exercise 153 (Page 269)

1. 250 kg m/s　2. 2.4 kg m/s
3. 64 kg　4. 22 500 kg m/s
5. 90 km/h　6. 2 m/s
7. 10.71 m/s　8. (a) 26.25 m/s　(b) 11.25 m/s

Exercise 154 (Page 271)

1. 8 kg m/s　2. 1.5 m/s
3. 1000 N　4. (a) 200 kg m/s　(b) 8 kN
5. 314.5 kN　6. (a) 1.5 kg m/s　(b) 150 N

Exercise 155 (Page 271)

Answers found from within the text of the chapter, pages 268 to 271.

Exercise 156 (Page 272)

1. (d)　2. (b)　3. (f)　4. (c)　5. (a)
6. (c)　7. (a)　8. (g)　9. (f)　10. (f)　11. (b)

Exercise 157 (Page 274)

1. 70 Nm　2. 160 mm

Exercise 158 (Page 275)

1. 339.3 kJ　2. 523.8 rev　3. 45 Nm　4. 13.75 rad/s
5. 16.76 MW　6. (a) 3.33 Nm　(b) 540 kJ

Exercise 159 (Page 277)

1. 272.4 Nm　　　　　　　2. 30 kg m^2
3. 502.65 kJ, 1934 rev/min　4. 16.36 J
5. (a) 2.857×10^{-3} kg m^2　(b) 24.48 J

Exercise 160 (Page 279)

1. 6.05 kW　2. 70.37%, 280 mm　3. 57.03%
4. 545.5 W　5. 4.905 MJ, 1 min 22 s

Exercise 161 (Page 279)

Answers found from within the text of the chapter, pages 273 to 279.

Exercise 162 (Page 279)

1. (d)　2. (b)　3. (c)　4. (a)　5. (c)　6. (d)
7. (a)　8. (b)　9. (c)　10. (d)　11. (a)　12. (c)

Exercise 163 (Page 283)

1. 50.0928 m　2. 6.45×10^{-6} K^{-1}
3. 20.4 mm　4. (a) 0.945 mm　(b) 0.045%
5. 0.45 mm　6. 47.26°C　7. 6.75 mm　8. 74 K

Exercise 164 (Page 286)

1. $2.584\,\text{mm}^2$ 2. $358.8\,\text{K}$

3. (a) $0.372\,\text{mm}$ (b) $280.5\,\text{mm}^2$ (c) $4207\,\text{mm}^3$

4. $29.7\,\text{mm}^3$ 5. $2.012\,\text{litres}$

6. $50.4\,\text{litres}$ 7. $0.03025\,\text{litres}$

Exercise 165 (Page 286)

Answers found from within the text of the chapter, pages 281 to 286.

Exercise 166 (Page 286)

1. (b) 2. (c) 3. (a) 4. (d) 5. (b)

6. (c) 7. (c) 8. (a) 9. (c) 10. (b)

Exercise 167 (Page 291)

1. $1270°C$

Exercise 168 (Page 292)

1. $203°C$

Exercise 169 (Page 295)

Answers found from within the text of the chapter, pages 288 to 295.

Exercise 170 (Page 296)

1. (c) 2. (b) 3. (d) 4. (b) 5. (i)

6. (a) 7. (e) 8. (d) 9. (e) or (f) 10. (k)

11. (b) 12. (g)

Index